PRAISE FOR SOCIAL F

"*Social Forestry* by Tomi Hazel Vaarde is a book of hope with the Earth and her forests does not have to be an extractive one leading to destruction. Through cooperating together we can regenerate our forests and rewild ourselves and the land, growing hope as ecosystems recover while empires crumble."
—Vandana Shiva, PhD, author of *Agroecology and Regenerative Agriculture: Sustainable Solutions for Hunger, Poverty, and Climate Change*

"Step into this manzanita-burning charcoal dirigible for the wildest ride you can imagine through Gaelic seasonal rhythms, Quaker ethics, medieval guild practice, permaculture insights, and deep-rooted tree goddess wisdom. Tomi Hazel will teach you how to manage towering bonfires, abundant gardens, forest gifts, feral decorating, land assessment, and communal love dynamics with a healthy serving of improv poetry and insight from days long ago. A true archetypal elder, Tomi Hazel validates your internal clock and doses out the medicine for your soul, your land, and your people. Do not miss your chance to experience this inimitable voice, guiding us toward a renewable, regenerative future."
—Jessica Carew Kraft, author of *Why We Need to Be Wild*

"Hazel is one of the few elders in the rewilding community. Their visionary approach to integrating humans back into the land has inspired thousands of people and given us hope for a transition to a vibrant, regenerative future. This book is the priceless culmination of their experiences, knowledge, stories, and passion gained and given throughout their lifetime. It will provide inspiration, insight, and direction for decades to come, right when we'll need it most."
—Peter Michael Bauer, host of the Rewilding Podcast

"It is clear that Hazel has developed a deep understanding—a knowledge, wisdom, and power—of how humans can be an integral part of the solution as we seek to bring about habitat restoration. Hazel stands in the new 'front line' in the battle to engage people with place, and *Social Forestry* will, I am sure, become a key text in the emerging literature on the theme. If we are to heal our relationship with our forests and woodlands, we would do well to dive deep into the writing of Hazel Vaarde."
—Alexander Langlands BA MA PhD PGCert FHEA FSA, Senior Lecturer - Uwch Ddarlithydd, History/Heritage - Hanes/Treftadaeth

"Inside this luminous guide, you will find practical placemaking advice, ancient lore, and a humor that shimmers. Receive these generous offerings—a lifetime of wisdom from an elder, a teacher of permaculture, and radical changemaker like no other—and you will be

transformed. Together we listen to the earth, we understand where we belong, and we find our way home again."
—Helena Norberg-Hodge, Director of Local Futures and author of
Ancient Futures: Learning from Ladakh

"Embracing past and present with a timeless eye on the future, the author weaves their life story into the story of chosen place, the Siskiyou Mountains of southern Oregon, walking with wisdom between the settler and Indigenous worlds that have shaped it. . . . Grounded in the material realities of land, people, and community to an uncommon degree, the author points ceaselessly toward the spirit in all things. For scholars of change, the bibliography alone is worth the price of admission. To find such a guide and visionary elder in the way of holistic thinking is a gift beyond measure."
—Peter Bane, author of *The Permaculture Handbook: Garden Farming for Town and Country* and Executive Director of Permaculture Institute of North America

"With *Social Forestry: Tending the Land as People of Place*, Tomi Hazel Vaarde takes us into what may seem to most of us to be an alternative universe, a world in which humans live as an integral part of a healthy ecosystem. It can serve well as a handbook for living in and with a forest, but it's much, much more. It offers, for most of us, the rarest of chances to step out of our doomed industrial culture for a few hours and learn what it would be like to live as a species that could thrive on into the indefinite future without degrading the ecosphere to the point of collapse."
—Stan Cox, author of *The Path to a Livable Future: A New Politics to Fight Climate Change, Racism, and the Next Pandemic*

"With *Social Forestry*, Tomi Hazel Vaarde gifts us with poetic musings and hands-on ideas for healing our relationships with the places we live through actions that help us better understand and restore our many imperiled landscapes. Vaarde shows us how Social Forestry is both a theory and a praxis anchored to the intentional work of 'culture-tending' that addresses our ecological grief by seeking local spaces to reignite our imaginations and act together in meaningful ways."
—Diana Negrín, Wixarika Research Center and UC Berkeley Geography Department

"Tomi Hazel Vaarde's new book is brilliant! And full of surprises. In these times of fear and isolation, their poetic trailblazing integrates deep ecology, the ethics of First-Nation cultures, small-is-beautiful bioregionalism, and 'mythic memory' to advocate the repair of the planet, forests, while rewilding our own damaged psyches."
—Chellis Glendinning, PhD, author of *My Name is Chellis and I'm in Recovery from Western Civilization*

SOCIAL FORESTRY

SOCIAL FORESTRY

Tending the Land as People of Place

Tomi Hazel Vaarde

Foreword by Starhawk

SYNERGETIC PRESS
SANTA FE • LONDON

Copyright © 2023 by Tomi Hazel Vaarde
All rights reserved.
Foreword copyright © 2023 by Starhawk
All rights reserved.

No part of this publication may be reproduced, stored in any retrieval system, or transmitted, in any form or by any means, electronic, mechanical, photocopying, recording, or otherwise without the prior permission of the publisher, except for the quotation of brief passages in reviews.

Published by Synergetic Press | 1 Bluebird Court, Santa Fe, New Mexico 87508 & 24 Old Gloucester St. London, WC1N 3AL, England

Library of Congress Control Number: 2022951377

ISBN 9781957869063 (paperback)
ISBN 9781957869070 (ebook)

Cover and interior design by Howie Severson Design
Managing Editor: Amanda Müller
Developmental Editor: Noelle Armstrong
Production Editor: Allison Felus
Printed in the United States of America

To
Elizabeth Beatty Januszkiewicz

TWO BABY SKUNKS

As learned from Tom Phelps at Camp Wakapominee, early 1960s, Adirondack Mountains.

ONCE THERE WERE TWO BABY SKUNKS.
ONE WAS NAMED IN, AND ONE WAS NAMED OUT.
SOMETIMES, OUT WAS IN AND IN WAS OUT.
OTHER TIMES, IN WAS IN AND OUT WAS OUT.
WELL, ONE DAY, IN WAS OUT, AND OUT WAS IN,
SO, THE MOTHER SKUNK SENT OUT OUT TO BRING IN IN.
OUT WENT OUT AND BROUGHT IN IN RIGHT AWAY!
THEN, MOTHER SKUNK ASKED OUT:
"HOW DID YOU DO THAT SO QUICKLY?"
AND OUT SAID, "SIMPLE, INSTINCT!"

This is a storytelling exercise.

The story is told with hand movements and facial expressions:
The left hand is IN and the right hand is OUT.

The hands flip back and forth separately, and together, side to side, following the action. The hands are raised on either side as MOTHER asks the question, and hang in the air at OUT's reply, feigning innocence with raised eyebrows.

CONTENTS

Foreword by Starhawk   xiii
How to Read Hazel by Megan Fehrman   xvii
Acknowledgments of Gratitude, a Poem   xix
Principles Clipboard   1
Introduction   13

Part I: Foundations   21
Chapter 1: Peoples of the Forests   23
Chapter 2: The Lineage of Social Forestry   39
Chapter 3: The Nest-Home of Social Forestry   61
Chapter 4: Relationships   83

Part II: In the Forest   107
Chapter 5: Forest Ecologies   109
Chapter 6: Forestry Work   151
Chapter 7: Fire   191
Chapter 8: Charcoal   217
Chapter 9: Treasures from Thickets   249
Chapter 10: Forest Shelters   283

Part III: Toward Cultures of Place   319
Chapter 11: Starting from Here   321
Chapter 12: Cultures in Transition   353
Chapter 13: A Place for Humans   379
Chapter 14: Social Order   397
Chapter 15: Carrying the Bundle   413

Part IV: Visioning   427
Chapter 16: A Year in Wagner County   429

Bibliography   459
Index   471

FOREWORD

A long time ago, in a conference far, far away, I wandered into a room where a storyteller was speaking. This storyteller, Hazel, was telling tales of humble wood cutters and charcoal burners living deep in the forests, and bringing a whole new lens to all the fairy tales of our childhoods. I was enthralled. I had been working with fairy tales for many years as a teacher and practitioner of earth-based spirituality, in ritual and in writing. I'd looked at folk tales and traditional stories as remnants of ancient pre-Christian traditions, as reservoirs of resistance to the imposition of capitalism, and as psychological archetypes, but I had never really thought of them as reflections of how people actually made a living in the woods. Hazel gave me a whole new way of looking at these heritage tales. What wood were those humble woodcutters cutting? What tools did they use? What skills were needed to survive in a little clearing in the forest? What economy, culture, and traditional knowledge do they represent?

Fast forward a decade or so. I'm now stewarding a 40-acre ranch in the coastal hills of Northern California. I'm also immersed in teaching permaculture: a system of ecological design that works with nature to meet human needs while regenerating the environment around us. At one of our Northern California convergences, I encounter Hazel, telling fascinating stories and holding a workshop on how to bring good fire back to the land.

Throughout the Pacific Northwest, climate change has exacerbated the dangers of wildfire. Fire has always been an overriding concern in these forests, but now extreme heat, a longer dry season, and the intense winds of climate change have amplified the danger to an extreme pitch. In fact, on the very night the convergence ends, a destructive wildfire sweeps through the hills of Sonoma County and into the city of Santa Rosa itself, destroying homes and taking more than 80 lives.

Hazel told enthralling tales of how a community might take up the land tending practice of intentional burning. For tens of thousands of years, the Indigenous peoples of the Pacific Northwest have managed the land with fire. They burned periodically to reduce fuel loads, keep the forests open, favor old growth and wildlife, and foster biodiversity. These forests have all coevolved with fire. They are meant to burn from time to time, but when that burning happened regularly, the fuel loads were kept low, the fires stayed close to the ground, and relatively cool and catastrophic crown fires were prevented.

Since that time, while wildfire, climate change, and the pandemic have all increased our challenges, a quiet revolution has been going on in the woods. Communities have begun to band together to protect our homes and wildlands by tending the forests and bringing back good fire. I continue to teach permaculture and regenerative land management. I am responsible for not just my own land but am also part of a forestry committee that manages 2,500 acres of our homeowners' association. So when I was offered the chance to write a foreword for Hazel's new book, I jumped at the chance to get an early reading. I've not been disappointed.

Social Forestry is a poetic and visionary exploration of how we might return to a state of balance with the land. We will never recapture the elegant relationship Indigenous people had reached before the settlers came. As Hazel points out many times, only Indigenous people can re-indigenize. But we can, whatever our heritage and background, learn to become peoples of place, to respond to the conditions of this moment, and draw on Traditional Ecological Knowledge, scientific knowledge, and our own observations and deep connection to the land to come into right relationship.

Hazel begins by outlining their underlying worldview and the principles of ecology and spiritual connection to the land that shapes their vision. They weave together anecdotes, stories, scientific principles, and poetry into a rich mix that hints at a spirituality fit for our times, one that does not require belief in dogma or miracles outside the realm of nature, but rather awakens our sense of awe and reverence for the everyday miracles embodied in the natural world: how leaves decay to feed the growth of the tree and the budding of new leaves, how the destruction of wildfire can renew the land, how nature provides in so many ways for our needs and pleasures.

Then Hazel gets down to specifics. The next section of the book goes deeply into the how-to of tending the woods, drawing on Hazel's many decades of rich experience living and working in the forests of the Northwest. From working with hand tools, to methods of thinning and pruning, to ways of using forest products for jewelry, for furniture, for natural building, and for biochar, Hazel shows us how they do it.

A whole section of the book is devoted to bringing back fire, as a community-led and deeply connected way of tending the land. As someone who spends a great deal of time with community projects to tend the woods, mitigate fire risks, and bring good fire back to the land, I found Hazel's visions and prescriptions affirming, inspiring, and sometimes challenging. How can we work with even more sensitivity to the ecology and spirit of the land? How could we get more products and multiple uses from the material we thin?

Hazel's articulation of principles that heads each chapter were especially useful. More than just a set of techniques, Hazel's principles help us expand our way of perceiving and understanding what's going on in the land. I found their list of 60 things to think about before cutting a tree especially useful, with delineation between projects that are quick fixes, retrofits, or long-term visions, and the different considerations for each.

Hazel goes on in subsequent chapters to discuss methods of natural building using materials from the land, and shares their own experiences creating cabins and forestry camps. They devote a section to other potential value-added products such as baskets, jewelry, furniture, and fencing from the by-products of tending the woods.

Finally, Hazel ends the book by presenting us with a vision of what a new culture of place might come to be. How can we live in right relationship with the land today? How do we honor Traditional Ecological Knowledge and Indigenous Ecological Knowledge; weave them together with modern science; work around the legalities, the structures, and the strictures of our current economy; and build toward a long term vision of people and land in harmony once again?

This last section of the book addresses "social permaculture," or the way we design and arrange our social relations to further the regeneration of the environment around us. Hazel presents their own unique visions for both short-term transition and long-term cultures of place, all of which are stimulating for thought, discussion, argument, and fruitful cocreation. I especially love Hazel's deeply nuanced discussion of gender and the potential for multiple gender roles in a culture. At a time when this topic has become deeply polarized, Hazel offers us a way to broaden our thinking and engage our imagination. Hazel also draws deeply on their own Quaker heritage, with its respect for the Inner Light within each person, its emphasis on simple living and social justice, and its practices of cooperation and peace. And, deeply in alignment with my own lifelong exploration of pre-Christian, earth-based cultures and religions, Hazel integrates the ancient Indigenous stories and customs from Europe, offering a way for those who come from settler peoples to reconnect with their own land-based roots, tales, and traditions.

From how to bundle buckbrush to how we might organize a rooted-in-place society, Hazel takes us on a journey that is theoretical, practical, spiritual, and visionary. This book will be invaluable to anyone living on the land, tending the forests, or simply looking for inspiration as to how we can adapt to climate change, mitigate its risks, and restore the balance. It will, of course, be especially relevant to anyone living in the Pacific Northwest, but its principles and vision can be adapted to many different climates and ecosystems. For homesteaders, permaculturists, foresters, land managers, climate activists, and policy makers, *Social Forestry* is a rich resource.

Storyteller, forester, scientist, poet, and practical mystic, Hazel shows us how we might regenerate both land and culture to become the earth's appreciative eyes, loving heart, and healing hands.

Starhawk
Cazadero, California
September 2022

HOW TO READ HAZEL

Reading Hazel is much like having a conversation with Hazel. A long, sometimes rambling, conversation that goes on for hours. Days. Years. But don't worry, it all comes back around and begins to make sense. As Hazel would say, "Fear not, Dear Reader, we are loopy." Your patience, attention, and curiosity will allow the ideas to gain depth, connect, and come back. You may have to read a certain passage not once, but twice. Possibly thrice. Like any self-respecting elder, this "Trickster Auntie" turned author will challenge you. Expect to struggle—you may even complain—and you will also learn.

It is likely that you have not read a book quite like this before. Hazel comes from a Hicksite Quaker background, based for generations in upstate New York on the Hudson River. Quaker Plain Speech is her first language—modern American English came second—so the sentence structures are different. You will find blunt and plain sentences, (some may even call them abrupt) that are poetic in their own way. Maybe you want to sing the text or commit the short phrases to memory, much like a bumper sticker.

To write in phrases is to demonstrate learning from a culture that is based in place and story. The phrases are shorthand and repeated so that you remember. You may notice many hyphens, especially in the first few chapters. Words connected by hyphens should be read as a whole, like long German words. There are also some short poem-like summations that are italicized and taken from the text, once again to emphasize what is important and to help you remember. You will also see italics when a term is introduced for the first time or when using a Latin name to identify a particular species and genus.

Chapter one will take you through the book and it will tease you. There are plenty of terms that are not defined yet, but they will be. The flow is being established here and the Principles Clipboard can be referred back to again and again as a map of where we are going and a review of where we have been. Keep going! You may just have your suspicions confirmed, new hypotheses generated, scenarios envisioned, and even receive some practical advice for small-scale forestry.

Much of Hazel's authority comes from experience, not only in the traditional culture that shaped their language and storytelling ability, work ethic, and a strong connection to Place, but also decades of experience working all over the world experimenting, refining, teaching, reading, and now writing about the principles of Social Forestry and how to tend the land as people of Place. Ultimately though, the true authority is the land where you live. Hazel will give you clues for the riddles, songs to sing, and advice to chant, but what is right for you, your community, and the forest you are a part of is not something that someone else can prescribe. Only through time, observing, tending, and collaborating with others in your Place can we repair, adapt, and belong.

MEGAN FEHRMAN
AUGUST 2022

ACKNOWLEDGMENTS OF GRATITUDE
A Poem

MATURITY

You are holding the grown-up in your hands.
This is the Elder Report. The elder thanks their Elders.
From many centuries, cultures, contexts, places. Hicksite, Abenaki, Mohawk,
The old country, Adirondack Mountains, three billion years old,
The Village of South Glens Falls, public schools,
Mohegan Council BSA, Troop Nine,
South Glens Falls Friends Meeting, New York Yearly Meeting,
Glens Falls Crandall Public Library.
All the hard stories of that Place, and cemeteries everywhere.

SEPARATION

Not just leaving, but going.
Thanks to the workers, warehouses, distribution to readers.
The book thanks the author, artists, and compilers.
The author is not the book.
All praise to Wisdom and Perspective,
All hail the books in the Bibliography.

DELIVERY

The Book Doulas gathered;
Such a process! Assembly!
All praise to the content collective,
And Siskiyou Permaculture.
All praise to Synergetic Press,
Our anticipation blooms astounded.
Thanks to the tree paper, and soybean ink,
Thanks to the presses, power plants, and workers.

LABOR

The manuscript became a draft.
The draft was revised. Illustrations emerged.
All praise to the editors, arrangers,
Noelle, Amanda, Allison, Jessica, Megan.
All praise to the producers,
Karen, Michael, Mulysa, Hazel,
with the help of many others,
Melanie, Hanya, Elizabeth, Doug, Tom, Vanessa, Joanna,
Isaak, Julia, Jennifer, Kjersti, Kylie, Rosemarie, Sue Ann,
Starhawk, Delia, Heron, Abby, Rose, and Peter.
And all our grandchildren, who distract us magnificently.

GESTATION

A very long one; time warped. The universe tickled and ticked.
All praise to the long run editors,
Gwion, editor emeritus, conspiring since 1969,
Four manuscripts: praise the valiant editors, Arcana, Miku, Jessica.
Larry, Tracie, and other early readers, PLAYA Writers' Residency at Summer Lake,
Wolf Gulch Ranch, Tom and Maud, Home and Place.
Endless thanks for the nurture.

CONCEPTION

The Nature Lodge, Camp Wakapominee, 1963,
We offered several merit badges, wild woods, whole wide wonders.
SUNY College of Forestry at Syracuse University
Took tree loving into laboratories and field stations,
Degrees 1969, and thanks to the State of New York.
The inspiration was fertilized at Laney College, Oakland, 1971,
Where the participants dictated the curriculum: Wild Edible Plants and Woodslore.
Permaculture and Eco-forestry followed,
Evolving through a rainbow of settings:
Design Associates Working with Nature, Berkeley,
Crater Lake National Park, Youth Conservation Corps,
Aprovecho, Cottage Grove, Ecoforestry Institute, Glendale,
Rogue Institute for Ecology and Economy, Ashland,

Heartwood Institute, Garberville, Lost Valley Education Center, Dexter,
Wilderness Charter School, Ashland, Ashland Ranger District, USFS.
Special thanks to the Permaculture Activist, 2005,
Publishing the article "Social Forestry in the Shasta Bioregion."
The Conception thanks the Inspiration,
Facilitators and enablers who nurtured the seeds, gathered the soil.
All praise to the emerging cultures of Place,
Communities, institutions, glorious groves, and celebrations,
Fairs, conferences, and many decades of
Delicious process.

INSPIRATION
Deep in Place and Story, up in the branches of a Sugar Maple,
The view is delightful. All praise to my families,
Fish, Ward, Durrin, Ross, Sisson, Smith, Southwick,
Schmidt, Liedecker, Ryan, Januszkiewicz, Barton.
The ancestors set up this manifestation which persevered.

PRINCIPLES CLIPBOARD

CHAPTER ONE

Peoples of the Forests

The basic goals and principles of Social Forestry for ecosystem restoration include this list. These ethical and precautionary principles will be visited again and again in the stories that elucidate each chapter.

The essential cultural principles of Social Forestry are

1. Working Together brings good fortune.
2. Drainage Basin Councils build and hold culture that reaches for appropriate local opportunities.
3. Cultural Habits emerge from cooperation with Nature in Place.
4. Let the land guide us: we sit in council with All Beings of Place, taking our humble tasks from complex cues.
5. This is a collective effort: we come into relationships that foster cooperation with Humans and the Others.
6. There is honor for everyone: All sorts of Humans have skills and propensities that are useful to the common effort. Every gift from the land is appreciated and celebrated.
7. We bring Human culture into the forest, and Forest offerings into our homes.

CHAPTER TWO

The Lineage of Social Forestry

The principles of what we learn from the stories include

8. The long-epics are Deep! Learn from these about cycles and the nature of time.
9. Our sources are various and need vetting. Narratives are seductive—we love a good story—yet the many dimensions of knowledge are a complexity, as are ecosystems. The whole story is beyond Human comprehension. Allow for humility in the face of immensity.
10. Traditional Ecological Knowledge (TEK) is a collection of stories and procedures that holds wisdom and pertinence for Place-based culture. TEK cannot be mined for narrow advantage; all TEK is conditional and precautionary. We cannot derive or discover knowledge without context. Some TEK survives in epics that have been moved through migration, yet still contain resonance in a newly settled Place.
11. Indigenous Ecological Knowledge (IEK) is ultimately place-specific. IEK does not transport well. IEK becomes unmoored from reference and pertinence outside of its Place. IEK is attached to a particular drainage basin; the landscape itself is doing the thinking and appropriate Human cultural alignment is only attained through long experience and effective memory. IEK must be respected as the sacred assembly of knowledge that fosters restorative tending.
12. Sacred knowledge is a spirit-spoken truth, corroborated by continuous revelation and cultural ceremonies of appreciation. Spirit is aside from our direct knowledge but proves true over and over, and is celebrated in powerful stories and teachings. There are clues to etheric and marvelous dimensions that appear to be emergent properties of complexity, seen as if in peripheral vision, at the edge of perception. This experience of the sublime is the most valuable and essential knowledge of Place and Being that a people can hold in Culture. Acknowledge and respect, learn from the clues of other cultures, but find Spirit where we are; it is present and luminous.

CHAPTER THREE

The Nest-Home of Social Forestry

Ah,
tis a twiggy-messy thing,
yet home

The Principles of Home include

13. Stay home and dress up. Be who we are where we are and celebrate it at home. Wear the appropriate fabrics from the local fiber-shed, made as durable and as lovely as can be. Miss the Place if you go away.
14. It takes a Culture of Place to manage continuous stewardship for ecosystem functions.
15. The Home-Place teaches us how to be. The way that we dance with our Place entrances and delights our Original Mind. We learn as children do.

CHAPTER FOUR

Relationships

The Principles of Relationship include

16. Respect leads to reciprocation; a high degree of implication leads to *balance-within-limits*.
17. We need a relationship to Place. Place needs us to come into the web of relationships that keeps All Beings present and doing well enough.
18. We are surrounded by dynamic natural dramas. The clown is following you. Learn to follow suggestions. Avoid temptation and know the differences between sympathy, empathy, and compassion.

CHAPTER FIVE
Forest Ecologies

The Principles of Forest Ecology include

19. A forest is an intertwined set of relationships anchored on tall woody perennials and stacked from deep roots to leafy crowns with layers of ecology. Save all the parts: we do not know how to assemble one from scratch. We can help forests get started or repaired.
20. Forest types are site, aspect, and climate specific. Forests migrate during climate changes. We can only tinker with these dynamics. Pay attention and look for emergent opportunities.
21. Nutrient cycles are local and elaborate in forests. This net of mutuality supports heavy carbon and water loads, modifies local climate, and allows maximum biodiversity with complex relationships.

CHAPTER SIX
Forestry Work

The Social Forestry Work Goals from Functional Principles include

22. Restore the wildlands: regenerate nutrient cycles and foster resilience and complexity.
23. Sequester carbon: hold carbon on-site in soils and vegetation. Live well within the solar carbon budget (net photosynthetic capture).
24. Bring back broadscale burning: use underburning (after clearing excess fuel) to recycle nutrients, stimulate carbon capture, and preserve biodiversity. Also called "cool burning."
25. Move local forests toward Old Growth: facilitate fire-tolerant stability and complexity by managing the understory.
26. Close the loops: nutrients, water, carbon, wildlife, and soil biota must not be exported or lost to erosion; loop, net, and sink. Beaver, Salmon, and Fire are our icons.
27. Hold water as high as possible in the drainage basin: keep it cool by shading the riparian corridors and building water tables upslope.

28. Bring back the Beaver, especially in the headwaters. Learn to live with their engineering; they create several layers of riparian sponges and sinks; they build for Salmon.
29. Perfect timing: become tuned to the land. The shape of Place holds the stories, and we can learn to come into alignment with potential magnification and elaboration.
30. Use the right hand tool for the job: sharp, well-tended tools are convivial. We work together safely and expediently while attuned to our culture and surroundings.

*These nine principles above
are our work goals,
now we will continue
with further social and cultural principles*

31. Restoration to what? The regenerated ecosystem functions will not be like any previous regime. Tending changes the Wild dynamics. Humans are implicated in co-evolution. Pay attention and stay nimble. Observe and interact.
32. Human cultural work-arrangements are reflective of the drainage-basin ecodynamics. We mimic a dance beyond our bipedal grace. If we cherish all the ways Humans can be useful, we can honor all skills with non-exclusive—but skill-focused—work-guilds that cooperate with the whole-regenerative-effort through drainage-basin spokescouncils.
33. Ceremony and ritual, guided by taboos and etiquette, keep local People of Place oriented. How we go about our tasks speaks to the sensitivity of our attention. Keeping an open-view of complexity benefits from layers-of-reminders, ornamenting our demeanor. Let us, at least, entertain All Sentient Beings.
34. Work skills are learned from example and honed through practice. Re-skilling is the work of manifesting the basic-hand-skills common to Human cultures but forgotten in modern-industrial-consumerism. We need to get back to muscle and attention knowledge (poise). Our bodies and our culture benefit from movement and gathering skills. Ergonomic muscle dance with sharp objects. Work together, but do not get distracted; avoid dangers.

CHAPTER SEVEN
Fire

The principles from working with Fire include

35. Fire is spirit and light. Fire is magical. Fire is powerful. Our Place and Culture are built on Fire and cooperate with Fire. Plan for catastrophe.

36. Ignition in dry-fine-fuels that climbs fuel-ladders to windy-dry forest-crowns can quickly explode. Usually, after ignition, the wild-fire moves northwest on the ground. Blow-up is a massive burn where whole slopes or valleys are vaporized and the air burns. Escape is difficult.

37. Cool Fire can be seen at the end of a wild-fire, as the weather cools and the burn drops to the ground, moving downhill. We can mimic its benefits and learn to work with moderate broadscale under-burns and meadow-burns.

As storytellers are wont,
first a story
to stage the set,
to be ready
for the details
of the practice.

CHAPTER EIGHT
Charcoal

The Charcoal principles include

38. When wood burns, the heat of combustion cooks the unburnt wood and distills out gasses that feed the flames. If we burn up all the wood, the remains are ashes. Ashes are mineral remains of wood. Wood loses (outgasses) carbon, water, nitrogen, phosphorus, sulfur, and other micronutrients, through combustion and pyrolysis.

    In most open fires, some components of wood do not directly burn and outgas as large or unstable molecules (carbon monoxide, hydrogen, methane, turpines—all together known as wood-gas or brown-gas). Wood that combusts completely becomes ash (minerals), water, carbon dioxide, nitrogen and sulfur oxides, and several other simple gasses.

    Pyrolysis is the distillation of wood. In an oxygen-control container (a single or double-retort kiln), combustion can be quenched just-in-time to capture charcoal. The flue-gasses can also be filtered and used as biofuel or condensed for varnishes and turpentine.

39. White smoke is water vapor. Brown smoke is complex gasses. Blue smoke is simple carbon burning. Smoke from an open fire has all these colors. Brown smoke is the most toxic, unhealthy to breathe, and contributes to atmospheric-greenhouse-effect-heating. Different colored flames are usually from different mineral compounds (imagine a Shaman, dusting Fire with color flashes) or mixes of complete and incomplete combustion.

40. Charcoal is stable, nearly pure carbon, the remains of partially combusted wood. The vessel and tube, and sometimes bark, structure is still visible. Charcoal contains less than half of the carbon found in unburnt wood. Charcoal is seldom consumed by Life, but offers a haven and refuge to soil micro-life with its range of tube sizes and textures.

CHAPTER NINE

Treasures from Thickets

The principles of Gifts and Yields include

41. The gift goes around and comes back. Opportunities are offered us that benefit from discretion, so that we are gifted-again. Reciprocation is expected and appreciated. We get what we need, without fear or worry, when we do not take too much. Host rituals of gratitude at every effort.

42. Skimming is accepting the gifts of Nature that we notice as part of our tending. When there are surpluses, and All Sentient Beings are content, we receive opportunities. By converting only by-products of ecorestoration to our cultural-resilience and participant simple-comfort, we bless the gifts with use and treasure the goods with care and maintenance. Skills are our contribution to transformative-work (offerings)—we carry the lump of clay to the wheel, and make a beautiful pot.

43. The Ultimate gifts of Nature, to her tenders, are Beauty, Serenity, and Wisdom. We receive these through forest-bathing, cooperative-work, and cultural-continuity. Our cultural-reciprocation is art through ornamentation, thrivance through Potlatch, and wondrous-appreciation through drumming, dance, and song.

CHAPTER TEN
Forest Shelters

The principles of Forest Shelters include

44. Staying warm in the winter, with a small heater and solar glazing, is easier than staying cool in the summer, which needs heavy curtains, mass to cool overnight, and perfect-timing. Mass is as essential as insulation. Small cabins and cottages, to retreat to during cold and hot extremes, can be built from on-site materials.
45. Most of the year we can live outdoors in open shelters. We need shade and rain-cover with wind-block on two or three sides. Most cooking can be done in open shelters. Many of our common needs can be met with shared, semi-open spaces: bath-house, tea-house, council-lodge, dance-hall.
46. Fresh air, of high quality, is essential to health. Airflow follows a set of patterns that move and capture air. Our shelters are located in landscape-scale air-shed flows that shape our lifeways. Midlatitude Humans have favored aspects: southeast, above fog, below snow.
47. We in the Pacific Northwest have year-round seasons of mild outdoor weather. The work of Social Forestry often involves moving camp and returning to the winter-village. Much of our up-drainage shelter is temporary. A hoop-of-shelters is a necklace on our drainage-basin. Well-made clothes, appropriate to the season, are our most important shelter. We Humans are naked without borrowed-coverings. I told you we were silly—who else has to do this?

CHAPTER ELEVEN

Starting from Here

The principles of Invisible Structures include

48. Invisible Structures are hard to see. We can understand them through naming and mapping. Natural patterns shape Invisible Structures, as well as Life.
49. Social arrangements are learned. We share these ghosts; they haunt us. The stories and ways we relate help us navigate challenges. Imagine a guided change.
50. Home ways need to fit Place. And Time. The celebration of our livelihood is adorned by our fitness in Time and Place. What we do now is connected to the past and the future. Live for the many generations to come; appreciate the ancestors.

CHAPTER TWELVE

Cultures in Transition

The principles of Transition include

51. Beware taking from Nature as if it is owed us. Avoid overfishing, overhunting, and overharvesting, of any sort. It is not virtuous to exploit Nature just because your ancestors were removed, placeless, and disenfranchised. Sit down and think.
52. All forms of social arrangements might best be modeled on natural patterns and built within the limits of Place. The local biological carrying-capacity determines our Culture. Celebrate what we have. Build a new tomorrow. Go Deep-Local.

CHAPTER THIRTEEN
A Place for Humans

The principles of Social Categories include

53. Different genders, sexualities, learning styles, physiologies, and passions prove useful to Social Forestry with support, guidance, and facilitation-without-compulsion. The wide variety of Human interests and abilities allows complex cooperation. Many of these tendencies may have coevolved with Horticulture, gifting us the social structures that support tending. Honor, grief, and praise hold us together. *Diversity is beautiful.*
54. Common traditional patterns of social relations seem to fit ecosystems, as we have ancestors that lived those ways. Even where social categories prove useful to holistic strategies, keeping perspective on unexpected troubles and shining Light, with advice and questions about rules and rigidity, will be our never-ending delight. *Pay attention, but don't buy it.*
55. Prejudice based on superficial reading of perceived markers is shallow and inappropriate. Pre-judging limits our relationships. We do it all the time for convenience, taking shortcuts, but do not forget how easy it is to make mistakes and how much we can grow through learning the stories that nourish our compassion. *Did we notice?*

CHAPTER FOURTEEN
Social Order

The principles of good-social-order include

56. Good-social-order does not have to be class based or hierarchical. It can be transparent and accessible. Good decisions get made by communities when the process matures and all bases get covered.
57. Shapes of social order become traditional when they serve well. Children learn the nuances early. Adopted members (*convinced-friends*) can seem inflexible as they try on what seems to be a set of rules. Experienced categories of participants carry weight. *Weighty Elders, Servant-clerk, Clown, Orator.*
58. Traditional Human social orders, at Home in Nature, are not predominantly egalitarian or individualistic. Membership and ceremonial positions carry intentions of service. Benefits accrue to the community-at-large. Individuals sacrifice (gift) to the whole. Social order carries the whole.

CHAPTER FIFTEEN
Carrying the Bundle

The principles of Ceremony include

59. Ceremony connects Humans to the Great Wonder. When we participate, we allow Spirit to whisper through us. Our daily busy-focus is diverted to enter the larger scope of Life. With our senses now open, we go forward to task with wider opportunities.
60. A ceremonial bundle is a collection of symbols and sacred objects, wrapped up for storage and carrying. The bundle connects us to the greater Creation through sympathetic stories, represented by the contents. Mnemonic cues orient us. Gathering rods into a brindle reminds us how we are All, tied together.

INTRODUCTION

WELCOME TO SOCIAL FORESTRY

This call is to do culture-tending. We will only learn the good ways by getting out there together and experiencing the wild. Then, as we re-wild ourselves and our relationships, we will find some relief from eco-grief, as empires crumble and ecosystems recover.

The whole planet needs some Humans to do headwater-forest repair work. The path to the real work on the land is a deep, well-worn evolutionary way of becoming more mature and wise Humans. This is a way of hope and beauty.

The mountain forests, meadows, and streams on the Pacific slope of North America would benefit from tending and ecosystem functions repair. Logging, grazing, road building, and mining have degraded these headwaters and ridgelines so severely that rain does not fall, runoff does not sink in, and the carrying capacity for native plants, animals, and peoples has failed to show resilience and fecundity. Forests are declining and catastrophes are increasing. What can we do?

Humans can be useful, their *disturbance regimes*[1] can facilitate ecosystem function enhancement. On the west slopes of Turtle Island, our compelling vision is the return of Salmon, Beaver, and Fire. This net of totemic icons implies *ecosystem functions*[2] that support immense complexity and enhance biodiversity. This is the Ecotopian[3] imperative: become People of Place and find a way to heal our relationships, for the benefit of All.

With familiarity and Place-knowledge,
we come together in council
and accept the ultimate welcome:
the restoration of kinship and webs of relations.
The deep feelings of belonging, included in council,
our membership in complexity.

---

1 Moving or adjusting elements of an ecosystem while tending or harvesting, disturbing as a type of tending.
2 The net benefits of complex landscape scale flows that deliver clean water, healthy air, biodiversity, and resilience to disruptions.
3 Ecologically oriented culture in Northern California and the Pacific Northwest.

The forests, woodlands, prairies, brush fields, stream sides, and ridgelines miss us. If you feel the call, start with *The Healing Magic of Forest Bathing* (Plevin 2019); meet the trees wide open, senses tingling. Move deeper in with hikes and treks, camps and trails.

Social Forestry as an intention-of-culture contains a whole set of options that we deploy as needed and appropriate. The important key here is *social*—we do this together. The great majority of detailed momentary decisions are made by guild-groups, drifting or swarming through edges and patches of *ecological mosaics*. The learned principles that could guide these decisions are so vast and numerous that too much thinking interferes with intuition and skilled practice. We also need to learn pattern-language, systems logic. The principles we have collected and experienced highlight the lessons of *Traditional Ecological Knowledge*, land-based knowledge kept in culture through stories.

Observation and noticing are fine-tuned so that mutual aid and translations flow, tickling out *emergent complexities* noticed on the margins of awareness. We pay attention to these essential clues, many of them learned from stories. "Then Coyote turned and looked back, a flash of green from under a log . . . " reminds us of fragile glow worms that puzzled us until we heard and then saw, to our delight. Markers of moisture and complex soil life.

We suggest that you find friends that live nearby, you-all might start culture repair with the practice of Salon, a traditional way to share and discuss, without argument or dialog, where participants speak to everyone gathered, and the flow is connective not contrary. As depth of community enriches through imagination and creativity allowed in Salon, you-all might move on to community inventory, re-skilling gatherings, group woods-work with local mentors, and eventually into passion-groupings we call guilds. As guilds learn to sit in spokescouncils[4], and neighborhoods learn to sit in drainage basin councils, wise-culture emerges—informed by accumulating knowledge of Place.

WHAT THIS BOOK INTENDS

Whatever the shape of our specific Place, there are Social Forestry ways to address concerns and necessities. We all need to learn a lot, and make commitments with others. But ultimately we will not learn from books, even this one—the real work and the details of process are Place-specific. Here we have assembled ideas and inspirations, stories and lists, to tickle the imagination, and act as checklists for council deliberations. Ultimately, we who take up the passion to change the paradigm will need the space, support, and opportunity to do the practice. We will need the guidance of local people, local biodiversity, and local conditions.

This book is not merely a recipe collection, where the reader can pick and choose their indulgences; rather, we want to suggest that all skills and opportunities are embedded in cultural contexts that shape action and involvement in complex ways that a book

---

4   A council of delegates from other councils.

cannot fully enfold. This book offers a sprinkle of stories and ideas, but we also need a flood of dedication and investment.

We hope our stories, examples, lists, and principles lend the reader and their community guidance and a lens for how to think about what can be done. The actual doing is far more contingent on real time and Place—we can only offer grist from what we have seen, heard, expected, and done. We leaven with suggestions and opinions, however, so excuse the sometimes directive (effervescent) tone, where we lay out sequences and linear dispositions. We beg your forgiveness; do not be misled, let your Place do the guiding. As Arne Naess, the Norwegian deep ecologist says, "Pay attention, but don't buy it!"

PRINCIPLES, POSTERS, AND POEMS

You, dear reader or study group, will find pop-outs and side boxes, with the stories and texts, that illustrate and help focus essential ideas and knowledge. These remind us of the greater whole context, and present comments that suggest other dimensions in the discussion.

Our stated Social Forestry Principles are clumped to fit the nature of the chapter subjects. These statements are essential poetic condensations of complexity. They help us learn to think in symbolic ways that are critical to making good decisions when working with Nature. Principles can be reminders of long stories and lessons, patterns and sequences, that we have precipitated into phrases, mentioned or repeated or made prominent during salon and council. Keeps us oriented. A sort of shorthand or gesture, referring to accumulated wisdom.

We use the stories in our chapters, like ornaments spun out from the principles. Nature and evolution have qualities and flows that can be observed and then stated as principles or patterns. Principles are about why, patterns shape how. A culture can create *Advices* through the consensus of experience and accumulate *Queries* that we ask ourselves as we live, work, and mature.

Principles can be seen in mapped relationships and suggested actions illustrated on posters that appear in the photo insert. We will refer to the posters using iconic glyphs that send you back to this insert, where you can view all of the posters together and scry some magic connections. Where we refer to a poster, the reader has the option of looking up the larger whole poster, or just remembering it while reading. In the best salon-style conversation-flows, interjections and pointed referrals act like web weaving, bringing the assembled community into group-mind space and time perspectives, to allow emergent properties of complexity to show us our best options and concerns.

Most of these posters are teaching and council devices, maps of systems views. They came to us through teaching systems courses and presenting at conferences, accumulating as inspired or needed, through almost two decades. These posters have been bundled to support this book—found art that nicely fits the fabric we drape to suggest the work.

Some posters are special-made to fit a text or explanation, such as the Poster of Posters. They can reveal a story pattern, offer a checklist of concerns, illustrate patterns and flows, or list critical information. Some posters are almost like board games; one can tell several stories by following various pathways through the entities, functions, flows, and relationships. Each poster announces a systems-mapping style that is appropriate to the subject. Each poster is dated and initialed by the crew that created them.

Social Forestry courses are taught with different sets of posters, clipped up like clothes, to support the various discussions and salons: a set for Water, a set for Fire, a set for nutrient flows. A deep and ambitious council session goes better when nested in a council lodge or sacred grove that surrounds the discussion with icons of epic knowledge. Pictographs and iconography, calligraphy and rock art, all efficiently prompt the symbolic thought flow of our dreams and imagination. Words and literature can be too narrow or distracting; to truly mature our understanding of natural complexity and our potential for wise behavior, we have to allow the symbols of deep evolution, of natural relationships, to reveal truths of Place.

30 Poster of Posters

Here you see the map of the maps, using our simplified poster icon-glyphs to show the web of relationships among subjects. The prominent posters of this book are Observe, Forest Care, Loop Law, and Now Next Forever. These four central posters propagate radially into sprays of detailed sets of posters. This epic array of information presents the context of Social Forestry and its part in the systems-of-systems universe of natural healing and thrivance.

TRANSITION TIMES

17 Now Next Forever

The Now Next Forever poster is a spiral story line. This poster is derived from an aural story that follows a pattern of revisiting sectors in a sequence of learnings and practices that build to Ultimate cultural maturity. This path, spiraling through sectors, puts a sequence to concerns, prioritizing a transition, step by step, through focusing now, to seeing ahead. We tell ourselves an epic story, while moving up the vision.

We use a triad—Quick Fix, Retrofit, Ultimate—to suggest priorities in sequence. A Quick Fix can be done immediately with materials readily at hand; Retrofit requires planning, education, imagination, and deep council and takes lots of investment; Ultimate is the visionary goal, the outcomes that should not be compromised by Quick Fixes and Retrofits, which contradict and shortchange the mature potential. The Ultimate, demanded of us by the patterns and principles of Nature, is revealed through careful observation, and deep traditional wisdom.

*To become People of Place
is to localize our support networks.*

Social Forestry is one part of the poster deck of concerns and ambitions. Even though restoring ecosystem functions is necessary for sustained inhabitation of drainage basins and tending the wild is a healing and fulfilling way to be Human, there are many other concerns, opportunities, and foundational supports we all need to facilitate in order to do the Social Forestry work. The poster Now Next Forever branches out of the Home sector, into Social Forestry posters, through the poster Adaptation.

 ## Adaptation

The poster Adaptation is a systems map style called a Venn Diagram. Three overlapping domains collect and sort activities and necessities into categories, to help us understand the relationships between different but compatible approaches to forested landscapes. After all these millennia of colonial forest destruction and exploitation, modern men made some philosophical adjustments, through the concerns of Conservation. It is useful to have a philosophical framework that leads us toward living within limits. Conservation, regeneration, and resilience are goals that suggest a set of principles and procedures that hold destructive colonization in check. Conservation is a Retrofit approach to forest care.

*What we need is Adaptation.
Conservation concerns smack of old school inertia and excuse.
Finding a wise way in the new world requires a paradigm shift.*

Horticulture, as opposed to Agriculture, is about tending, magnifying, and enhancing complexity, with appropriate tools and attitudes. This suggests important skills and sensitivities learned in gardens and edges near home, before we are invited to extensively disturb Nature. This is the most productive and intensive ecosystem modeling agenda of small village culture. The work of reciprocal engagement with whole drainage basins through tending uses the skills and tools of Horticulture. There are many books in the Bibliography that are great resources for learning these arts.

Social Forestry is a sector on the Adaptation poster to remind us of the opportunities and priorities of drainage basin–based Cultures of Place. We will unfold this sector with other posters throughout this romp of a workbook. The Social Forestry sector branches into four other posters: 7. Forest Terms, 8. Prescription, 9. Social Forestry Transition, and 10. Social Forestry Terms, which explore the concepts we use to tend the forest.

Social Forestry is landscape-scale tending that needs cultural-scale cooperation and coordination. Forests, woodlands, brush fields, and riparian strips can benefit from pruning, thinning, planting, and under-burning. Cultures of Place gain food, materials, and medicine through eco-restoration processes.

The shorthand phrase for the shape of skimming while tending is "The Four S's": Sequence the work through council, Stage the work with cooperative coordination, Sort the materials into craft categories, and Stash the bundles for winter fireside arts.

We will next visit the poster Social Forestry Transition, or "Soc For" Transition, its nickname.

Social Forestry Transition

This system poster style is a bubble diagram. The entity bubbles are generated through an ideas process, such as brainstorming, and the mapping of concerns and relationships (shown on the spokes and cloud) are worked out by discussion, experience, and research.

The transition to whole landscape wild tending may go through social-cultural shifts in relationships we inherited from the imperial colonization of globalism. These old forms of social trading and overextended supports may not fit in the Ultimate vision. As much as we might want to jump into Ultimate, dreaming of the future, we need to do the hard and messy work of transition (Retrofit) to get there.

For example, the Wilderness Charter School at Ashland High School used this poster to orient their curriculum and practice the field work of roadside stewardship in forests above the city. Much of the phrase-tags emanating from the entity bubbles are key to Retrofit. The functional words on the connecting lines may be temporary arrangements, and the overweening systems challenges we certainly hope resolve themselves, without compromising the Ultimate relocalization.

29 Loop Law

Loop Law poster and Observe poster are foundational and all encompassing, working in tandem with the principles lists and arrangements that map out "How Things Work." They are the cheerleaders, buggy flags, solid truths with wide confirmation.

The Loop Law poster reminds us that cycles, not quests or expeditions, are the way of life on this planet. Dance the circle, walk the pilgrimage, around and around.

1 Observe

Observe posters keep us oriented. Pay attention and notice how things work. Prolonged and thoughtful observation deepens our humility and teaches valuable lessons.

Measure twice, cut once.
Think before you leap.
Be concerned for others.

Somewhere in the mix of posters you may recognize Place-specific pathways, traipsing into the Ultimate. And we need all the posters, arrayed among the prominent posters to orient ourselves to do the real work, where we are.

POEMS AND STORIES

The ornamental and referential poems and stories strung out through this book are rest opportunities from the density of discussion and principle. Particularly pithy parts of the text are italicized and pushed to the right margin. At the end of the book there is a long visionary tale transcribed from aural presentation and edited to be more literate. Here and there are small sketches to spice things up and show patterns.

We trust these pathways lead you into the woods.

As all wild tending is Place-specific in its knowledge and practices, we will continue our Introduction to Social Forestry with an introduction to the Siskiyou Mountains in Southern Oregon where the experiences and examples revealed in the book were born. And first, as at the start of a proper council lodge meeting, we speak words of gratitude and acknowledgement to All Sentient Beings, the Place itself, and the First Nations who preceded and informed our local version of Social Forestry.

PART I

Foundations

The essential cultural principles of Social Forestry are

1.
Working Together brings good fortune.

2.
Drainage Basin Councils build and hold culture
that reaches for appropriate local opportunities.

3.
Cultural Habits emerge from cooperation with Nature in Place.

4.
Let the land guide us: we sit in council with All Beings
of Place, taking our humble tasks from complex cues.

5.
This is a collective effort: we come into relationships
that foster cooperation with Humans and the Others.

6.
There is honor for everyone: All sorts of Humans have skills
and propensities that are useful to the common effort. Every gift
from the land is appreciated and celebrated.

7.
We bring Human culture into the forest,
and Forest offerings into our homes.

CHAPTER 1
Peoples of the Forests

A BRIEF TOUR OF THE SISKIYOUS

West from the Pacific Cascade volcanic mountain range, which runs north and south, is a tangle of geology pasted on the advancing North American tectonic plate. This complexity of metamorphic, granitic, and uplifted ocean bottom is often referred to as "The Klamath Knot."

The Bear Valley corridor leading southeast, past Ashland to Siskiyou Pass, connects the north and south sections of the Chinook Trading Route. This braid of trails and camps snakes over this highest of passes near Pilot Rock, an ancient volcanic dike. This storied trail system runs from Canada deep into Mexico.

The Siskiyous are probably named by Russian miners after women's breasts, as the Tetons, from French trappers, are so named in Wyoming.

The central, east west running, and highest Siskiyou range is also one of the rare high elevation corridors going east west from the Pacific coast, leading across the north south Coast Ranges, volcanic interior north south Cascade Range, all the way to the Great Basin of rifts and block ranges further east. The drainages on the north slopes of this central range feed the Dragonfly's River, which collects the Applegate River at Grants Pass and then cuts through the coast range, gathering up the Illinois River, in a spectacular canyon, on to the sea.

The Klamath River drainage, on the south slopes, runs from deep in the Great Basin, through the Cascade Range and down a complex set of canyons and geological fault lines to the sea, collecting the Scott River and the Trinity River on its way. The Sacramento River, south of Mount Shasta and the Columbia River farther north (and the home of the

Chinook First Nation) also manage to cut through the Pacific coast's great, long mountain ranges, from the interior all the way to the sea.

Here in that mosaic of steep mountain drainages and perched basins have settled many different First Nations speaking versions of almost all the language groups found scattered across the whole continent. These well defined landscape features protected many small villages and clans, connected by river travel and mountain pass trails to neighbors, intermarriage, and trade. The mosaic of language groups isolated in their small domains is more dense than anywhere else on the North American continent.

PEOPLES OF THE RIVERS

The Klamath River is named after the peoples of the upper Klamath Basin. Following the flow, the Upriver Karok, then the Downriver Karuk, meet the Hoopa where the Trinity River joins the Klamath River, where the Yurok carry the Salmon tending out to the Pacific Ocean. The Shasta people carry the Shasta River and the Scott River stewardship. The Tolowa people kept the Smith River drainage sacred. The Rogue River is named from a pejorative slur against the Dhalgelma peoples, who care for the whole river to the sea. The proper name is "Dragonfly's River." Their upriver relatives are called Latgawa. The Dakubetede Athabaskans hold the Applegate River down to Dragonfly's River, and the Tolowa and Dhalgelma peoples tend the Illinois River (named by settlers after a First Nation from the mid-continent).

We here now, admit that we are settlers,
On the unceded territory of First Nations.
Where treaties, retaining sovereignty,
forced on survivors by the US Army,
at the time of their removal to reservations,
were never ratified by the US Senate.

We now need never forget to recognize
that First Nations retain traditional rights of access
to cultural uses of their Places.

Almost all the river names were brought here by the new settlers. Let us never forget the First Nations who tended these rivers and uplands. And these river corridors were crossed by many travelers from other tribes, following the ancient trade routes. The Siskiyous are a very complex web of multi-ethnicity. And this has been going on a very long time.

The ancestors of these small tribes arrived here more than ten thousand years ago. The complex jumble of ranges and valleys also became connected long ago by the Chinook Trading Route, running north and south, while the rivers drain mostly west.

The valleys, running through and out from the Siskiyou Mountains, harbored many small villages, with language dialect clans clustered inside drainages and basins. Lots of small cultures with different language roots living just over a mountain range from the dialect variations spread up the rivers.

The newest peoples to arrive in these river basins are mostly European settlers. Compared to the thousands of years the people they displaced spent tending and loving these special places, they just got here. The "new settlers" have a lot to learn. This is not an easy place to thrive and the knowledges and practices best suited to persisting for so long, alongside the stories that keep this Place-based wisdom embedded in cultural ways of being, are not always honored and valued by these newly arrived settlers.

The new settlers with their extractive and genocidal impacts, came from the east over the long Pacific Coast ranges, across the Great Basin deserts. They did not understand what they found and did not appreciate the delicate balance of tending on this wild mosaic of complexity. The First Nations people had learned many important lessons of persistence and resilience.

In the flow of stories and peoples, traveling through or trying to stay, there would be great healing and wisdom growth in learning about this Place and the Original Peoples. As generators of local new Place-based dedication and culture growth, we want to hear the long story. There is so much to hear about. So much to learn to expect from the larger environs, and the seasons of disaster and abundance. As all North American Indigenous people have taught (Kimmerer, 2013), we should always start any feast or ceremony with gratitude and greetings to the First Nations and all sentient beings of this spirit filled land.

After making our intentions clear, new settlers have a lot to learn. This new storytelling walks us through a scenario of coming home. We always start with acknowledgement and thanksgiving. This needs to be tradition for all local councils. After we thank All Sentient Beings, all the implicate ecology that holds us, let us speak gratitude and greetings to the First Nations past and present.

A complete thanksgiving, a recitation of gratitude, should take days. May we find the way again soon to have this time sense, to adopt an understanding of interconnectedness and co-creation with all our co-inhabitants, to state our intentions of staying and belonging to Place.

Not many new settlers have really dedicated themselves to learning how to be and thrive with this amazing tapestry. Most new settlers have tried to bring their ways from far away places and paste them here without discretion or finesse. We, and our allies, intend to learn to cherish and tend to our surrounds. This is a spiritual and practical task of cultural growth and evolution. We dedicated new settlers will only build our we-tending by becoming People of Place.

WE, OF MANY-ME'S

Social Forestry is participatory, interactive, and observant. We do it together. We practice skills that allow work which reciprocates and appreciates, inspired by principles, which we learned from stories, and then experience in Nature, reinforcing Traditional Ecological Wisdom. This cooperative, coordinated flow of sequence, staging, sorting and stashing, requires People of Place who are sensitive, engaged, and dedicated to local tending, stewardship, and ecosystem functions.

This is work that refreshes. The planet needs some Humans to step into the space and ways familiar to our mythic memory, with the intention of healing ourselves, local culture, and interactive biodiversity. This participation is natural to us, we can thrive simply and elegantly on less stuff and with more heart. Get and keep ready for the work through Forest Bathing.

As we join up and mushroom into mutual aid pods, looking at the work, and considering the immensity, we dig in. The headwaters call us. The trees sing to us. We can reciprocate. Talk back.

THERE IS NO RECIPE

This can't be done mindlessly. Attention is required. Growth and orientation is promised. Local solutions come from prolonged observation, ongoing knowledge gains, cultural process practice, and feedback from initial disturbances.

Tales of Traditional Ecological Knowledge can be imported with new settlers, perhaps allowing some precautionary tentativeness. Local knowledge of Place is gained within personal and social orientations such as stories, ritual, and passages or initiations. Living through the aggravation and shock of paradigm changes demands some comfort. The good news is that old news, the long story—with continuing and evolving revelations and social tools for change— lead us to point back to totem clans and Social Forestry guilds, dancing the inter-species entanglement with Place-specific shapes of etiquette and taboos, noticed and learned, emergent from Nature. We-all get belonging, inclusion, support, and guidance from elders with soul.

*Wild tending
re-engages our natural propensities best
through work together for All.
We can get some relief
from our eco-grief and shock,
through immersion in Nature.*

DOWNTOWN AS WELL

We suggest that occasional, seasonal, guild work can be practiced almost anywhere. In the Pacific Northwest urban woodlands, steep slopes, riparian corridors, untended parklands, and city forests are all candidates for stewardship. If neighborhoods coordinate with city planning, opportunities should rise for more shade, fuel reduction, invasive species harvesting, food forest plantings, and trail systems to reduce erosion, limit trashing, open wildlife corridors, and provide access to ongoing ecological restoration. We may even be able to accommodate semi-itinerant seasonal workers on drainage-basin scale hoop-camp circuits.

The Hoop, a semi-nomadic seasonal sequence followed by landscape tending cultures, should be explained with reference to its Indigenous origins in the Great Basin of the US. When tending is spread out over large landscapes and takes place in seasons and conditions along changing elevations, a semi-nomadic sequence of camps going up country in summer and back down to winter camp allows the tending culture to support its needs while keeping the ecosystem functions tuned up and resilient. Materials and foods gathered, processed, and carried back to the home village keep everyone busy and happy during the winter storms.

The city neighborhood becomes drainage-basin savvy and re-orients to whole-system council considerations. The community learns to cooperate, coordinate, and celebrate together. The landscape-scale goals of the return of Salmon, Beaver, and Fire apply to the whole region and inspire local stewardship. House-holding needs to be the new economy, reducing consumption, recycling energies and materials, and making do with flair. As we learn to cherish local resources and vernacular styles, we will celebrate ornamentation, ceremony, mentoring, elder oversight, and community/personal passages.

ACKNOWLEDGEMENTS ABOUT PLACE

One of the slogans for the intentions of settling in a newfound Home is "Indiginate to Build a New Culture." Here in the Siskiyous, we hope that allies will join us to support the still present First Nations in eco-restoration and the return of Salmon, Beaver, and cultural Fire to our beautiful mountains and river basins, so that all Beings of Place can thrive. The following comment recognizes and celebrates the First Nation Peoples of the Siskiyou Mountains. This is an example of following a *Basic Call to Consciousness* (Mohawk 1978), an acknowledgement of Place and the long story.

Only First Nations peoples get to re-indiginate.
They return to the homeland that they know so well.
New settlers need to indiginate to become People of Place.

WE START THE PRACTICE, AND CARRY THE VISION

Social Forestry gathers the vision, the people, and all participants of Place together. People of Place learn to steward a drainage basin to ensure the return of Salmon, Beaver, Fire, and thus resilience and persistence. All the Non-Human-Beings, across the landscape, come into relationship with Human tending.

This is inherently social.

Practicing place-specific Horticulture, guided by Indigenous Ecological Knowledge, is a way of coming into the *right-relationship*. We reciprocate the gifts of Life humbly, while we acquire a vast skill set and deep subtle knowledge.

Social Forestry cultures learn from Indigenous cultures (First Nations). Creation stories, trickster stories, epic tales, and deep memory is an ultimate gift if we are so blessed. We also learn from careful observation, assembled books and hard-copy resources, apprenticeships, mentoring, community salon and council, working together, direct feedback from Nature, patterns and systems savvy, reading the landscape, deep-ecology ethics, art, dance, song, epics, science fiction, and celebration.

So much to learn, but this is ultimate dedication to Place-work for all our sakes. Our communal intentions and mutual support nets will get us going. All of our disturbances, ameliorations, and culture changes that People of Place employ are site-specific. We learn from the land.

Social Forestry is the culture of forest people. The forest is the source of multiple benefits. A society that understands the value of the forest will be persistent and sustainable. The culture that embraces *Social Forestry* lives in close identity with the local slopes, glens, and drainages: a Place.

Although the examples and stories used are centered in Ecotopia, the Pacific slopes, west of the north-south tall mountain ranges that contain the maritime influences and shadow the dry Great Basin—all on the edge of the North American continent—we trust you will find resonance and modeling that proves useful where the forests of any context persist.

Social Forestry is a cultural fabric woven in Place. The ecologically managed forest delivers surpluses of seasonal products to a culture that uses those benefits in daily life. The community of Humans finds the variable work and materials—and the turn of the year—to be an organizing flow for festivals, rituals, and household practices. This natural economic life is as diverse as the ecological mosaic of the forest. The ethics and principles of the culture support a continuous, ongoing interaction with the forest.

One can see long-established Social Forestries in traditional villages around the world that have forested drainage basins. Seasonal celebrations and product rich lives have thrived over generations from deep time. The forestry practices are learned in context and taught by stories and apprenticeship. Both the Human culture and the forest context are thus stable, resilient, and healthy.

Intertwined.

To be Human is to participate. We are social animals just like so many other cultures, Wild and domesticated. We naturally yearn to be in relationships with all our neighbors. The Forest and the shape of Place hold our stories, the epic tales of consequence and entanglement. As we manage to live in cultures of Place, we easily come into a mesh with gifts of each season. We Humans have been co-evolving with Forests for millions of years and have deep remembered patterns of proper Being. The Forests and Beings of Place miss us and welcome us with our good hearted attempts at cooperation and our silly entertaining ways. If we open up the space, the neighbors come around to see what we are doing and comment on our style!

Social Forestry, done well and attentively, holds the potential to heal the landscape and our relations to Place. As part of our tending, we will find opportunities to "skim" natural processes without betraying our principles and best intentions. We can live beautifully, simply, and well, sharing the bounty of a vibrant drainage basin.

We can restore the ecosystem functions and restore our spiritual and social cultures.

The challenge is the Transition. Modern colonialist consumers are blind to their condition. Times are changing and much needs sorting out. The truth of the extractive empire, for a start. Most settlers are displaced people themselves, moving into occupied territories and displacing the original people with inappropriate means of extraction for industrial speculation. And that is the shorthand summary: Displaced folks displacing other folks without learning where they had arrived and how to behave properly in this new, strange stolen land. In some cases, the locals actually offered to teach cooperation and limits but the driving force of Empire predominated, destroyed, and took dominion over the spoils.

This requirement of humble submission to limits is tough on the opportunistic, peripatetic Humans. But the gift of language and culture brought us ways to cooperate and share. An ordered local culture of settlers can grow into the wisdom of careful tending. The process of becoming a people of Place is not impossible. New settler cultures that show restraint, caution, and humility can eventually come into supportive mutual aid with nearby remaining First Nations and nearby drainage basin councils. New cultures do not re-indigenate, that honorific is reserved for First Nations folks. Settlers, at best, are absorbed by Place and seem to disappear, melting into the ways of Place. Think big, we are talking Ultimate Balance here.

*The gifts of long suffering and memory,
ecological etiquette and taboos,
reinforced by elders and clowns,
keep the circle dance turning.*

The collapse of complex societies is a repetitive pattern in the epic tales. A concentrating and extractive empire eventually implodes under the weight of postponed infrastructure repairs, speculative exuberance, unbalanced debt bubbles, and excessive complication. The global consumer-industrial spasm is doomed. The present appetite tries to swallow two earths constantly while perched on only one. The future of institutional and corporate collapse and chaos is messy. How else does an empire fall? There are social and relational choices that we can make, and dedicate ourselves to the repair and restoration of forested drainage basins.

The very well-used traditional way to survive and rebuild is called re-localization. The goal is to dig in and dig out, surf the random salvage opportunities, and dodge the waves of disaster. The ecological term for Places that support survival and persistence is *refugia*. With a bit of old held wisdom, and the clear intention of being useful, we will find wonder in the work of Social Forestry.

THE CONDITIONAL TENSE
A local homily during a plague

Here we are without much certainty. More observation is recommended.

And we are writing in English, an authoritarian mess. Let's do our best to condition our discussions.

We can make tentative observations about what we are experiencing but we should offer those as gifts to our community.

Political and systems observations are appropriate as long as we place them in the context of discussion. Can we do this politely?

The poem we chant before lighting fires in Social Forestry is this:

> *We are idiots!*
> *We do not know what we are doing!*
> *Please forgive us.*
> *We are doing the best we can.*
> *We accept feedback.*

I am finding that if I do not send out a frantic e-mail and let it compost, it transforms quickly and is not recognizable in two days. We are experiencing a storm of useless information. And when we want to blame someone, or want to demand that someone else does something to fix it, we should be careful. We need to change ourselves.

The big overly complex systems are failing in so many ways. This we should not be surprised by. They were over extended and false and rigged all along. We might want to study David Fleming's *Lean Logic* (2016), which suggests that we can live better simpler lives. I am feeling and experiencing a phase change. By systems thinking, we cannot predict the future, especially from our position in this fast changing now.

Nonetheless I want to predict *relocalization*. This is serious stuff. The most important work we can do right now is local and social. Let us find ways to support each other. And plant big gardens.

I am surmising that spending a lot of time complaining about big bureaucracies and political positions will not get us very far unless we are really lacking entertainment.

(continued)

This is the crazy time. We are living in a psychic storm. We do not have the science about this pandemic that we need to make our plans.

And we are isolated and confused. At least I am willing to readily admit such. But we do, as culture change artists, have the systems training and design practice to pay attention without jumping to conclusions. Let us be kind to one another. Compassion is a good perspective. Sympathy and empathy is real and very natural. But keeping our own center and staying in open observation is key to being useful to our communities.

I trust this homily will be helpful. I am hopeful that we are all prepared to assess reality carefully. Take a deep breath and go for a walk if you can, in a forest.

The spring flowers are lovely and we are still getting some rain.

All the best to you and yours!
Hazel of Little Wolf Gulch

TRANSITION TIMES

In times of chaotic change, the myth of the Tower of Babel comes to mind. The languages diversify and fragment as the tower falls. As the Quakers say, *"The book, it will perish, and the steeple will fall, but the truth will be there at the end of it all."* Just as we now ground ourselves in local truths, for reasons of persistence and preservation of wisdom, so have remote communities in the past. *Ecosteries*, monastic tenders, are one model of redemptive tending. These refugia of knowledge can be an intentional community of preservation and practice, a *life boat* for turbulent times.

Worldwide communications between localized efforts for global mutual support and connectivity should be facilitated to coordinate the planetary efforts that restore ecosystem functions and mollify this looming sixth great extinction. There is worldwide work to do in forests. If we can connect several dimensions of Human culture and weave a web of common effort, we have an opportunity for healing.

The local transition efforts of retro-feudal manor farms, hoop culture camps, drainage basin councils, and community inventory can be supported by federal stewardship contracts on the commons, caretaker cottager cooperatives, inter-agency science and modeling, non-profit research and advocacy, international earth repair congresses, and educational campuses re-oriented toward greater coordination and a sharing of common goals between departments. Although these overwhelming imperial chimera may evaporate, leaving us all to our own efforts, keeping appropriate relationships on eerie, wide, and global scales can help our local truths find confirmation and reflection.

Subsistence cottagers, nomadic graziers, and Indigenous horticulturalists have been removed from the commons, in waves, for centuries (Maitland 2012, Berman, 2000, Martin, 1982). In these times, healthy villages that had protected and maintained natural resources and did not contribute to the commodities-based global economy, have disappeared. National parks remove and exclude local, traditional people, their cultural activities, and access to their sacred sites.

With a bit of enlightened subsidy, or at least no taxation, knowledgeable local stewards who are allowed to remain will deliver ecosystem services and contribute to global eco-restoration. Governments need to focus on supporting re-localization and de-growth, especially in the face of climate weirding. So perhaps the time has come to turn around a great disgrace and destruction.

Support Indigenous tribes and resettle wasted drainages.

Modern ecological sciences have taught us how the forest works. Can we learn to apply the same principles to our economic and community lives? Can we learn from the forest? The possibilities are both culturally enchanting and rationally intriguing as we consider the challenges of supply chain collapses and the decline of easily available industrial feed stocks.

Economic globalism is failing: re-localize and become People of Place!

SOCIOLOGY IN THE FOREST

Social Forestry explores the shapes of relationships. With experience in the woods, access to deep narratives of old forest cultures, and our visionary scenarios, we nurture the entanglement of mutuality in Forests. This is peoples moving toward the future, working from ecological, spiritual, and sociological principles to imagine and remember new/old ways of living with and tending complex forested landscapes.

Most upland drainage basins have been seriously degraded by industrial extraction, the removal and destruction of wildlife, and exclusion of First Nations' *Traditional Ecological Knowledge*. As every Place is special and specific, we will present paths and recipes as examples of how to get there from here. We will sketch and imagine the potential, reporting on the experiments that promise healing, restoration, and cooperation, to rebuild natural resilience through nurturing complexity.

The Human social opportunities and traditions of working and living in the forest have co-evolved with changing landscapes in a multi-dimensional context of time and place. As we learn about these traditional and re-invented ways, moderns can rediscover Culture and Place and come again into immersive and reciprocal relationships with Nature.

First and foremost, we do forestry and Horticulture in groups of cooperators. To do this socially means to avoid both industrial methods (loud machines) and looking exclusively for a single product or raw commodity through extraction. This also means avoiding ways of working that separate us from our natural cohorts and from sensitive consideration of all the dimensions.

*On the land,
all senses open.*

As Social Foresters, we are returning to social relations with the natural entities that comprise a forested Place. Lots to learn from careful observation and from working together to celebrate social skills and cultural ways, such as craft specialty guilds, clan and family alliances, deep remembering, and visionary story telling. Collective decision making, spokescouncils, etiquette, and taboos are the shapes of good Human social order. Ecology is social, relationships are the web that holds and nourishes us all.

This is a way to escape the hyper-individual "self-improvements" and the compulsive consuming that "globalism" has infected us with, and subjected us to. Instead, we can reach to find comfort in dancing endless cycles, and with just enough Human artifacts. How much material consumption do we really need to feel engaged and useful? How can we work together to do eco-restoration in ways that also supply our basic needs (not wants)? Can we find humility through tending the complexity of forests where we will be learning to return to the seasonal cycles of traditional Horticulture?

We re-inhabit our deep ancestral ways, with homes full of hand processed forestry products. Our seasons are marked by festivals and ceremonies that rekindle our relationship with the land. The cycles of the sun, stars, and moon keep us oriented, to expect the next activities. We will find our yearly hoop trek route, that takes us out and up, to seasonal camps with various work and celebrations—all appropriate to the whole drainage basin—so that Fire, Beaver, and Salmon return. We go on pilgrimages around our home watersheds, tracing the ridgelines and around the sacred mountain, to rekindle our dedication to Place and wonder.

Let us clump up and hold the shapes of natural systems, in the light of close-held stories, so that we can practice good timing and useful disturbance. There is a lot to learn about the "Other" and to come into a reciprocation that feeds souls. We thus allow children, through nature immersion, to have the set of experiences, rights-of-passage, and symbolic repertoire that lead to adult considerations of social-ecological-wholeness and time (Shepard 1998).

PRESENT CHALLENGES OF TRANSITION

We raise hope that in the short term, drainage basin councils will deflect any threatened privatizations of some of the few remaining Commons: roadside and public forests—state, federal, and county. If local non-profit cooperatives, bundling together emergent cultural entities, can successfully bid and contract for long-term stewardship, we can then secure ecosystem functions that impact whole river systems and drainages. We are more likely to persist in our restoration efforts if we can dodge accelerated extraction facilitated by industrial greed and the worship of growth without limits.

This is whole systems thinking, reaching for cultural co-evolution beyond modern consumptive lifestyles. Perhaps covenant-contract based conservation areas or perpetual public trusts can be explored that also recognize Indigenous peoples' original rights, traditional place-based cultural practices, and *TEK*. Perhaps local cooperative re-organization can begin with the re-establishment of the best first treaties between settlers from colonial empires and First Nation peoples, with the offering of mutual aid treaties. Will drainage basin councils facilitate the emergent new Peoples of Place?

Stay tuned.

De-growth or post-growth thinking moves beyond many environmentalists' ideas of "sustainable growth." Some de-growth theorists still imagine a lot of leisure and fair distribution of existing goods and services, as if industrialism and globalization can continue in some form. This politically expedient expectation seems to ignore the need for restoration processes that heal. The harms and displacements accumulated through colonial industrial extraction of raw commodities and the extensive transportation and pollution necessary to "rationalize" financial capitalism are in dire need of mitigation. Fluid capital and mobile labor, forever? Really?

Who actually enjoys relegation to redundancy, forced migration, or internal displacement, not having a home? Humans like to think we are useful and engaged with each other in common goals. This Social Forestry offers the joyful, and mindful, work of simple house-holding economies and the collective (Something for everyone to do!) seasonal work of drainage basin and woodlands stewardship. We re-discover the creative vernacular grounded in local values, using the surplus materials and flows of the restored landscape's resilience and fecundity.

Industrial commodity extraction for mercantile consumption by fragmented and alienated individuals is based and founded on constant and forced economic growth. This false accounting has led to *Peak Everything* and the terrible pollution and degradation of the planetary net of interactive ecosystems. Gaea may reject Humans as a failed species.

To repair and restore essential ecosystem functions—such as healthy water, air, and soils—that support downstream cities, we need to close loops, recycle nutrients, and avoid mistakes in the headwaters. Use precautionary principles! Move immediately to de-growth, steady state economies, and cooperative planning to preserve and enhance the Ultimate Commons: forests, river basins, carbon based soils, air sheds, and the Horticulture of persistence. We do have some support from modern science, especially the fields of Ecology and Anthropology, with resources in the bibliography. This is a social challenge more than another mathematical or mechanical opportunity. Local peoples, especially forested drainage basin dwellers, are key to re-establishing the anchors of fertility and facilitating water table recharge across extensive geographies.

SOMEWHAT TRANSPARENT ASSUMPTIONS

We do not intend to promise more opportunities for the monetization of Natural Capital, the depletion of the true sources of the wealth concentrations that built modern civilization. Our Social Forestry is meant to supply the replacements for capital flows with material, social, and crafts flows. We celebrate frugal house-holding. In these following chapters we explore the visions and goals of re-localization, tending the commons.

Thus, we assume the need for steady-state economic policies and not-for-profit business arrangements. Trade and gifting of value added specialty products (such as tinctures, baskets and fiber crafts) unfold when community reliance is shifted from global to local. When we use the term *Transition*, we are suggesting lower cash flows, home based economies, and managing for the highest average health of the biggest pool of diverse participants possible. Our systems-thinking ultimate goal is resilience through the multi-dimensional complex entanglement of relationships.

We expect that Humans will recover interest in rewarding physical labor that is skillful and varied, with ergonomic as well as psychological benefits. We expect the community to find work and value for a wide range of personalities and propensities. We hope that families and villages arrange useful work for all ages and all sorts of relations through the Council of All Beings.

So far, in the late imperial era, sustainability only seems to mean new profit sources and continued concentration of power. Massive personal and family debt and insecurity have crippled truth and vision, limiting our common prospects. To step away from the great speed up and rentier skimming, we need the gumption of positive visioning and remembering and we need to appreciate the real natural world.

*That takes education
and access to practical information.*

The extra-basin export of the value-added materials of eco-restoration are limited in scope and vary by landscape and Human cultural skill. The goals of resilience through fecundity manifest incrementally, and we need accurate monitoring to stay within bounds. There are limits. After so much damage, lacking balanced reciprocation, we can use the concept of reparations, to show humility and stay sober in our budgeting.

Eco-restorative emergent culture is not only rural, it also happens in urban areas. Even brown-fields (abandoned industrial wastelands) can be remediated. Fungal and bacterial inoculation (Stamets, 2000) leads to the return of trees and wildlife. We need wide understanding of what can help Humans and culture move on from here. Landscape-scale scenarios suggest that we need everyone educated about natural processes and we need place-appropriate cultures of tending that support each other, along with trade wisdom and specialties. The transition first-adopters are counter-culture folks and social activists, who are already alienated from modernity and are willing and motivated by love and careful study, to make the personal and *Ecotopian* investments in positive and inclusive change.

The audience for these messages is everyone.

The principles of what we learn from the stories include

8.
The long-epics are Deep! Learn from these about cycles and the nature of time.

9.
Our sources are various and need vetting. Narratives are seductive—we love a good story—yet the many dimensions of knowledge are a complexity, as are ecosystems. The whole story is beyond Human comprehension. Allow for humility in the face of immensity.

10.
Traditional Ecological Knowledge (TEK) is a collection of stories and procedures that holds wisdom and pertinence for Place-based culture. TEK cannot be mined for narrow advantage; all TEK is conditional and precautionary. We cannot derive or discover knowledge without context. Some TEK survives in epics that have been moved through migration, yet still contain resonance in a newly settled Place.

11.
Indigenous Ecological Knowledge (IEK) is ultimately place-specific. IEK does not transport well. IEK becomes unmoored from reference and pertinence outside of its Place. IEK is attached to a particular drainage basin; the landscape itself is doing the thinking and appropriate Human cultural alignment is only attained through long experience and effective memory. IEK must be respected as the sacred assembly of knowledge that fosters restorative tending.

12.
Sacred knowledge is a spirit-spoken truth, corroborated by continuous revelation and cultural ceremonies of appreciation. Spirit is aside from our direct knowledge but proves true over and over, and is celebrated in powerful stories and teachings. There are clues to etheric and marvelous dimensions that appear to be emergent properties of complexity, seen as if in peripheral vision, at the edge of perception. This experience of the sublime is the most valuable and essential knowledge of Place and Being that a people can hold in Culture. Acknowledge and respect, learn from the clues of other cultures, but find Spirit where we are; it is present and luminous.

CHAPTER 2
The Lineage of Social Forestry

NOTES BEFORE THE EXPOSITION

We readers drift so easily into utilitarian considerations. The linear hypnosis of the alphabet channels our attention. Opportunistic impulse has kept the species going, at best with some sense of stewardship, yet that is not the whole of social relations. Inclusion of the many dimensions of wholeness is elsewhere, held in the wider context, while we follow the sentences of text. Storytelling in an embellished council lodge with special effects gets closer to thanksgiving than cocooned book time. So "Bear" with us as we reinforce overly narrow narratives in order to expose threads of lineage.

Consideration also means the care of "Others." A humble survival, especially during Transitions, while helping foster thrivance at the edge of chaos, is our properly aligned Ultimate Intention. This has been shown to be best done in the community of Place, supported by the practical reminders of ritual, while we widen our attention and clear our clutter to achieve presence and compassion.

Read these stories please, in order to understand subsequent stories. Here are our fragment-threads from the greater tapestry that we can only imagine.

THE PREDECESSORS

There is an academic study of Social Forestry that focuses on protecting the relationship of rural villages to the forested drainages that supply water, fuel, and cultural materials, to support the local needs of marginal farming. There are programs in South Asia and Africa that study ways to facilitate traditional persistence in spite of the demands of corporate and national ambitions.

Community Forestry is a term that refers to communication and coordination between rural residents, corporate logging, and government agencies (Charney 2007, Lenentine 2006). Forestry is a term that has come to mean "getting out the cut": a focus on timber to the detriment of other landscape values. Social Forestry in Nepal, India, and Kenya in the late twentieth century looked at reconsidering forest in urban areas, on temple grounds, and near villages to work toward supporting local culture and ecosystem services (Dove 1995). The main activity still seemed to be timber products—replanting forests with exotic species such as Eucalyptus and Pines in plantation groves was a compromise between rebuilding forests for watershed functions and producing commodity wood products for export and industrial feedstocks.

Social Forestry in poorer countries focuses on reducing fuel wood cutting. Deforestation is the ongoing result of landless poor farmers desperate for cooking and heating fuel. Improved cooking stoves, such as rocket stoves, reduce fuel wood cutting and help preserve forests (Hyde 2000).

Ecoforestry is a term that refers to attempts made to mitigate timber extraction with adjustments that intend to preserve remnant biodiversity and ecosystem functions (Pilarski 1994). This includes leaving stream side buffers to reduce water flow impacts, a few seed trees standing in barren clearcuts, and replanting industrial forest ground with more complex species matrixes. The goal remains timber production but with more ecologically aware methods (Morsbach 2002).

Analogue Forestry, developed in Sri Lanka in the 1990s, combines Permaculture principles and cultural forest values to increase biodiversity and re-establish multiple product cultures. A link was forged between temple ground native plant and animal refugia, and barren industrial forest plantations, using epiphytic plants to attract native birds and insects and the native seeds and inoculants that can restart native forests under the nurse crop cover of industrial forest species. Once native forests have begun to recover, the exotic plantation trees can be removed.

All these alternative forestries were still focused on timber production. The social component seems to have been used to soften the impacts of ongoing timber extraction. In the Pacific Northwest community, forestry-type efforts retreated in the face of funding cuts after extensive timber harvesting moved to Asian and African clearcutting; the social and ecological mitigation values lost support. Environmental preservation non-profits continue to receive funding and government policy support to try to protect remaining remote forests. However, because Social Forestry multiple product forest tending does not fit into the monocultural commodity extraction supply lines, there has been less and less interest in supporting local communities and their watershed forests.

Often, a traditional people becomes separated from access to accustomed use of wide lands, by government redistribution of extraction rights for timber, minerals, and water, towards export-commodity corporations. This fracture can lead to the failure of traditional settlements and add to the flight of refugees to urban areas and resettlement camps. With the loss of traditional ecological management, downstream effects bring loss of quality drinking water, increased flooding, accumulated chemical toxicity, catastrophic fire, and compromised air quality to urban areas. Short-term extracted values compromise the long-term regional environmental qualities. The counter-balance is the work of Transition and mitigation with ameliorative planetary consequences.

Similarly, there is a movement to motivate and facilitate Ecosystem Restoration and Global Repair through non-governmental organizations and local non-profit efforts. This wave of activism and good intentions is necessarily overseen by local drainage basin cultures and is being fostered by communications/interconnections and available volunteer help from urban areas.

Meanwhile, all displaced Peoples have ancestral knowledge of how it was when the forests were still our original homes and we lived in small villages, with mixed seasonal activities. These roots of Social Forestry are fragmentary, but some American settler communities have preserved intriguing remnants of forest cultures. A wide range of influences have come to North America from various worldwide forests. We have a selection of these displaced stories to cling to as we learn what to do, where we are now. Ultimately all settlers need to learn the stories of what happened on these emptied and stolen lands.

THE REPORTS OF FIRST NATIONS HORTICULTURE ARRIVE

Where we work in Southern Oregon, we're on the Pacific Rim of the North American continent. If you look at a map of Indigenous languages on the West Coast and compare it to the rest of the continent, the granular language root complex of the West Coast suggests there have been escapees from older cultures, landing on the West Coast over tens of thousands of years. We have a lot of mountainous drainage basins to settle in. Over immense time, this incredible mosaic of cultures co-evolved with the forests and ecosystems of the Pacific slope to such a degree that plants changed their habits to better thrive with human management.

It's a fantastic flow of coevolution, and the major tools they used were baskets, stone tools, fire, and digging sticks. The digging stick was a fire hardened, Mountain Mahogany, or some other dense rosewood branch or stump sprout, that was used to turn over turf and pop bulbs out of the ground. That sort of cultivation is turn over and back, not mixing or inversion. The best coevolution example is *Brodiaea*—related to Onions and Garlic.

Some *Brodiaea* bulbs became accustomed to vegetative reproduction (facilitated by digging sticks) to the point where their seeds are not readily viable. All those little bulblets around the main bulb would stay in the turf when the digging stick turned over the sod and popped out the fat center bulb. Now the bulbs of *Brodiaea* are dropping deeper and deeper into the soil and fat bulbs are hard to find, because there's no human intensification. Geophyte is the inclusive term for all bulb plants, many of whom are essential to Indigenous diets.

Proving that tended ethnobotanical resources have adapted to their tending is very difficult and would require DNA and archeological studies. The hypothesis generated by these proposed studies would still need extensive experimentation as epigenetic changes are also difficult to quantify. We are left to our surmising, to do the best we can with tending.

When we look at any situation that includes "wild" ethnobotanical species, such as bulbs, we are trying to learn how to proceed from here. Some speculation is warranted. Holistic management demands wide considerations, and our cultural and personal insights are worthy of council consideration. The vast and inclusive mind of Nature works through connections—we Humans may be in line to transmit emergent ideas that turn out to be very useful. We have had a hard time researching, thinking, testing, and arguing about evidence in reductive ways; the linear hierarchy seems dysfunctional. There are other ways. Observation, noticing, mapping, singing, and dancing, waiting for some light to shine.

We suspect that *Brodiaea congestum*, a small tasty Lily family bulb, shows some co-evolutionary adaptation through tending by digging sticks. We now, centuries after the loss of active tending, seldom find masses of bulbs near the surface. We search for seedlings with no luck. Most bulbs we have dug up were too deep for sticks, embedded down in our hard adobe clay. The best mats we have harvested were found on side cast bulldozer soil, downslope of twenty year old fire breaks. Disturbed ground, scraped and turned over. And somehow, there is this bounty.

This suspicion is tickled by reading and hearing about flipping over thin sods and shallow soils with the digging stick, pulling out the fat bulbs and leaving the bulblets tucked back in under a re-flipped sod. Disturbed ground-kept-fluffy favors bulblets (vegetative reproduction) over seeds perhaps. The Lily family seeds have short storage lives in general and their seeds need special care in keeping.

Then we need to consider the other bulb tenders. Gophers move a lot of bulbs and dig storage tunnels. Thus, some uneaten bulbs push up from the subsoil. And the newly arrived, introduced Turkeys are taking almost all the near-surface bulbs. Both of these may be in high populations for lack of hunting. When we speculate, some understandings come from complex noticing; the mind feeds back some thread or clue we can take to council.

This landscape on the West Coast was inhabited by many small cultures practicing a local Social Forestry, a complex dance within a tending ecological mosaic, ranging up and down elevations and river corridors, with all kinds of products and traditions. We are blessed to have some reports at last that give us good insights to becoming good stewards of Place. The early ethnobotany investigations were mostly by male academic anthropologists doing fieldwork. The lists they made do not have cultural information, they are simply a list of plants used. Only a peek into ancient ways of living.

There are some diaries and journals that contain a spattering of social observations but they are the notes of folks looking through European cultural lenses and need interpretation and skepticism. In the late 20th imperial century, the Boaz branch of anthropology[1] matured and started to be interested in what fragments remained of tending and processing cultural materials and foods. This eventually included women academics who talked to Indigenous women, through which different stories began to emerge.

In the late 20th century and early 21st some women, including Indigenous women, made it through the doctoral dissertation defense and did documented field work that is revealing a lot of ignored memory. We have seen a rediscovered treasury of ideas and specific examples of landscape-scale Horticulture and broadscale burning, the traditional provenance of women, in these recent contributions.

One of the most valuable readings I have recently encountered is Nancy Turner's *Keeping It Living*. The anthropological essays in this book, submitted by various contributors, report on broadscale burning and what was seen by early colonists. When some European colonists got to the West Coast and looked around, they had no idea what they were looking at. They decided that the Indigenous people were lazy, had so much food, they didn't do anything on the land, and therefore the land was available for colonization.

There's imperial law about this and a body of literature where White People declared that the Indians were not using the land—look at it! Empty and under-exploited. Obviously the Indians were not using the land. So they wrote off all their rights and customs. Turner goes back and renegotiates the treaties from that time— which were not treaties, but exploitative surrender terms—laying out their flaws. This book was published in 2005, the same year of publication as *Tending the Wild*—a wonderful tour of what was probably going on before the European invasions of California—by M. Kat Anderson, and refers to Anderson's earlier work quite a bit. M. Kat Anderson's work is anecdotal; she documented people she was living with. She writes, for example, about growing a bow out of a juniper tree and other sophisticated tending.

Nancy Turner, through research over decades, found that intensification of semi-wild production was always marked. Every single patch of enhanced yield: roots, herbs, berries, nuts, and Cedar trees were all marked territories, named and owned, recorded in oral history. It was a complete and continuous record of stewardship that is being reassembled

---

1   Early 20th century academic interest in capturing disappearing First Nations.

from oral history and is now the basis of First Nations, especially on Vancouver Island, getting back eco-restoration responsibilities and opportunities; getting back into a usufruct relationship with their ancestral landscape.

We're paying attention at last to how they were doing fertilization, transplanting, weeding, and seeding, using digging sticks, cultivation, and pruning; how every tended Place had a name, had an ownership. Not ownership like we think of in Europe but designated harvesting and stewardship responsibility.

This advanced Horticulture creates nutritional balance in a high Salmon diet, with plants like *Oxalis* roots (Sour Weed) and *Potentilla* roots providing certain amino acids and vitamins that are lacking in a seafood-only diet.

I would also recommend *Indians, Fire, and the Land* (Boyd 1999), a more expensive book from 1999 that contains a lot of social record—how we work together as a Human culture that persists, traditional knowledge, and the misunderstandings of early European visitors who witnessed the remnants of deep and old burning practices. This is what we have for a peek at broad-scale landscape burning.

Robin Wall Kimmerer, of Algonquian culture, added *Braiding Sweetgrass* (2013) and reports on work she did at SUNY College of Environmental Sciences, formerly the College of Forestry (from which this author holds a Bachelor of Science degree).

Robin showed that many semi-wild ethnobotanical gardens, such as Sweetgrass meadows, are accustomed to disturbance and harvesting, but wane and fail to thrive when ignored or "protected." That thesis was not easy for her to defend: "Is this Science?" she was asked. Tao Orion has added *Beyond the War on Invasive Plants* (2015) to show a better way to think about conservation, restoration, our relationship to Nature, and the mistakes that have been made. All these women are reminding us of our universal deep ancestry of Social Horticulture and the lovely, productive relationships elaborated on a natural template with an unending glory in view that haunts our long Human memories.

NOTES ON TEK, IEK, AND TSK

These ideas are abstract map/poems at the meta-level of systems thinking. We are thinking about thinking about systems.

These terms are about traditional knowledge. When we say science, we refer to observation skills and acquired knowledge, accumulated knowledge, confirmed through focused experiments. My friend and mentor for traditional knowledge on the Pacific slope is Dennis Martinez with Indigenous People's Restoration Network, IPRN.org. There, one finds articles by Dennis, talking about indigenous knowledge and Social Forestry. Within the last ten years, he taught a course with M. Kat Anderson at Occidental Arts and Ecology Center.

DENNIS MARTINEZ

Dennis Martinez and I are old friends. We have worked together in many venues, teaching Permaculture Design Courses, attending eco-restoration conferences, talking to environmentalists, and connecting rural and tribal cultures to a wider range of forestry perspectives. Dennis worked with Design Associates Working with Nature (D.A.W.N.), Earth First, and government agencies to advance the use of native grasses and traditional forest practices. Dennis was on the interview committee of tribal elders that I presented to while applying for the Agriculture Instructor position at D-Q University, Davis, California, in 1985.

His work with tribes and government agencies helped develop stewardship agreements and adaptive management areas, including near the Sinkyone Wilderness in southern Humboldt County, California, and Star Gulch near the upper Applegate River in Southern Oregon. Dennis has worked with multiple tribes in Hawaii, Oregon, California and many other places. The return of the Dhalgelma Salmon ceremony to rivers in Southern Oregon is one of the remarkable efforts he has facilitated. He has taught many communities and agencies in the arts of cultural burning and native grasses seeding (Martinez 1993).

Commercial industrial technology does not recognize traditional knowledge, because complex consideration throws a monkey wrench in the gears, slows them down big time. If we do not account for whole systems, scientifically, for the damage being done, consumer globalism is not sustainable.

IEK is Indigenous Ecological Knowledge and is precisely local, sacred, and special. It should not be compromised and cannot be exported. All traditional knowledge is Social, and keeps a people together, through Wisdom stories. Deep memory is valued, as it carries portrayals of whole systems, to help us remember what we may be missing.

Traditional Ecological Knowledge (TEK)—a collection of stories that hold complex experience in cultural memory, and can be carried by migrants—or Indigenous Ecological Knowledge (IEK)—which is very local, are both epics, embedded in vernacular languages and customs—are both anthropological terms. Traditional Scientific Knowledge (TSK), is also sometimes mentioned, to value the tried and true methods of conserving wisdom for all peoples who are embedded in Place-ness and the Long Story.

TEK (traditional ecological knowledge) is a story-based collection of knowledge that contrasts with experiment-based modern science, which does not appreciate anecdotal observations no matter how recorded and repeated. The traditional part means empirical knowledge is held in lesson-stories, based on long-time actual experience. The ecological part means nested in relationships, not fragmented and separate. Cartesian and Newtonian science demands proof-in-numbers and declares all else unimportant. This western shortsightedness has brought much harm. Combine this convenient rationalization with bureaucratic inertia, and global finance is a runaway train wreck.

TEK that is still held in persistent cultures is special and sacred. An outsider would not understand all the references and reminders. Children in these cultures hear lots of stories as they grow and this builds a repertoire of memorized references. Simple phrases spoken by elders in council trigger complex memories and allow a more whole approach to culture. TEK needs respect and not appropriation. We can learn from other cultures but we need to acknowledge our sources, and learn how similar lessons from many cultures can guide re-localization.

Indigenous Ecological Knowledge (IEK) is Place based and does not travel well. Any IEK that is still extant needs be kept secret and cherished. This advanced level of knowledge can be acknowledged as possible and doable. We can be inspired to reach for stories from our settlement-restoration work, and we can thereby learn to move towards the humility of Place-wisdom.

Traditional Ecological Knowledge (TEK) is a collection of wisdom, not a liturgy of religion. Many of the stories do have a spirit-full resonance and teach us wider attentions. Accumulated practical observations get generalized symbolically, in stories that remind and review the best of a collective memory of Place. This begets precautionary cultural principles, long-view conservation. Similar principles are held in world-wide religious texts, as well. Some text is inherited from deep tribal experience, preserved in oral history, and then incorporated into written and Natural law. Even the dogma of convenient politics retains some wisdom such as, "A rock wall without mortar cannot be taxed," due to its temporary nature. This residue gives imperial orders and policies some grounding and seeming reasonableness.

There are practical applications to learn, and ways of listening, to understand Traditional Ecological Knowledge. TEK is often revealed in metaphorical stories, collections of Horticultural and Place-based experience. TEK is the general term for kept Wisdom and much TEK is preserved in fragmented snippets, and has been carried far from its origin lands. All cultures have some ecological savvy kept in stories and sayings. Read on, kind tenders. And build your list of references to explore and newly met kin to greet.

THE WORLD-WIDE PATTERNS

As empires fall, illiteracy rises. This vacuum allows and encourages re-localization. Oral epics, with specific eco-wisdom, re-emerges but only thrives if the accumulated systemic imperial-degradation of basic carrying-capacity and resulting refugee desperation does not aggravate any attempts at settled local persistence. We need a break, at least for a while, of orientation and culture building. Much collective wisdom is about survival through times of big change.

The persistent stories of scattered and displaced, but still coherent, cultures able to re-adjust to new local-survival imperatives and regimes are carefully preserved epics that bundle stories, with a cast of symbolic characters and generalities that keep identity, language, and names. The repeated telling of these stories relieves some of the cognitive dissonance from loss-of-Place.

Both religion and traditional knowledge use symbolic references. TEK, at best, is held intact and uncorrupted through careful ceremonial memorization, but still offers the opportunity for reinterpretation. Dogma resists the heresies of *continuing revelations* and the *testing of scriptures.* Among polytheistic and animistic Horticultural, or tending, hunter-shepherd-trader /nomad-folks, the stories and songs are vivid and colorful. Are their magical expectations the over-estimation of subjectivity?

Survival still depends on recognizing practical patterns from memory, so Humans learn from memorable stories. When we write these stories down, they fade. We Humans coevolved with memory cultures, not alphabets. The most compelling memory aids that travel well at all turn out to be ideograms and symbolic glyphs, reckoning back to real life: tapestries, scrolls, bark paintings, and architecture. Memorization training in early modern times was in edifices, The Pantheon for the Greeks and cathedrals for the Christians.

The very best mnemonic template of culture is the landscape itself. The shape of the land and the prominent features, or the Places of events (mythical and memorized) are celebrated with stories. The very pebbles can be said to be shaped by the ancient ones, the mountain a creation spirit come to rest. Or a trickster got caught and stuck for our benefit; we learn what to avoid, just look over there!

A few displaced tribal peoples have managed re-envisioning phases, a series of council fires (Martin 1982), where the stories are re-imagined with new symbols, such as Hemp, Horse, and steel, for the Sioux (Farb 1968). And thus new ways of working with Nature can be integrated. These fresh and newly appropriate re-arrangements can generate symbolic pattern-stories that find nearly universal planetary application. All Species and All Places; when they resonate together, the new epic can become as resonant as religion, for the purposes of capturing and spreading memory. That Human appreciation of resonance spreads evangelically, among other vulnerable displaced folks, who are eager for a colorful new/old cultural basket and a Place to live in peace.

Social Forestry is an opportunity for modern seekers to return to the forest, with new eyes and expectations. Good science and visions of *good-social-order* offer a template for re-organizing our expectations and hopes: The Return of Fire, Beaver, and Salmon.

EUROPEAN IMPERIAL FORESTRY

Working with trees and forests has deep cultural roots and the local details vary by context. The world still holds forest-dwelling peoples with complex relationships and stories, under stress from imperialism, but as yet tangled in their forests. Even colonialist (and deracinated) European settlers have thick poetic bundles of forest-associated language. On many levels all Humans can relate.

Well before the Romans, the forced demand for timber rose from militarism: ships, wagons, wheels, forts, charcoal, mining, etc. The many other products of the forest that were used locally were degraded by focused log extraction. Many drainage basin villages were removed along with their forests, the people conscripted, enslaved, or destroyed by imperial orders: "Remove them!". After the forests are removed, there may be centuries of semi-nomadic grazing as the rivers dry up, thus the deserts of North Africa and Arabia. Much later, even the Scot's highland crofters get removed, in order to run nothing but sheep to feed and clothe armies, navies, and to export Humans for industrial labor.

And the slave trade removed and enslaved captives as a commodity of globalism, still necessary for imperial ambitions at the beginning of the industrial revolution, until fossil fuels and mechanization displaced muscle and brain. The city dwelling privileged classes justified their accumulation and subjugation without noticing the brutality to Nature. Steam power accelerated saw log extraction: steam donkey skidders, narrow gauge rails, big sawmills, and steam ships. There went the coastal forests of the New World.

The New York College of Forestry was founded by Gifford Pinchot in the first decade of the 20th century. He had been trained in Germany and already had experimented with industrial forestry. When he got to the Adirondacks, with the vast forests still there, he carefully studied clear-felling, now called clear-cutting, and replanting. He found the trajectory of site quality and log production was always down with each subsequent cut. The industrial forestry model that most fits machine efficiency demands is a disaster. The college went on to find, or invent, ecological justifications for industrial log extraction so that colonization for imperial extraction could proceed.

The federal agencies of North America facilitated "resource development"—which was actually removal—by subsidizing railroads and giving away First Nations' lands. Go ahead and read all the West Coast treaties that were never approved by the Senate, while the Indigenous refugees (an oxymoron?) were corralled. So it has gone for many empires over ten thousand years. Eventually they all fail. Empires are a sort of social/trade spasm and like eco-spasms (an opportunistic explosion of one species) they overwhelm or "overshoot" (Catton 1982) the carrying-capacity and reach crises of "bottleneck" (Catton 2009).

The "settlement" of the western basins and ranges was labor intensive. There was a great migration of tools, machines, and Humans to build the infrastructure and process the "raw materials." We are constrained in our language to even begin to represent the vast scale of industrialization. There has been a series of resource rushes and financial collapses, leaving ghost towns and desertified mountain ranges.

In the Little Applegate valley, first it was the Beaver, extracted by Ogden and company by the ten thousands. Then gold, and with that, timber, water, grass, and other minerals and aggregates. The water tables and bulb meadows were extirpated, gone. The Dakubetede, Athabaskan speakers, longtime residents, were massacred, battled, and removed. Then, after the gold rush, with run-down ranches and their seasonal logging, mining and grazing, came the extraction of value with the gambling of mortgage failures, farming challenges, drought, and rural social isolation. Books have been written about these western Human experiences with frontier expansion (Stegner 1971, Austin 1988).

The decline of easy pickings, and now, the final rush to extract the very souls of Place and its citizens via relocation, retraining, downsizing and debt, became most evident in the early 1970s. After that, wages went down and small towns, small farms, and small local manufacturing collapsed, while small cities, with older sawmills, went through industrial abandonment. The new highways and centralized mills, geared toward international markets, abandoned what local forestry culture and knowledge had accumulated. Port cities and transportation hubs bounced back with gentrification, spreading development, creating a city/wild-lands interface, and increasing wildfire suppression—and subsequently, excess fuel explosion (excess standing dead and crowded forest biomass) from lack of tending and care.

Here on the Pacific Rim, chain saws and bulldozers really came on from WWII. Clearcutting, on a giant machine scale, did not get going until the late 1970s. The whole technology of cutting, and yarding, and loading, and milling, logs became very efficient, meaning less Humans required, and thus the impoverishment of local economies. This has led to a loss of local knowledge; hunting and fishing licenses are way down. There are less folks—yes, it was mostly men—walking around with "eyes on the land." The state actually wishes to depopulate the rural areas further so that there are cleaner roadside tourism vistas and less local resistance to industrial resource extraction, just out of sight.

There were a lot of experienced older saw-log-focused foresters that did not like what this meant. Gone are the sustainable yields with good monitoring. Trees are seen as stale assets that need liquidation, to keep the financial gambling going. Labor and capital need be mobile, while all resistance to empire is local. So the remote corporate owners rotate the middle management to prevent organized local resistance. These political tactics do not serve persistence and resilience, as we reach multiple environmental limits.

RE-EMERGENCE OF LAND ETHIC

Thoreau's journals in the early 19th century and Aldo Leopold's *Sand County Almanac* (1968) in the mid 20th have tracked a thread of land stewardship deep in the American myth. Those men and a few women (Austin, Alcott, Gilmore) reached for a connection with Place, even really big Places. Wallace Stegner, starting with the University of Iown Creative Writing program, guided that thread through to Stanford University, several novels, and a string of famous students: Edward Abbey, Ken Kesey, Doug Peacock, Terry Tempest Williams, and Wendell Berry. All of these writers explore our relationships to Place. This is the literary canon of the modern Western environmental movement.

Our earliest written sources of anthropological observations of Indigenous peoples are limited and biased. The first European contacts with the Indigenous peoples of Turtle Island were traders, trappers, and missionaries. We mine their journals to interpret what they were actually seeing, rather than what they thought they saw and wrote down. This is some help in historic reconstruction. Then came a series of anthropologists and artist/writers, in the late 19th century, who recorded their own filtered impressions. The earliest Ethnobotany notes are merely lists of plants. We can still hear stories and jokes about "what to do when we see an anthropologist coming" at pow-wows. Often, the old tribal village peoples sent these pretentious tourists to the town fool, who would gift them with reams of idiocy.

Reaching back to aristocratic European land romance, there is a string of preservationism and landscape pretensions (parks and preserves, castles and waterfalls) that led to a modem wilderness ethic championed by John Muir and eventually the Sierra Club in California. This brought preservation and limits to despoliation, through declarations of protection and non-interference. The onslaught of industrial demands was closing in on the last remnants of wildness.

Conservationists who invented and promoted the National Parks movement worldwide removed and excluded Indigenous peoples from their ancient, sovereign lands. These early conservationists did not understand the value of cultural tending and the strong place-based spirituality that shaped what was seen as "wild" (and needing preservation). This is still harming both places and peoples of place.

The environmental movement that followed with the wilderness preservation successes and the passage of environmental regulations in the early '70s was heroic and fierce but did not slow down the beast much. Colonialism continues onward, with more financialized instruments of torture. Most of these modern enclosures expelled and refused traditional First Nation practices on these lands. Parks and preserves are a worldwide pattern sponsored by paternalistic pretensions with academic grant lubrication.

It has taken another movement, the recognition of the rights of Indigenous peoples worldwide, to bring into better focus the abandonment of stewardship and the necessity of living with limits. The environmentalists have slowly come around to understanding the importance of Human tending of the so-called wildness, and the limits of the definition of wilderness and national parks that exclude traditional cultural tending. Conservation is compatible with Horticulture and Social Forestry.

6 Adaptation

MODERN SOCIAL FORESTRY

The emerging fabric of Ecotopian Social Forestry is woven from a convergence of story lines. A conventional forestry college education, following a childhood of land-based practices, is exported to the Pacific Northwest and meets stories about Coyote the Trickster and Creator and a scattering of old persistent cultures, as heard around campfires and in plays and literature (Doty, 2016), new settler savvy, and back to the land counter-culture. Weave in ecological education teaching opportunities at national parks, junior colleges, and overseas institutes, with Indigenous connections and permaculture design mind and it starts to look like something. Definitely syncretic, cobbled together, but with some traction in face of the disasters of imperialism. So hang onto your hairdo! We got some reading for you. Hope we get out into the landscape and do some good, soon.

In the mid '80s Guy Baldwin, Michael Crofoot, and us-all talked extensively in Davis, California (near the mouth of the Sacramento River), about prospects for Social Forestry in the Shasta bio-region. We imagined a "necklace" of mid-elevation forest properties strung around the headwaters of the Sacramento River. These could be occasional access areas without permanent development. An itinerant crew would rotate through the properties with the seasonal work called for by the specific circumstances.

Ultimately, these discussions led Michael to publish a pamphlet about inoculation strategies for nursery and farm and Guy (then publisher of the *Permaculture Activist*) to write a seminal article about seasonal and long-term opportunities for Social Forestry (Volume IV, No. 3, August 1988). Both went on to harvest nitrogen-fixing (N+) shrub seeds for trade with New Zealand where overgrazing is resulting in erosion for lack of native pioneer species. The diversity of N+ species around Shasta is very high. The lack of Pleistocene, species-reducing glaciation and the abundance of unstable soils (from fresh geological instability) offering pioneer sites has nurtured the co-evolution of shrubs and nodule forming bacteria. *Ceanothus integerrimus* seeds seemed to be the first alternative forest product to focus on.

Guy managed to settle in a foothills olive orchard and Michael went off to New Zealand. I went back to Ashland after leaving D-Q University (1986) and finished my first book, *Greenward Ho!* (1990). D-Q University is one of the first multi-tribal institutions of higher learning on the planet. It was formed through cooperation between the American Indian Movement (formed on Alcatraz Island) and La Raza (the urban movement of Latino peoples).

In the early '90s, I sat in with the Ashland Watershed Stewardship Alliance about watershed restoration and management possibilities. The city is a partner in a 1929 memorandum-of-understanding (MOU) that allows coordination with US Forest Service management of the almost 10,000 acres in the Ashland Creek watershed. I became interested in the relationships between the city, the university, the federal bureaucracies, and the citizens. Although I submitted a Social Forestry proposal to the city and sat with the USFS Ranger in the University Provost's office, where we traded document bundles toward systems theory education, the political process of arranging coordination was slow, and everyone wanted to know where the grants and subsidies were so that a whole systems approach could be studied and implemented. At AWSA, unemployed, newly immigrated scientists looking for monitoring jobs and grants had more patience than I for endless meetings.

Better luck was had with the Siskiyou Permaculture Resources Group (SPRG, "*sprig*,"), our local club, with the apprenticeship program at Bush Street, called Tom's Garden Cottage (TGC) and with the Wilderness Charter School (WCS) at the Ashland High School. These were more grass-roots, citizen-led, efforts. SPRG applied to the Ashland Ranger District and took on a stewardship arrangement for up to 8 miles of Upper Tolman Creek Road south of and above Ashland. The apprentices, six at peak, learned woodcrafts, built prototypes at Bush Street, did assembly at a local birdhouse and feeder factory, and worked the Tolman roadside with the WCS. The WCS took on Permaculture as a core curriculum—which lasted thirteen years until budget cuts closed it—and developed the straw-bale classroom grounds. The design process enfolded a vernacular aesthetic of fencing and natural-building using fire-reduction roadside harvests of Douglas Fir poles and willow wands from the Tolman roadside project. A Southern Oregon University intern documented Permaculture at the Charter School (Runge 2002).

***

In the late '90s, I installed a semi-formal garden design at 117 High Street in Ashland, Oregon, with woven Scouler Willow gates and fences and Madrone and Douglas Fir arbors amongst other garden elements, such as a recirculating waterfall. High Street was toured twice by the Oregon State University Master Gardeners Program.

A lot and cottage nearby on Bush Street, two tenths of an acre near downtown, was eco-developed for seven years and is documented by a master apprentice's thesis (*TGC*, Jacob Squirrel 1999), a half hour made-for-TV high school produced video, and a slide show for teaching. Also in the late '90s, I documented 80 acres on Elk Creek, above Trail, for government forestry subsidy, with an ecological forestry plan containing Social Forestry elements (*Elk, Dancing!* 1999). Organic farms in southwest Oregon have increased their use of windbreaks, woodlots, and hedgerows. The Southern Oregon Woodcrafters Guild has annual shows with many native woods cut, turned, joined and polished.

ORGANIZATIONS, EXPERIMENTS, AND ADVENTURES IN ASHLAND

Meanwhile, several local non-profit entities have emerged. Lomakatsi Restoration Forestry (founded 1995) has done significant tree and shrub plantings and has done industrial forest work in slash-and-burn fuel reduction, as necessary emergency procedures, subsidized by the government. They also had a natural-built housing demonstration on a covenanted land trust. Their tightly knit crew does public fund raising and more recently have taken on projects with federal agencies, city watershed management, and First Nations. The ongoing training in Indigenous fire management and prescribed burning is heroic and inspiring. They demonstrate big success, by still being together and busy after more than two decades.

The wilderness preservationist non-profits have, in the past, dismissed utilitarian and Indigenous interests and protested all harvesting, but have come around in the last few decades. The interesting, but now defunct, Rogue Institute for Ecology and Economy made a brave attempt to organize and publicize green certification, endangered species management, forest industry displaced-worker retraining, and alternative forest products. Mixed, hardwood-log truck loads of Black and White Oak, Big Leaf Maple, Red Alder, Madrone, and others, were sorted into single-species I-5 long-distance truck loads to special mills up north. The lack of a good sorting-yard system and efficient brokerage proved economically crippling.

Manzanita bird perches, made by local entrepreneurs in Grants Pass from twisty-smooth red branches, were distributed nationwide to zoos and collectors. Manzanita is one of the overgrown brushland species that need to be thinned before we can reintroduce cultural fire practices. Portable computer-aided sawmills were demonstrated, a replacement for long distance hauling to industrial-scale mills upstate. Locally processed flooring from various species is brokered in-region with green certification. This angle of economic opportunity is fraught with compromises and hurry-ups that prevent perfect timing and integrated stewardship under the control of local people. But the term "alternative forestry products" is supposed to present less damaging extractions than clear cut logging.

Early in this century there was a stir of interest in business approaches to Social Forestry. First a team associated with the University of San Francisco tried to organize a group of local Social Foresters to manage industrial fuel-reduction crews in Southern Oregon—we refused. There was federal money promised and we might have found economic rewards through federal subsidy. This new business-model eventually ended up hiring displaced workers at low wages, undercutting already-existing local crews.

Then there were meetings with a lawyer and a CEO-in-waiting to form a company called Willow Works and finance a forestry with multiple products forestry and venture capital. Willow furniture, basketry, fencing, and privacy screens could be offered, with other products to-be-developed. However, much upfront research and business planning is necessary, especially if one does not have an immediate money maker, to fund business expansion. The CEO quickly pulled out.

This and the experience of the Rogue Institute of Ecology and Economy show that commercial and public pressure continues to demand jobs and the extraction of cheap, raw materials. The land is supposed to produce profits, and the locals should be happy to work hard for cheap. Multiple products are expected to enrich the brokers for their information work, and nonprofits asking for grants should show jobs creation or they will not continue to receive support from foundations.

This Willow Works Project Field Work Flow and Products table shows a sequence of work in the forest over two years. To set something like this in motion is logistically daunting. Parts may stand for a while as separate businesses, but not sustainably. To marshal such a system, we would best reach for unanimous consent through watershed council and then be assured of the dedication of all parties. All entities continue to meet in council to share information and administer certification of sustainable practices. Local culture celebrates seasonal specialties and vernacular aesthetics: the beauty, color, pattern, and tastes of the place where we live.

This table illustrates the products and timing involved in manifesting these possibilities on the forestry side. Workers in the woods deliver harvested materials seasonally to craftspeople and small manufacturers. To work on the social side, business and education must inform and support both consumers and workers.

The sociopolitical realm is presently oriented towards conservative rights of primogeniture, as in first-come first-serve water rights and the sanctity of private property to

```
Willow Works Project Field Workflow And Products
Notes by Tom Ward 29n September 2002, Ashland Oregon

PROJECT AREA—Private and public lands in the "INTERFACE", especially along roads
and firebreaks

PROJECT GOAL—Stable and productive "old grpwth" canopy or overstory, able to be
underburned and/or intensively managed

     WORKFLOW . . . . . . . . . . . . . . . . . . . . . . . . . . . PRODUCTS

A. INITIAL ENTRY—after scoping and flagging (first early winter)
     1. remove ladder fuels—high pruning
     2. lop and scatter fine dry fuels
     3. thin Douglas fir stakes . . . . . . . . . . . . . . . . . . . SAILS, SHORES FOR HURDLES
     4. prep hardwood stools for coppice
     5. select small hardwood logs . . . . . . . . . . . . . . FURNITURE, BUILDING MATERIALS
     6. shred greenwood . . . . . . . . . . . . . . . . . . . . . . COMPOST
     7. select mushroom inoculation logs . . . . . . . . . MUSHROOM INOCULATION LOGS
     8. select culls from 4,5 and 7 . . . . . . . . . . . . . . CHARCOAL
     9. inventory threatened & endangered species . . SEEDS

B. SECOND ENTRY (first summer and fall)
     1. inventory pharmeceuticals . . . . . . . . . . . . . . . HERBS
     2. thin coppice sprouts . . . . . . . . . . . . . . . . . . . BASKETS, FENCES
     3. build trails, sort pads, camps, ponds . . . . . . . INFRASTRUCTURE
     4. harvest from mushroom logs . . . . . . . . . . . . . MUSHROOMS

C. THIRD ENTRY (second early winter)
     1. thin understory trees . . . . . . . . . . . . . . . . . . . SMALL DIMENSIONAL LUMBER
     2. shred green slash . . . . . . . . . . . . . . . . . . . . . COMPOST
     3. seed natives for future crops . . . . . . . . . . . . . EROSION CONTROL
     4. wildcraft various species . . . . . . . . . . . . . . . . HERBS, BARKS, SEEDS
     5. select culls from onsite milling . . . . . . . . . . . CHARCOAL, COMPOST

D. FOURTH ENTRY (second summer and fall)
     1. ongoing coppice sprout harvest . . . . . . . . . . . BASKETS, FENCES, FURNITURE
        . . . . . . . . . . . . . . . . . . . . . . . . . . . . . . . . . . . . ESSENCES, COMPOST, BROWSERS
     2. single tree select logging . . . . . . . . . . . . . . . TIMBER, ONSITE MILLING LUMBER
     3. wildcraft various species . . . . . . . . . . . . . . . . MULTIPLE PRODUCTS
```

the detriment of general community health. A helpful transition might be to declare an ecological-opportunity zone, and enable them with eco-rational covenant development codes, tax-free trade, local currencies, catastrophic health care coverage, social-arrangement tolerance, and local oversight of forest planning. As so much of the Shasta region is in public domain or absentee landlord ownership, the initial arrangements might be by stewardship contracts with rights and restrictions enumerated.

A slow and piecemeal buildup of local entrepreneurial businesses is proceeding and this could benefit from a more coordinated economic development. The State of Oregon has a buy-in-state website to connect commerce. A Shasta region organization could facilitate trade in, and certification of, ecologically harvested multiple products. For Social Forestry to be better commercially established around Shasta, a cooperative effort will be necessary. The inter-institutional meetings so far have only opened the discussion.

The next modern-style step would be a conference with a call for papers and documented experience. An arena for such a conference could be prepared with the display of maps, posters, and storyboards to support a planning process. Good facilitation is increasingly available, as folks learn whole-systems approaches. The examples of various NGOs and activists such as Saul Alinsky, for community assessment; Tom Atlee, for community decision making and democracy; World Wildlife Fund, already established in the region and useful for ecological assessment; and international community-forestry interests would be available and new grant supported efforts are springing up like mushrooms. Indigenous peoples offer tremendous perspective, knowledge and example. They can speak from experience. The empire is not friendly.

The challenge is to positively envision a Social Forestry appropriate in present and future times. Our Quick-fixes and Retrofits should not compromise Ultimate Intentions. We must study the traditional human forest cultures and use ecological and whole systems sciences to map the possibilities. Then we can inspire the various modern institutions and interests to arrange a new/old forestry that feeds the needs of the forest and the human culture.

This social forestry bubble diagram shows possible resource streams between FOREST and TOWN and associated educational, commercial, cultural and woods-craft entities. This is a map of possibilities. Also see the poster, likewise titled "Social Forestry".

9 Social Forestry Transition

BEST LESSONS OF THOSE TIMES

For 13 years I taught at Ashland High School in their Wilderness Charter School (WCS) where I was the permaculture guru. Through this experience, many of these kids changed their majors (towards biology and planning) from what they previously wanted to do in college. They came back home afterwards because they wanted to be at home. The students and staff built an incredible-edible campus with a strawbale classroom and permaculture gardens. Some of the fencing and arbor materials that were used at the school

THE LINEAGE OF SOCIAL FORESTRY 57

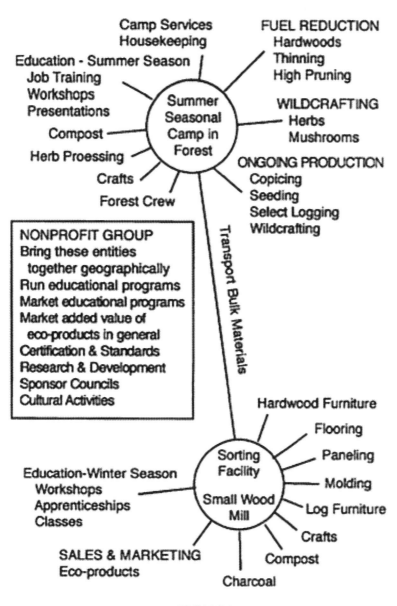

came from fuel hazard reduction work in the forest. We went out in the forest, got materials, brought them back and made things that got used every day, and thus came to be People of Place.

The entire road system of the federal forests and rangelands has already been through an ecological assessment. The road profile from side to side is called the road prism. It's on a line across the road from the top of the cut to make the road to the bottom of the fill dumped to support the road. The federal roads in local forests have a road profile that varies from 50 ft. to 300 ft. The roadside contained in the prism is a piece of the commons which can be immediately managed for stewardship through local ranger districts and local BLM districts on an extraction permit basis.

Much of the materials in the prism are available by permit to be harvested with no fancy paperwork. Permits are required to haul materials out of the forest. Roadside plant clearance (which used to be chemically controlled with herbicides), fuel hazard reduction, sight and view along road, campground cleanups, culvert clearing and other light road maintenance has been cut from the federal budget. So there's an opening right now for negotiating Social Forestry stewardship on a million miles of road sides.

The WCS asked for a project on eight miles of Tolman Creek Road, which happened to be, for the best part, on a north slope set of keylines (benches that drain sideways) with crossing creeks and springs. We coppiced Willow. In this case it was Scouler's Willow and we took poles, exotic (non-native) medicinal herbs, seeds, nursery stock, basketry withies, and removed exotic roots. The federal government only had three categories of materials that they'd give permits for: floral materials (that's herbal medicine), grape stakes (rods and withies?) and small poles that are up to 6" diameter breast high (dbh). Creatively, we qualified all our material harvests. We were in turn providing roadside maintenance and fuel reduction to the commons.

The Ashland Watershed Stewardship alliance had declared that anything over 17" dbh is old-growth in order to establish protocols and create a benchmark. We can high prune big trees to fire proof them. Bigger than six inches is the definition of commercial timber and we can't touch 6-17 inch dbh, even if it's on the prism.

Up until all the changes following 9/11, 2001, I also had a client market for roadside materials. I was designing ornamental permaculture gardens in Ashland, for people that had just moved there, using fuel hazard reduction materials, such as those harvested by the Wilderness Charter School, and creating a vernacular aesthetic, showing how one town can be visually different from another town.

Loose dogs and little boys with air guns had been outlawed and now there are bears and deer in town. We can hardly garden at all. There could be a City wildlife manager, a city hunter, because some of the deer are injured or runts and we could end up with beefier bucks downtown if we just do a little management. The wild population of

Mountain Lions has rebounded and deer are seeking refuge in towns that seem safer than the wildlands.

We had a learning curve with roadside Willow coppicing[2] because Elk moved in and mowed the coppice sprouts. We then put brush mulch over the stools (the stumps left to sprout when you cut down the overgrown thicket). The regrowth produces lots of thin new long withies. These can be harvested and made into all kinds of useful tools and structures for living. One can find illustrations of hurdles (woven modular walls), bent-wood chairs, tool handles, baskets and all kinds of cultural paraphernalia in old English coppice craft books. Indeed, years later a survey of the abandoned project showed regular yearly harvesting of the willow we had coppiced, some folks noticed our bounty but only skimmed it for their own needs. It takes a culture of Place to do lasting stewardship.

All this commercial interest is a Transition tactic. What else can we do while we prepare for collapse? What we really need is settled cultures of Place, able to be nimble and effective with 'the least work for the most effect". Much more Social Forestry has unfolded since these adventures near town. In this next chapter we will look at Staying Home. The lessons we learned on the street are going to be useful in future negotiations as we can cite our experience. What did and didn't work, and can we derive some principles?

---

2 Managing a tree stump for sprouts on a rotation of cutting and growing.

The Principles of Home include

13.
Stay home and dress up. Be who we are where we are and celebrate it at home. Wear the appropriate fabrics from the local fiber-shed, made as durable and as lovely as can be.
Miss the Place if you go away.

14.
It takes a Culture of Place to manage continuous stewardship for ecosystem functions.

15.
The Home-Place teaches us how to be. The way that we dance with our Place entrances and delights our Original Mind.
We learn as children do.

CHAPTER 3
The Nest-Home of Social Forestry

Ah,
'tis a twiggy-messy thing,
yet home

In this chapter I am introducing Quaker terminology. We Quakers actually call ourselves "Friends." Quaker was initially spoken by others as a slur. These are my people, The Society of Friends. I offer these stories as examples of cultural adaptation to Place and Right Livelihood. You will learn a lot more about the ways of Friends in Chapter 4, Relationships.

INDIGINATE OR RE-LOCALIZE? BECOMING PEOPLE OF PLACE

Some intact tribal folks cannot "re-indigenate" because the land that spawned their stories is destroyed or they were forced out. Only folks still living on traditional lands who somehow become distracted from their relations, can re-indigenate. Traditional healing ceremonies in many Cultures of Place tell the names, the connections, to the distracted one, in case the lost one might remember how to come back to home.

I speak as a bridger, a teller of intersectional stories. I have deep roots in traditional Cultures of Place. As a refugee to the Pacific Northwest, I have walked the Klamath Knot for a half century.

How do newcomers bridge
to become locals?

*Time and contacts, stories and visits,
reading and listening.
Learn all the birds
and their songs.
Miss the place
if you ever go away.*

The Northwest Pacific running rivers, although draining very complex geologies, carry critical and essential nutrient loops that run up and down rivers and keep the forests green and the biodiversity high. Tens of thousands of years of First Nations tending kept some balance of ecological resilience with human intervention. This continued all through epic changes of climate and geology, floods and volcanism. These First Nations called and convened repeated long councils, and adjusted. They held populations that lived within local carrying-capacity.

This got changed. Industrial extractions and genocidal and ecocidal colonialism wreak havoc. Now we have landscapes that carry artificially supported over-populations of clueless box-dwellers. And the biological foundations of essential ecosystem functions have broken—All Beings, fragmented and alienated, just like the modern Human cultures. Living on Earth looks to become even more artificial and what Life is left to feed the robotic demand for feedstocks? Please confirm that the laws of physics still hold; if it can't go on it will not go on. Modern Humans are a species in danger of extinction. Their cultural and legal structures do not benefit from local control, the laws are international and global. This self-replicating systems-spasm is eating two earths at a time and this cannot go on.

Ecotopian (back-to-the-land) culture has dug in, in some places. There is resistance and repair. Attempts at energy-addiction healing continue. Some of these ritual-based future-oriented dances toward local sufficiency and resilience have traction. Small amounts of on-the-ground work are being practiced to the best of our collective abilities. This may seem silly, in the face of the empire, but this is the work of drainage basin repair and enhancement. The sort that Humans have done everywhere in some times past and do in some times present, the way the species ought to, and co-evolved to, tend the Wild.

To facilitate the return of Fire, Salmon, and Beaver, we all have similar work to do, in various drainage basins. The return of the stories of interconnection and reciprocation weld us to the web of complexity and fit us into the cycles of change. Some work to do is obvious: shade the riparian stem-flows, put old roads to bed, soak in winter storm drainage, reforest the ridge lines.

*Long trajectories.
Only Human cultures of Place
can do it well enough.*

There is work to do wherever we are. Cities can do this too. Lots of untended landscape out there. And yet the most challenging task of modern humans is to submit to the good order of cooperative culture and mature through community justice and watershed council. Hyper-individualistic consumers (acquisitive dilettantes with severe nature-deficit stress) come on to the hoop slowly, with much resistance. They do notice emergency situations that motivate radical change. To try to build Culture of Place, that preserves principles and patterns on a socially and geologically appropriate time scale, is wildly optimistic. Circumstances demand it.

The Salmon is the traditional continental-scale nutrient loop on these west flowing drainages. The complex runoffs of soil, carbon, and water-soluble nutrients are mixed in the rivers and ocean and return as upriver floods of concentrated bio-attached nutrient packages. A Salmon run is a sort of giant bathing of goodness, that spreads uphill with animal dispersal, against the rains and soil erosion that slumps down, to fill up our hungers. What a wonderful event is the return of the silver wriggling hordes.

With rivers draining to the coast, on the ranges that drop to the Pacific Ocean, the upslope forests capture these nutrients, holding them back, through carbon attachment and fungal re-distribution. North facing slopes can be enormous carbon sinks, with layers of giant logs in various stages of moldering decay. As we go up-river through the wall of mountains as few rivers do, into drier basins and even deserts, fire becomes the local recycler by returning nutrients to the soil that were trapped in standing fuel loads.

Cool fire is best, fire that jumps around and stays on the ground, under the overstory canopy. We do not want catastrophic volatilization, where the whole forest burns, and all the mobile nutrients outgas. Whole different circumstances dictate different work and various cultural arrangements, and that builds skills fit to the local specifics.

21 Just Sink Carbon

Return the Beaver to the high mountain meadows, to foster wide thickets of wet brush and marsh, hold back the winter waters, and widen the bands of complexity. Start up high, holding the nutrients in resilient broad adaptations, meadows, thickets, and forested ridges and slopes. Then the late summer stem flows, kept cool by shady stream-sides, carry the Salmon smolt back to ocean and chitin feasting, from sea crustaceans. With their phosphorus-based exoskeletons, this Shrimp and Krill diet bulks up the Salmon, who bring that wonder back upstream to keep the cycle turning. Without phosphorus, plant flowers fail to mature and set seed, which means terminal meristems fail to grow and extend.

This is true of both the wet side and the dry side of the Pacific ranges. Great basin loops may not reach through to the ocean but they have their own local loops to learn and augment. The variety of soils and vegetation over vast reaches of continental proportion leads us to endless learning and practice opportunities.

*The focus can be on Home,
and still we need the big picture.
Pay attention.*

24 Soil Water Rock

This poster—Loop Law—shows a simplified garden loop for phosphorous and other nutrients. Net, sink, and loop is our motto. Nutrient cycling and storage is managed by catching (net), stashing (sink), and keeping local (loop).

29 Loop Law

The patterns of tending will vary with place. Cultural ways likewise. The control of erosion, the repair of bad forestry, the return of fire to savannas and forest floors, are traditionally coordinated by semi-nomadic hoops of activity. The landscape inspires local Human culture, ornamentation, work, and skills. The widely noticed rituals, worldwide, of gratitude and reciprocation, seem to nudge Humans toward best practices.

Only through patient council and good social order can we end up cooperating at the level of tending that is locally appropriate. These known social ways are the practical path, we hope, as Human cultures, to re-orient to the work of maintaining carrying-capacity, and restoring ecosystem functions.

All these grandiose-sounding, over-generalized, scientific concepts only have real meaning in the context of Place grounding. When a local cooperator repeats an often heard phrase as a reminder, nearby listeners will have a rich set of memories flood in. In new settler cultures this usually takes generations, but there will eventually be remembered-stories to reference forever. Culture building needs to be locally appropriate and yet borrow from the collective planetary experience. This invitation is for All Peoples: wherever we are: there is work for all, many skills are useful.

Are we willing and ready to step up,
embrace our Place and Peoples,
and do the work together?

DIMENSIONS, DIRECTIONS, AND CALENDARS

Places have edges, not boundaries. And the Human experience of time and space is shaped by remote horizons. These edges are mysterious. They defy our *skrying*, our search for meaning in patterns. As David Abram teaches us (1996), the edges of the sky, the planetary horizons, are veils. They pass things away, and they pass things in. Our local world seems separate from those worlds across the far horizons.

Likewise, the horizon we stand on, the ground, the soil, the bedrock, is beyond our ken. We dig there to find the past, and end up trying to interpret from fragments and clues. The air itself is a horizon, in that it is invisible, except when revealed by smoke or clouds. The breeze whispers something to one ear, touches the hairs on our arm, and the message seems important, but soft and effervescent. These three fuzzy edges keep us guessing.

The Sun, and the Moon, and the Planets, travel the sky and cross the ecliptic horizon, with a procession of arrangements. The show in the sky is mesmerizing, and many tales have been told, to follow the parade of signification. Humans in-Place orient themselves with the *calling of the directions*. Six or more directions are personified and spoken to, with welcoming and thanking, acknowledging their essential natures, by citing the praise story for that direction. This basic call to orientation is common in many cultures. And the four dimensional remembering of this set of key symbols then surrounds All Present and sets the stage for breathing together in ritual, before sitting down to council.

Proper good-order in council also takes time to praise All Beings. The rituals of thanksgiving go on for a while, but they bring us into the holograph of relationships, so that all present are humbled and full of gratitude. Getting *set for meeting* is well known to be necessary to useful *corporate* (whole cultural-body present) considerations.

> *Corporate amongst Friends*
> *means in-community,*
> *as with corporate worship and decisions.*

Humans are special for being able to have compassion for other Humans and All Beings, to show hospitality, no matter what the clan or family. We have also, in shamanistic rituals, been able to learn to be more fully Human by extending our natural compassion to All Beings (Snyder 1990).

Thus, apparently inanimate forces and objects come into relationship with People of Place. Not only can we talk to them (as we have named them, though we may refer indirectly to be polite) but we can borrow their symbolic meanings as tools in our thinking and re-membering. When People of Place find themselves in the nest of their real experience, they have the tools to consider the work at large: whole-landscape stewardship.

If any of us, as re-localized folks, are going to avoid bad mistakes, and make useful messes (*disturbance-regimes*), we will need a lot of humility in the face of complexity with All Beings, and our Human Friends, and *The Others*. Try as we might, we will never think like a Bear, or a Stone, or Grass. But we can engage, with expressed gratitude and intentional reciprocation.

In All Species Council *others* can be mimed, danced, and interpreted by Humans, to reach deeper and wider relations, something John Seed and Joanna Macy have explored. After some orientation by elders, a volunteer participant tries very hard to think like a tree and has a Human interpreter to speak for "tree" in council, as the volunteer dances "tall standing one" wisdom, very slow. All Present listen, feel, and watch very carefully, for the trans-species channeling. Later, gratitude can be offered with giant puppets and outrageous costumes, as we walk through neighborhoods to re-mind us, in an All Species Parade.

At times, fully immersed in Nature, one can get a sense of the Greater Mind. It is as if the landscape is thinking and *all sentient beings* are participatory. The landscape dreams in natural images, symbols that are laden with implications. Similarly, our Human psyche thinks in symbols, as Carl Jung has taught us. Several ancient cultural systems of symbols are used to help experience complexity and relationships more directly, more intuitively, by leading us to see patterns that perhaps we hadn't been conscious of.

The I-Ching, with the hexagrams, and the Tarot, with its set of cards, are the two familiar symbol sets used to see into complex questions. The common set of symbols to Human dreaming and Nature dreaming is best learned directly from Nature or through deep meditation (the *Original Mind*, Snyder 1990). Paul Shepard, in *Nature and Madness* (1998), suggests that the most formative years for nature immersion and natural relationship learning are three to nine. When children are allowed a full exploration experience they receive (awaken) deeply co-evolved archetype re-enforcement.

Personal Human maturation proceeds through stages that are celebrated with *rites of passage*. When the patterns of clan totems, stages of maturity, earned names, *original instructions* (TEK), and iconic representations of the cultural epic converge, passage is enabled. Adults can also renew the connection with the Greater Mind, through Nature immersion and *forest bathing*. Many Human cultures have large sets of recorded symbols, collected through all antiquity. The empathic observation here is that animal brains think in symbols not words, just as we do. Humans have found their own complex languages, but *All Sentient Beings have Spirit*, and they participate in wondrous ways. All Beings have community, language, and culture, if we would notice.

With Place-bonded adults, council will be less stymied in deliberations. Our Human community process, with our children, is Culture being built and reinforced. Besides child free-time in the woods, water, and fields, children can be taken on pilgrimages, up a creek, or into a forest, or up a hill, to special Places, where year after year they hear the same stories of Place and rocks, animals, plants, and wonder. With every season, every year, that special walk, with the whole community, ignites Love-of-Place.

Longer adult pilgrimages may circle a sacred mountain, or walk to the end of the Earth, and always follow a story-line (Chatwin 2016). Everyday pilgrims can walk a maze, or tour a cathedral of Trees, a jumble of Rock People, or a temple of Human-revered icons.

As we work and live in close association with the Nature of Place, our adventures often remind us of stories we have heard. Then, we see there is ongoing learning. Continuing revelation enfolds us endlessly. Coming back into relationships with all of Nature and Forest, with Beaver and Salmon, Fire and Compost, will generate celebration and persistence as we learn, again, to be social cooperators. As we do the hard work of becoming Human, in community with All Beings. This is what the *social* of *Social Forestry* is all about.

TEN STEPS TO BECOMING PEOPLE OF PLACE

Everything is changing, what to do?
Become People of Place.
The most enticing future beckons.

1. State the intention: "We are becoming People of Place." Sing the new songs. Usher in the focus, dedication, and decision with ceremony.

2. Learn the names and stories of All Beings of Place. Get to know the storylines, sacred shapes, and participate in the dances. Speak and sing gratitude to All Our Relations. Talk back, give gently hugs and kisses. Acknowledge our humility, sit down quiet. Discover new stories to explain the amazing events we witness. Notice there are many stories without Humans at the center.

3. Learn what happened where you are now. Who and what was removed, or introduced, or just arrived. Learn where your people came from and why they had to move. How did you end up here? What can you learn from your ancestors? How can we heal the angst of colonialism and the old deep traumas of forced removals and enslavement? Do the research, take in the reality. Offer reparations where useful and as available. What do we have to offer to Place?

4. Stay home and dress up! Wear the best clothes we can get. Local fiber-shed materials, local spinners and weavers, clothes made to last. Well kept fabrics, that brush off and seldom need laundry-care. Just enough second-skin covering to be comfortable and appropriate. Easily adjusted layers for inside shelters and outdoors. The way that we live in our place-inspired homes, will be evident in the clothes, ornaments, head-scarfs and hats, that indicate who we are, what we do, and where we are. House-holding is the ultimate economy.

5. Travel outside your drainage basin seldom. Make big plans to return safely home after such a momentous excursion. Mostly travel locally in loops and circuits. Learn the Drainage Basin Pilgrimage route (the watershed ridgeline) and start preparing to complete the sacred journey, walking the edge without leaving Home. Tend the Hoop, do the seasonal ecosystem function tune-ups. Get to know your Human neighbors. Work together to return Beaver, Salmon, and Fire to our drainage basins.

6. Reduce Energy Import Dependence. Increase graceful simplicity. Love your chores. Reciprocate the gifts our ancestors consumed with understanding, education, and actions that heal while we elegantly subsist, building the bio-diverse abundance of place-based resilience. Net, sink, and close the nutrient loops. Sequester carbon. Build water tables.

7. Consider what ecological-restoration might mean. Restoration to what? Do we know what we are doing? Pay attention and do not jump to conclusions. Notice where disturbance and intervention leads to magnification and not degradation. Life will have its way. Big changes demand adjustments. Nimble tending can support some stability. Cultural, clan and family maturity depends on coming into alignment with Spirit and traditional wisdom. Deep past informs deep local evolution.

8. Support Biodiversity. Learn what can be done. What can we do to help increase bio-diversity, resilience, and fertility (fecundity)? Praise edges, mosaics, and disturbances, where they shine. All the while persisting as People of Place, subsisting well enough. Cycles of cycles, nutrients in motion, changes and drifts. The highest average health for the greatest diversity of participants. Relations between clans and drainage basins are modeled on arrangements in Nature. As we learn and grow wiser, we are humbled by complexity.

9. Read the Treaties. Find the original texts and notes. Even though the empire did not ratify all these documents, we can learn what was said and not said. Restore accustomed access and tending options through community education. Support tending of Place through renewal of traditions and return of honor to Indigenous Ecological Knowledge. Recognize and learn the names of sovereign First Nations. Study the language group and traditional territory maps. Try to understand how peoples might live here, within natural limits.

10. Establish Drainage Basin Councils. After a new settler culture learns to work with their local neighbors, and connections are established with nearby drainages, new agreements of mutual aid and cooperation can be negotiated through exchanges of culture and diplomacy. These arrangements could allow regional coordination and benefits. Any new People of Place needs to recognize the sovereignty of First Nations. Deep time allows healing and thrivance. With the help of All Sentient Beings, our council processes will hold enough wisdom to do the real living of People of Place.

MY QUAKER VILLAGE BACKSTORY

Taylor Thoms, retired from being head of the USGS and Princeton University Department of Geology, sat old and blind in the late 1970s, wrapped in a blanket, in a rocking chair near Ashland, Oregon and described letters he sent to "important" leaders demanding plans for major water works to support the relocation of urban masses who should be resettled on homesteads and small farms. This return to the land would be supported by hand tool kits and video tools for documentation. He insisted that women's councils need to control warehousing and security in general. Many wise, well-studied generalists and academic department heads were reaching similar conclusions back then, after retirement and considered-thinking about long experience.

My meetings with Taylor were overseen by his traditional black dressed plain-clothed wife. I was a friend of their daughter and granddaughter. My own traditional Quaker upbringing gifted me with some context for his mentoring. He was not the first elder Friend who took me into confidence and advice. We will later explore some Friendly procedures, such as queries and advices, and the good order of Friends, found in the various Faith and Practice documents of the historic multitude of local meetings (North Pacific Yearly Meeting 2018).

Here we are talking about Home, so this is my narrative of Place. My mother is from the Fish clan of upper Hudson River Quaker farmers. The Quaker farmer thread weaves back to northern Europe and "the removals," the eviction of settled village peoples as the emerging capitalist state "enclosed" the land and claimed the profits through ownership. The end of the commons provided a working class to mercantile capitalism and armies of extractive colonialism. Long settled cultures, thus displaced, confused, and disoriented, mixed and exchanged stories as they migrated and then displaced other cultures overseas.

With intermarriages over centuries, the Fish line mixed with other early settlers. There is a constellation of villages where we have history. The homestead farming built Place wisdom and infrastructure. This basin of mixed soils and plenty of running water was productive using methods brought over in the early 17th century by a mix of northern Europeans who came from somewhat similar geology across the North Atlantic. After the English Revolution of the early 17th century, many immigrants claiming to be Quakers moved to the English colonies (Christopher Hill 1991). They brought with them languages and practices of a different landscape.

My Quaker Sisson and Fish lines of descent came over with the early Puritans, but were persecuted as witches and heretics. The Sissons helped foment King Philip's War in the late 1600s and were forced out into Rhode Island and eventually up the Hudson estuary to settle against the Adirondacks.

The wisdom and woodslore learned from Indigenous First Nations happened centuries ago and cannot be traced back to their sources. We assume that the woodslore, often learned in summer camps or on nature walks, came to us from the Abenaki people, who most likely tended the food forests of the southeastern Adirondacks and the Green Mountains of Vermont. The Ward and Fish family teachings always credited, "like the Indians did," after these lessons. Even sitting quiet in Meeting for Worship was reinforced by repeating that phrase. The direct Mohawk wisdom came to South Glens Falls Meeting through Thomas Cornielisson, a visiting pastor in the early 1950s.

My father's last name is Ward. So the Fish woman married the Ward. I heard some jokes growing up. The marriage agreement included that the children would be raised Quaker. The Wards also can trace a long story in upstate New York. Both clans have lineage threads that go back to the Early Dutch colony of Fort Orange (near Albany, New York). The accents, attitude, and physical appearance of these old settler descendants is Dutch more than any of the other influences.

My family stories take place in the upper Hudson valley and mountainous Adirondack headwaters that early in colonization saw several wars march through. The colonial wars against the First Nations, the French and Indian War against the English, the Revolutionary War of the colonists against the English, and the War of 1812 against the English Empire. The Civil War brought the Underground Railroad through my childhood village. My settler ancestors were run over or chased by armies again and again. The choice of parking at the crossroads is good for multicultural experience but hard on persistent peaceful farming.

Growing up in South Glens Falls brought these threads together with a strong sense of Place. Both my parents were storytellers. Dad taught history and woodslore from Algonquin traditions. Mom taught the long story of farming villages. Together, we explored of woodlands, creeks, lakes and mountains with plant and place-names that contained stories. Local public education filled in more stories. Museums, parks, battlegrounds, cemeteries, monuments and old buildings, bridges and canals. And my family has only been around those parts for four centuries. I inherited a treasure-trove of place-ness. I have managed to take pilgrimages back to the homeland. Lots of relations live there.

Our village culture contained herbalists, wildflower gardeners, backyard cows, fence-lines with native plums and berries, gardens and hedgerows, fruit trees and huge street trees. Wild animal signs could be found in the backyard. The Hudson River canyon above most of the factories was at the end of our street with the village beach. We used old fat tired bicycles to find places to explore and catch fish to eat or wild edible plants. I sold vegetables from my gardens from age five until going to college.

The seasonal festivals of old Europe were still carrying some old customs and accrued some new place-based practices. May Day meant May baskets left anonymously (we had hedges to hide in). Easter did egg hunts and early wildflowers. Summer had shelling and shucking bees. Halloween followed "cabbage night" where unharvested produce got redistributed creatively. Christmas and New Years were festooned with native plant wreaths and Balsam Fir trees. The manger was embedded in forest boughs.

I was apprenticed out at age 10 to my God-grandmother, Mae Barton, who taught me about wild gardens, plant propagation, and proper cups of tea (Ward 1990). I was sad to lose this idyllic fantasy when, after I left for college and the West Coast, the Glens Falls area descended into a toxic industrial sacrifice zone. The Hudson River is full of chemicals seeping from old factory sites.

My father and maternal grandfather met early mental decline from long employment in a heavy metal pigment factory complex, "The Imperial!" (originally the Imperial Wallpaper Company). South Glens Falls became a dumping ground for PCBs and TCEs tank trucked up river from New Jersey. The village water system was poisoned and the garden village culture timed out. It was as if I was raised in the 19th century and suddenly found myself in the 21st.

THE LITTLE APPLEGATE

The Little Applegate Valley (LAV), where I have made my home, has been through a lot since the Dakubetede, Athabaskan speakers, were removed and Beaver before that. The Human social fabric exploded through the gold rush with many thousands of miners camping, tearing apart the stream beds and tunneling along quartz seams. There was even a Chinese community who built the China Ditch and were not treated as badly as elsewhere. A silent film was made of downtown Sterlingville, of which no trace remains. Buncom, now a ghost town, was a small city. Even Crump had thousands of miners shacked-up all over the place, and several small businesses.

After the gold rush of the mid to late 1800s, the settlers who stayed were the ones who had started farms and ranches to feed the miners. These few white families filed homestead claims as well as mining claims and held grazing permits. There were several small sawmills in the valley and almost every ranch did some of everything to extract value from the landscape.

The farmed soils were poor to begin with and only got worse, the mining ditches, later used for irrigation, became less reliable, and the forests close-by did not grow back very fast. This is a dry valley, with steep slopes and metamorphic soils, which are alkaline and dominated by magnesium, and has not been friendly to settlement based on extraction. Many ranches have gone bankrupt more than once.

Over time, layers of settlement have included old families, new money, a few miners, firewood cutters, hunters, off road gladiators and counter-culture folks. Starting in the early 1970s, back-to-the-land folks, hippies, found abandoned homesteads for relatively cheap, with no building codes and reliable water. Since the late 1970s a drought has set in.

Tree ring records in the Sierras show 60-year drought cycles and the LAV has been through droughts and good years since the 1850s. The new Ecotopian settlers have brought environmentalism and farm-rescue into the story and resisted the latest extractive efforts of timber corporations. The industrial scale clearcutting that became the norm after 1980, in the PNW, is even more disastrous on our brittle landscape.

Meanwhile, in the late 1980s, small organic farms started to operate and the first renewal of community organizing, other than forest defense, bloomed. By the early 2000s land speculation for country homes had ballooned and now, plenty of fancy cars are on the roads.

The hippies somehow brought gentrification in their wake.

The Grange movement had its day during the 20th century, initially founded in 1867. Now new non-profits and cooperative marketing have entered the community. There is a certain frontier-flavor of culture here. The LAV of the late 19th and 20th centuries never had churches; there were dance halls and schools. The ranching neighbors always have cooperated on cattle roundups and barn raisings and next-door kids have married each other, forming ties between ranching families.

The newest layer of settlement by small farmers has brought more Ecotopians to the valley and we now have re-emergent culture. A critical mass of young people have settled or visited to allow cultural institutions to rise parallel to the more business-like nonprofits. A competition, of sorts, began between farms after small musical bands showed up at Siskiyou Cooperative CSA farm celebrations. This gelled as the Battle of the Barn Bands. We have had at least four of these hosted by different farms, sometimes even in a barn. A private community center started to host a yearly social event called Cabaret, where folks put on skits and acts, playing with themes wide and challenging. A local wildcrafting winery hosts tastings with entertainment and catered food.

Reading groups and clothing exchanges, seasonal themed parties and volleyball at the local county park materialized. Some of the farms built businesses such as a multi-farm CSA, bakery, brewery, daycare, massage, a couple of creameries, a couple of small sawmills, and seed saving. Farms and businesses learned to cooperate and got acquainted. Several farms hosted educational events and courses. Environmental protection work has continued, and now we are figuring out how to live and work with the cannabis industry.

The LAV reached a level of population and commitment that suggested even more integration and so the idea of community inventory and mapping as a shared group effort rose up.

There is a thread of continuity in the story of the last 170 years in our valley. There is rarely any government presence—those institutions are remote, as we are remote. There is barely ever any law enforcement. Too far from the coffee urn? Too small a tax base? We as a community of one hundred have mostly been on our own. We have succeeded and failed on our own wits. Through all of this the land has taught us. We have learned from Place. Some subsistence arrangements have remained and some have been left behind. The Dakubetede were right to call themselves "the people of the beautiful valley." We are along for the ride because we love being here.

COMING INTO HOME AT WOLF GULCH AND LITTLE WOLF GULCH

When the project at Wolf Gulch opened up in 1999, I was entranced with the pocket desert forest mosaic. My new friends, Maud and Tom, wanted to find a farm that no one thought could be farmed. When we toured Wolf Gulch, and it was so run down, I did notice the tall Star Thistle and basic irrigation layout. There was only a small spring deep in the downcut gulch and a slow well and the irrigation water had been turned off in 1962. You can see the results on YouTube if the satellites are still up.

After we did all the earthworks and infrastructure upgrades to support a seed growing farm, I looked at the adjacent 40-acre Little Wolf Gulch that we had labeled "wilderness." I asked if I could negotiate a life-lease and establish a forestry camp and start tending the wilderness. Twenty years later we have lots of stories to tell. And we learned to burn safely. Feels like home.

Little Wolf Gulch (LWG) is a 40-acre section of Wolf Gulch Ranch in the Little Applegate Valley of Southern Oregon. Wolf Gulch Ranch includes Wolf Gulch Farm where a 20-acre community supported agriculture and seed growing business has been ongoing since 2000. The community of seasonal workers and residents varies between 8 and 20 people, a small hamlet. The ranch is located in a pocket desert, one of the driest places in western Oregon, that sees as little as 10 inches of rain a year and as much as 30 inches. There are three steep gulches that cross the ranch: Wolf Gulch itself, in the middle, has thirteen head gullies. The east gulch is almost always dry, and is undeveloped, with a small drainage basin. Little Wolf Gulch is on the west and has four head gullies that start on the ridge of Little Bear Butte.

A gulch is a steep, dry valley, that has no dependable running water.

LITTLE STORY

The modest little story of Hazel's life goes like this: "Hazel has a little cabin in Little Wolf Gulch just below Little Bear Butte and just above the Little Applegate River and we have a little bit of fun there but we do not brag about it." We ask visitors to not wax too glorious when reporting their experience; perhaps it would be better to tell about all the challenges?

Very high wildfire potential. Rattlesnakes, Ticks, and Poison Oak. Mountain Lions and perhaps Wolves already. Isolation, peace and quiet, which is hard on some people. No cell phone service. Difficult to find as even GPS does not work here: one needs to read maps and follow directions. The canyon is a "low signal environment." Dirt road with blind hairpin curves. Hunters, loggers, pot growers and radical hippies. Hard working ranchers and farmers.

When we found this derelict ranch in the late 1990s it was a mess, with a view. The gulches all run to the south and the view is of the Siskiyou Crest where the Pacific Crest Trail runs along a 6,000-foot ridge line with 7,000 foot mountains; Dutchman's Peak is due south. The Powells and their extended family wanted to find a farm that no one thought could be successful. After Tom and Maud had been to Ladakh in northwest India, they were impressed with difficult but farmed landscapes. We like to modestly suggest that there is nothing at Wolf Gulch Ranch that anyone thought was valuable: no water, no timber, no gold, no aggregates, lots of star thistle, and rundown buildings.

The ranch was first homesteaded in the 1880s and the Sterling Ditch, a mining canal, that was constructed on the range above the ranch provided irrigation water to all three gulches until 1962, when the diversion was closed. After the end of irrigation water the ranch was largely abandoned and we bought it for less than the price of any other properties that we looked at. After a lot of planning, heavy equipment work, and a big new house construction project, we were able to start farming with water from a spring in Wolf Gulch and we got a winter diversion permit to fill our ponds.

After we got the farm going, Hazel negotiated for a life lease in LWG. This forty acre section of the ranch is classified as wilderness on the ranch plan. We have developed a net of trails and a small forestry camp to support the farm and reduce fire hazards. LWG is a fire laboratory: the Cantrell Gulch fire burned half of LWG in 1987 and we have since restored and under-burned acres of remnant Oak/Pine savannah. This landscape was burned regularly, every three to five years, by the Dakubetede First Nations people, who were removed during the gold rush of the mid 1800s. Tree rings show fires since the removals on a 20-year cycle. There is a lot of remnant ethnobotany still persistent, especially geophytes and basketry stools. This landscape is fire-explosive with thick chaparral, grown up since regular broadscale burning ended with the removal of First Nations people. We have two charcoal kilns and multiple burn pile sites.

LWG is also a laboratory for simple living and post fossil-carbon lifestyles. The vision is that small organic farms in the larger Applegate Valley could have small forestry operations that provide biochar while doing fuel reduction to protect the farms. Alternative forestry products can be marketed to support the forest workers. Oregon has a history of laws and principles that support forestry such as "access to economic opportunities," "temporary housing for seasonal workers," and "work camps for forestry activities". We, at Wolf Gulch Ranch, demonstrate ways to support people and farms, while we restore periodic "cool burning" fire to the landscape, hopefully to avoid catastrophic, hot wildfires. We demonstrate how we might have Humans living again in close relationship to Place while doing restoration and stabilization work.

The daunting future of climate change and the backlog of excess fuel on this landscape suggest the benefits of a re-localization culture. We have been teaching about this and envisioning through Siskiyou Permaculture with our Permaculture Design Course (PDC) and our advanced Permaculture courses including Social Forestry, Optical Surveying, and Ethnobotany.

We have also been developing value added products and tools that support de-consumerization, systems of living that do not use imported-energy. This is ambitious and not very lucrative, but does get Humans back on the land and in close contact with Nature and healthy living. So far we have produced several types and grades of charcoal, firewood, building poles, baskets, natural buildings from on-site materials, and decorative products for ornamentation that celebrate Place.

*Vernacular living,
appropriate to what we have,
where we are.*

A DAY IN THE LIFE OF LITTLE WOLF GULCH

After an intern completes the application process and goes through the get-to-know-you trial and orientation, the intern sets up house holding. We have found that the chores of "simple living" take about four hours a day and so we are lucky to get four to six hours a day to work on projects. During breakfast, we discuss the plans for the day after we have assessed the weather and opportunities.

If there are one or two interns they can stay in LWG. Three to six interns can form their own temporary "family" at the other camp facilities in Wolf Gulch. Hazel needs some private time, don't we all? A critical mass of interns, forming their own community, gives Hazel some space and time off. Meals are a good time for planning, reports, and sharing, with at least one meal a day as a community meal. Meal planning and preparation are part of the daily chores. We mostly eat meals that take some preparation ahead of time: Perhaps we have to soak something, or grind something, or cook for a while, or wildcraft some greens. The cabins in LWG and the classroom in Wolf Gulch have wood stoves and we try to avoid using propane or butane stoves. In most weather we can cook on charcoal stoves in open shelters.

Drinking water in this desert forest is part of our challenge. Wildlife has access to tree holes and the river in the canyon. We do bring in some spring water from off ranch but mostly we use rain water from barrels and tanks and mineral rich spring water, stored in a tank at LWG, from the few months that the gulch spring runs (March through June, in a good year).

Housing codes for small unpermitted buildings do not allow indoor plumbing; all our water is carried into the cabins in buckets and jugs. The tank-stored water needs to be filtered or boiled or both. Bathing is done in warm weather with solar-shower bags, and in cold weather with sponge-baths, inside the cabins. There is a sauna one mile away from LWG and we have started yarding materials to build a sauna nearer by.

In our Permaculture Design Course we talk about CLP: convenience, license and privilege. These are attitudes that need examination to prepare for the post fossil-fuel future. The industrial expectations of moderns demand shortcuts in order to meet the hurry-up demands of financial capital and global trade, but these expectations of endless growth are killing the biosphere. The very old Zen saying "chop wood, carry water" is very much our practice at LWG. This at best can lead to enlightenment.

Meanwhile we have some attitude adjustments to practice. When one is truly present, and aware of what it takes to live a good life, slowing down and paying attention, have their own rewards. We try to avoid waste. We only go to town for special extras twice a month, which takes some planning. When we need something we do not just go to a store, we need to figure out how to do without, repair a tool, search our stashes, or use local materials.

Communications out here can be a challenge for folks who are used to checking their screens constantly. We are "off the grid" at LWG. There is no cell phone reception. We have a satellite dish for internet powered by solar. Since we are using stand alone small solar-electric systems, we do not have surplus power. That means we use very small area lights for reading and we only check e-mail once or twice a day, depending on cloud cover.

Our deep-cell 90-pound batteries have to be carried in by wheelbarrow, and that means we do not have huge power storage reserves. All our cabins and the classroom have lights, but they are small and we have to budget our power uses. That includes re-charging devices. This is righteous simplicity as well as learning to budget.

There is a landline at the main barn and house in Wolf Gulch but that, as we have already noted, is a mile away, uphill, and private. The closest cell phone reception is eight miles away by road and that means driving and using fossil-fuel or an all-day bicycle trip.

The rural mail boxes are two miles away and down canyon, which is an athletic bike ride, perhaps to find someone has already picked it up, then returning to the farm in searcht. We tend to get our mail in town from a post office box twice a month!

We prefer to have interns stay at least two weeks, in order to get into the routine and become comfortable with all the new skills of off-the-grid living. Once an intern is accustomed to our ways, they are always welcome to return for a visit, as they already know how to fit in and have a good time. Sometimes we have special two-week intern programs and those are more intensive. New folks have to get with the program fast. We prefer that interns have already been to Wolf Gulch for a course, so that they have some idea what to expect.

The toilet systems are called "dry," because they do not flush. That means Human manure composting and urine composting. We also compost all our food scraps. We do not want to tempt wildlife with messes, especially the Bears and Raccoons, so we clean up fast and we take out any smelly garbage and packaging that we cannot compost (see "Wild Animals Tell Us" in the next chapter). There is a lot to learn about not interfering with wildlife while camping in the wilderness. Avoid temptation is the motto! Clean-culture requires some learning. . Wash hands with minimum use of water! Avoid waste and garbage messes!

Then there are the forestry-work projects. We emphasize safe and skilled use of hand tools. We also emphasize attention to detail. We are doing complex forestry here and that means sorting the wood we are cutting into several categories. We are also using a lot of fire, so we have to be prepared to avoid any escapes, the ranch needs protection. We need to avoid any injuries and expect interns to be responsible for their own safety! Our work is not industrial or continuous and unvaried; we take lots of breaks to discuss what we see going on and to plan our next moves. There is a lot to learn about identifying species, protecting sensitive ecological features, understanding the effects of our work and finessing perfect timing.

The social aspect of our forestry includes getting along with each other and learning to coordinate our activities, as well as getting along with Nature and noticing subtleties. The forest misses Humans and modern Humans are mostly disconnected from the wildness that we evolved with for so many eons. This is a big part of Social Forestry; we all have a lot to learn about getting along with the landscape and each other. Hyper-individualistic moderns are insensitive to *others* and self-centered in their priorities.

Social Forestry widens our priorities and emphasizes reciprocation and gratitude. We will be suggesting a lot of reading, in order to orient and integrate our attentions and values. The direct experience of Nature talking back to us is a powerful re-connection with our deep heritage. Reading the land and learning to talk back to—and listen to—all our Wild neighbors is polite and kind. We hope that everyone who spends some time in this special place-ness grows into maturity, with full hearts, ears, breath, touch, and taste.

THE ANNUAL SOCIAL FORESTRY COURSE

In midwinter, before anything starts to sprout and while most animals are hibernating and the tree sap is still down in the roots, we schedule a week-long intensive to introduce folks to the broad subject of Social Forestry. Wolf Gulch Ranch, with its sun facing drainages perched above the noisy river in the canyon, dries out quickly on some winter days.

The east–west running Siskiyou Crest ten miles south of us blocks wet storms from the south but sometimes captures slow rains from the west. We try to guess at weather windows by following the moon through the stars. If we get what we expect, there is a window during the course where the fine fuels are dry enough to have a day or two of broadscale field and forest underburning.

Young folks these days are looking for hand skills, a chance to do cool burning, charcoal making, and community decision making. We can't do it all in a week but we take the temporary community of winter camp through some adventures and stories. The course always fills up with early registrations.

This is the most recent iteration of skills and knowledge that Hazel has been teaching since the early 1960s. First, it was the Nature Lodge at Camp Wakapominee, then a natural field studies bachelor's degree. When the Vietnam War drove Hazel and friends to

emigrate to California, we were confused and distracted for sure. Blacklisting from medical relief work and prejudice against conscientious objectors blocked graduate school or any teaching opportunities.

The emergent Ecotopian counterculture spawned organic farms, farm to city natural food stores, distribution systems, free clinics, free universities, and collective efforts at social change. These kept us busy but economically times were tough.

OK, NOW WE GET A PERSONAL STORY, A NARRATIVE OF THE NATURAL

It was late summer 1970, when a friend put up a course proposal for "Wild Edible Plants and Woodslore" at Laney College in downtown Oakland and 36 people signed up. I was amazed to be told (having no clue what was going on) that I was now a college instructor! I was taken to the Dean of the Experimental College and was asked, "Do you have any college degrees?" I had hard copies thanks to having done student work-study in the registrar's office at college. I then tried to say something more, thinking this was a test, and was told to shut up and show up tomorrow morning in the designated room. There they were, 36 young folks with some diversity and I asked, "What do you want from me?".

They brainstormed a list of outdoor skills and I took notes. The course took off and I was relieved to be doing something I loved. The next semester we had 120 sign ups, nine breakout labs (natural fibers, edible wild plants, organic gardening, plant identification, ethnobotany, herbal medicine, natural foods, mushrooms, and "survival skills") with paid assistants, and buses to wherever. By then, I had soaked up a lot of California Natural

History and Ethnography— the used bookstores near UC Berkeley were awesome—and I added a West Coast wing to my college book collection.

The four years at Laney College landed me on the West Coast. I was becoming local. But settling in a home drainage took a lot of trials and relocations. I moved to Southern Oregon and sampled several of its diverse provinces. I started freelance herb walks and short courses, spending two summers as an Outdoor Education Specialist with the US Park Service.

Then in 1982 I was invited by friends at Tilth in Seattle to participate in the first Permaculture Design Course on the West Coast. There, I found a basket of design approaches that wove together my skills and interests. My time with permaculture teaching and organizing has taken me to various corners of the Earth and allowed me to interact with the remnants of local and Indigenous cultures. I have been welcome in diverse local cultures because I come from a local culture and can relate.

Can you follow this nest weaving? Messy and sideways for sure, but we come out where we started with woodslore and farming. Social Forestry has always been there, we are just trying to open it up and tickle it a bit. Let's see what we can find in this twiggy-ness.

The Principles of Relationship include

16.
Respect leads to reciprocation; a high degree of
implication leads to *balance-within-limits*.

17.
We need a relationship to Place. Place needs us to
come into the web of relationships that keeps
All Beings present and doing well enough.

18.
We are surrounded by dynamic natural dramas.
The clown is following you. Learn to follow suggestions.
Avoid temptation and know the differences between
sympathy, empathy, and compassion.

CHAPTER 4
Relationships

The social in Social Forestry is all about relationships. The interaction and exchange between Humans, and between species, Place, and All Beings, weaves a web of entanglements. Much of this complexity is beyond our Human ability to map or hold-in-mind as individuals, although it is possible to imagine collectively, through ritual, to approach/become One with All. Some of the iconic ancestral eco-tapestries are celebrated in traditional stories, and some new emerging relationships are discovered through dreams and revelations, reported from pilgrimages and vision quests.

As we pay attention to Place, as we begin to use all our senses—when we can start to hear what the Rock People, the Tree People, and other non-Human peoples have to say—there will be ongoing revelations and realizations. That is attention to being social; we listen to and interact politely with our neighbors. If we treat all relationships with respect, we can learn how to move in ways that support the Web of Life.

THE STATE OF MODERN RELATIONS

Tremendous stress has been put on our thin planet's skin with the widespread establishment of civilization. Now, recovery and regeneration require a return to tending the remaining forests, shrub and grasslands, rivers, lakes, seas, swamps, and marshes. The benefits of carbon sequestration, aquifer recharge, soil biota enhancement, and protection from aquatic pollution, all imply the need for "more eyes on the land" as says Wes Jackson.

Kings first removed European traditional landed-folks from their drainage basins to meet the imperial demands of mobile labor and general extraction (Maitland 2012).

Now, with massive industrialization, urbanization, and chemical agriculture, much of the landscape has been abandoned and degraded. A new Retro-feudalism with small farms and Social Forestry may ironically be called for, to soften the excesses of the ongoing addiction to fossil fuels. The real work is social, re-inventing relationships, so that we are cooperators and not oppressed by rigid hierarchies: local solutions need Place-appropriate social arrangements. We need to recognize that All Beings have cultures and languages.

Thus we become Naturalists, Herbologists, Bodgers, Charcoaliers, Sawyers, Rangers, Mothers, Weavers of baskets, fiber-twisters, and all-around-trouble. Who knows when and how these ways-of-the-woods will become widespread again? Most social and ecological scenarios of persistence and resilience emphasize re-localization: net-carbon-neutral energy use, living within resource limits, habitually embedded in Nature.

Our Ecotopian vision
sees a mosaic
of complex perennial Horticultures,
interwoven with
shifting and shimmering
Wild-ness.

ECOLOGICAL RELATIONSHIPS

Let's review the vocabulary of ecological relationships. Not all inter-species interactions are narrowly beneficial or detrimental. Some are seemingly neutral (probably common, just un-studied). Yet the existence of the arrangement itself contributes to the greater web. *Commensalism* or *Mutualism* conveys a neutral expectation of useful exchange, but these are actually generic terms that can be further specified.

Commensalism refers to co-inhabitants, or to different species that share the same physical or trophic space. **Mutualism** refers to species that have intertwined sharing that does not exclusively benefit only one partner.

Lichen on tree bark, or epiphytes perched on branches, are *commensalites* with the tree. The benefits to the tree are subtle, perhaps habitat for insects. A Mustard flower and a Solitary Bee have a mutualistic exchange, with neither partner harmed or exploited. A gifting economy is mutual aid.

Symbiosis implies a tight relationship with mutual benefits which can be highly specialized through long co-evolution.

Lichens are a symbiotic union between a fungus and an algae. The fungal hyphae penetrate the mat-embedded algae cells, and there is an exchange of minerals and water for sugars. The spores of the lichen are complex and carry the relationship with them. Many trees and perennial shrubs and grasses, perhaps all plants with roots, have symbiotic relationships with fungi.

> *We've all heard of
> the fungus and the algae
> that had a lichen for each other?*

Mycorrhizal fungi, such as Boletes and Truffles, allow Pine tree seeds to germinate, sometimes under harsh challenges, and extend the root system of the young tree, in exchange for sugars and complex enzymes from photosynthesis in the tree's needles.

Deer find Truffles by their awesome perfume, and then spread their spores in pellets through the savanna, inoculating new tree seedlings. A song can be sung about how All Beings on Earth are woven together within a mutually balancing dance called Gaea. The term Symbiosis has been adopted by many cultural artists and activists.

5 Animals in Ecosystems

Parasitism suggests a non-mutual exploitation, but even this can be subtly mutual, as the parasite might convey immunity to certain diseases, or at least keep the host alive as long as possible.

It is said of alkaline deserts that "the parasites have parasites on their parasites." The brittle environment of the desert favors certain highly stacked relationship solutions to persistence. These systems of adaptation can be brittle in themselves by being over-specialized. Parasites can be beneficial to ecosystems while taking advantage of host species.

Ecospasms can express themselves as epidemics with high carrying-capacity degradations, such as Voles on drained muck lands (Ashworth 1979). An explosion of populations.

Or they can be a nutrient transmutation force, such as some bark beetle infestations in already stressed and overstocked Lodgepole Pine stands. After the snow load lays down the tall thin trunks or a wild fire converts the standing snags, a new forest with more berries and species complexity re-covers the site, until the young Pines grow thick enough to close their canopy and exclude sunlight from the remnant understory layer, at which point the stand could be said to be overstocked, senescent, and vulnerable to catastrophic conversion.

IMPLICATE ORDER, THE RELATIONSHIPS OF RELATIONSHIPS

Implication is a qualitative mapped measure of the functional complexity of species, in a web of relationships.

Usually, this is different from charting the associates of *keystone species*, such as predators. This is more horizontal, as in "the high level of implication for the White Oak." Implication can refer to a whole ecosystem's complexity, as a collection of *highly implicate species*. The White Oaks are storied with this web-ness.

The petals of a Rose,
in the unopened flower bud,
are enmeshed in each other;
*the **implicate** folding of relationships.*

A **Keystone species** is one that, if it fails, could be the beginning of ecosystem collapse or decline. A **Highly Implicate Species**, because of its abundance of relationships, has all the resilience, built-in redundancy, and access to accumulated experience that being well-connected implies. See "Acorn Woman and her Consorts" below. An **Indicator Species** mimics or represents the overall health of an ecosystem.

*By watching the **indicator species**,*
we have a measure of complexity,
a window.

For example, Humans have come to depend on Oak/Pine savannas all across Earth. Not only have Humans learned to find nourishment in the mix of bulbs, berries, nuts, and animals, but they have learned reciprocation through Cool Burning (see Chapter 7) and disturbance regimes: a digging stick, or a hunting burn-flush, or a downhill, night-time slow-fire, or a tree-fall, or dig-pit, or sod-flip—all these are disturbances, where Humans mix things up and make messes.

Perhaps folks, with appreciation for what they have wrought, might offer some reciprocations with seeds spread, or sod turned back over, or cleaning up dead branches with long poles, shaking up the Oak crowns, dropping the dead wood (in winter). Perhaps this all leads to an under-burn, before shaking again (the next fall), for mast. Multiple ecosystem functions are tickled.

*The relatively gentle smoke
of an under-burn,
infiltrates Oak crowns
with complicated fallout.
Who comes around soon after?
Flocks of small birds?*

Another way that Humans reciprocate is with attention, song, story, and ritual. All of which slows down the Humans, to gather their presence and observation powers, but also recognizes Kinship and says so. "Talk back" is what Abrams suggests in *The Spell of the Sensuous* (1996). Often enough, Nature talks back to Humans in especially propitious moments of encounter, delivering seeming miracles. Humility, on the part of Humans, is appreciated by the Consorts of Oak, but some messes and silly behavior is also good enough to be some giving-back. Thus Humans become implicated with Nature. Different ecosystems blend different cultural practices.

This is why *implications* are more encompassing than simple food webs and trophic cascades; more attributes of relationship are accounted for. Linear mapping of perceived cause and effect does not lend justice to the more multi-dimensional aspects of interrelated Nature.

Another favorite relationship idea is that Human horticulturalists have been very useful co-evolvers, where their disturbances and harvests have encouraged active nutrient loops and provided disturbed edges.

The concept of **ecological meta-stability** is that pocket/patch patterns (ecological mosaics) and periodic disturbances that create new edges support biodiversity by maintaining some stock of pioneer and ready-repair species.

Thereafter, with a minor or major disaster, the landscape-scale recovery is relatively quick. A continuous, closed-canopy mature forest without any openings, or big tree falls, or landslides, or burns, may not have enough seed bank to recover after a stand-replacement fire, one that burns up the whole forest.

Thus I believe that the highest biodiversity on Earth, ever, was found within the last few thousands of years and the highest biodiversity per area was found on human-managed landscapes (as hypothesized by Kimmerer 2013, and Anderson 2005). Species, such as *Brodiaea* and some other *geophytes*, seemed to have changed their habits, with more small bulb reproduction and less seed viability. The digging stick, over tens of thousands of years, rewarded break-away bulblet propensities. This is Horticulture, not agriculture. In the last ten thousand years, power-driven agriculture has massively simplified species relationships and showed up on all continents. Agriculture comes with armies and commits massive uniform disruptions, not as friendly to complexity as the distributed disruptions of Horticulture.

> **Geophytes**
> are bulbs and roots dug for food;
> "a plant below the ground."

As we further discuss below and in Chapter 13, craft guilds, drainage basin councils, and ceremonial societies facilitate complex, landscape-wide Horticulture, by practicing traditional relationship and skill sets that coordinate sequence, with perfect timing. When we contrast Horticulture with agriculture (Hemenway, 2009); we could use the terms Wildcraft/Bushcraft, or Wild Horticulture, or Deep Local Horticulture as synonyms but not "hunting and gathering" which implies only taking, not reciprocation. This slur has been used as justification for the removal and genocide of First Nations peoples.

Our Social Forestry/Horticulture imagines accumulating multi-dimensional knowledge of highly implicated ecological matrixes, and the procedures that have best outcomes. We have some clues on how to learn to do this from TSK, TEK, and IEK. We have some good science, and need more, to answer questions. And here, in this book, we present the visionary work from many quarters, to lead our imagination to re-localize and become People of Place.

Humans have relationships, both past and future. Through language, symbols, art, dance, and ritual theater/council, Humans dance through the ages. Humans, as a species, persist best through *Culture*, as long as they remember the Long Story, and as long as they have the opportunity to do the ecological repair and gather the community focus to do the visioning dance.

Always coming into relationship.

ACORN WOMAN

Acorn Woman is the PNW guardian spirit of the White Oak tree, an ecologically key species of the forest in our region, especially in the Oak/Pine savannah. The White Oak has a high degree of connectivity and relationship in ecosystems, which is called *implication*, a poetic word in the English language with lots of different "implications." White Oak, Acorn Woman, the Triple Goddess, the White Goddess, Diana the Huntress, all of these names are associated with the White Oak across different cultures, in different places, showing us the central ecological importance of the White Oak tree. Because she's such a strong woman, she has many consorts, the spirit/animal/plant/fungi eco-guilds of the White Oak. We're now going to meet her friends and talk about their relationships, to her and to each other, and learn about our growing Human relationships to Place.

There are many White Oak species in the genus *Quercus*. The Oak genus can be divided into at least four groups: White Oaks, Black Oaks, Red Oaks and Live Oaks. Oaks are found in many ecological situations all around the northern hemisphere of Planet Earth. White Oaks can be found in low lying swamps and growing as thickets on 7,000-foot (2,000 meters) mountain ranges.

Perhaps because of her many consorts, she thrives in a wide variety of habitats. The White Oak acorns have fewer tannins than other groups and the oil content of the acorns is high. The acorns are easy to leach, to remove the bitter tannins, but do not last in storage as they go rancid. Some local legends say, "We only eat those acorns in heavy mast years, when there is plenty for All Beings. We leave all in a lean year for our relatives that do not have our hands. We can store and leach other acorns."

The White Oaks are consistently personified by Human cultures. They are associated with goddesses and nature spirits. There are compelling stories told of her, that the children hear again and again. It is obvious, on careful observation, that the White Oak is the tree species with the highest degree of implication on many landscapes that support Human cultures. Thus, she is held in high regard and is protected from harm through taboos and etiquettes.

The wood of White Oaks is dense and heavy. It is fibrous and does not split easily. The oldest tree rings in the center of limbs and trunks (the growth rings that show fat spring wood tubes grading to tiny summer wood tubes) support the deposit and crystallization of minerals. These old water vessels, superseded by new wood closer to the bark, come to be blocked by net like tyloses inside the abandoned water vessels. This makes White Oak wood a superior barrel stave wood, as it does not weep, or slowly leak, as Black Oak staves do.

White Oak center wood turns black with this mineral resistance against heart rot. As the trees age, they are also injured by passing fires, climbing animals, high winds, and falling snags. This allows some of her cohort fungi to get established in the inactive heart wood. The trees become hollow but the wood around the hollows hardens wonderfully, thanks to tyloses. This way, she survives decay and continues to "rent out rooms"; she is a gracious host to all sorts of cavity dwellers and lives a long and, sort of, stable life (Marionchild 2015).

Oak/Pine savannas ring the globe at the mid-latitudes. These are grasslands with scattered clumps of berry bushes, occasional big nut Pines, lots of herbs and roots, shoots, and a changing and flowing contingent of herd animals, predators, ground dwellers, and birds. Well tended, a Pine/Oak savanna is among the most productive landscapes for feeding, clothing, and sheltering Human cultures. Thus she is called Acorn Woman in many places, the Triple Goddess in Europe and North Africa, and various goddess names in northern Asia. Everyone notices, if they stick around long enough, that the White Oak is the Queen Goddess.

SHE'S BEEN HERE A LONG TIME, IN SOUTHERN OREGON.

One can pick up petrified White Oak on the south-aspect slopes of Grizzly Peak, north of Ashland, that is millions of years old and apparently the same species living there now. One can sit under her and be sitting on the bones of her ancestors. Makes the Human species look like a flash in the pan, just passing through. With such a long and glorious epic story, of course she has a lot of friends. After all, she offers food, fiber, and shelter to many animal species, not just Humans.

Many cultures have seen this poetically, that there are stories to be told about White Oak Woman, Acorn Woman, and her renters. That way, by learning the stories, we'll move a little closer to where we are, and who's here, and what our relationships are to Place. This is ultimately about Place. Stories orient us.

THE ORCHESTRA ASSEMBLES

The cohorts of Acorn Woman that dwell in the ground, let us praise them.
The Bull-snake, the Racer, the Garter, and the Viper.
The Alligator Lizard, the Blue Belly and the Skink.
The Ground Squirrel, Mole, Gopher and Mice.
Mushroom mycorrhiza, root hairs, and bacteria.
Morel, Bluet, Bolete, Amanita.
Brodiaea, Fritillary, Sego Lily, Lomatium,
Green Thistle, Oshala, Yampa.
Grasses, Mustards, Mints, Capers.
Monardella, Clarkia, Collinsia, Yerba Buena.
The special worms, all blue and pink.
All dwell together, some eat others, others are eaten.

The cohorts of Acorn Woman under her boughs, let us praise them.
Bearberry, Deerbrush, Snowberry, and Honeysuckle.
Ceanothus, Manzanita, Yerba Santa, and Silk Tassel.
Mountain Mahogany and Lacquer Bush.
Spoon Willow, Dogwood, Mock Orange.
All these stand up and feed many.

The cohorts, up in her crown, towering and binding, let us praise them.
The Grape and the Honeysuckle, Old Man's Beard, Mistletoe,
The Jeffrey Pine, the Ponderosa, Grey, and Sugar.
The Ash, the Maple, the Black Oak, and the Alder.
From creekside to ridge, the trees stand with her.

The cohorts, the four legged, the ones passing by, let us praise them.
The Deer and the Elk, Bear and Grey Squirrel, Chicory.
The Mountain Lion and Wolf, Bobcat and Coyote, Fox.
The Hare, Skunk, Badger, Ring Tailed Cat.
The Woodrat, Mouse, Shrew, and Raccoon.
All these pass by and nibble or pounce.

The cohorts of Acorn Woman, flying by, on wings, let us praise them.
Flutter-bys, Tortoiseshell, Duskywing, Mourning Cloak, Sister,
Coppers, Skippers, Whites, Sulfurs, and Swallowtails.
Wasps, big and small, Beetles, Moths, Bees, Hornet,
Aphid, Scale, Stick, Cicada, Red Bug, Praying Mantis.
The birds that search and hop and dive and swoop,
Quail, Grouse, Turkey, Flicker,
Nuthatch, Creeper, Hairy, Downy, Acorn, and Pileated peckers,
Bushtit, Chickadee, Vireo, Goldfinch, Warbler, Sparrow,
Thrush, Bluebird, Flycatcher, Junco, Kingbird, Cowbird,
Nighthawk, Swallow, Swift, and Bat,
Scrub Jay and Stellar, Raven and Crows,
Sparrow Hawk, Red-tail, Coopers, Sharp-shin, and Eagle,
Osprey, Blue Heron, Dipit, Duck and Goose.
Some live here, some are just passing by, roosts galore.

And Humans, the tenders, the watchers, the bringers of fire and song,
We praise them All. All things have Spirit. All our Relations.

INSPIRATION

Take a deep breath.
See all the lichens hanging down?
The air is good here.
Can you hear the stories in the breeze?
The clicking of insects in her hair?
I bet you feel welcome in Oak woodlands.
Think of your ancestors.

The White Oak was central to the stories of the old-old Humans of Europe—tree pantheons of goddesses, the sacred groves, and led the Cultures of Place. The Mistletoe was cut with the golden sickle and to harm her was to suffer punishment by custom. Travelers who took shelter in hollow Oaks had visions and dreams that changed cultures. Europe has seen waves of trans-migration. Lots of stories lie in the ground. These old-old Humans were redheads (a rare recessive trait), they were amazing cave painters, they buried their dead in graves full of flowers. They knew a lot about glaciers and bogs. They were displaced by dark folks in the west and Pony and Cow herders in the east. The Pony folks spread west and became the Gaels. To call them Celts was meant to be an insult. The Gaels were interpenetrated by empires from the south and raiders from the north. The languages mixed together. Many accumulated stories were kept as songs and epic poems. See Graves, 1948, and Matthews, 1991. The Triple Goddess of the Gaels, associated with the European White Oak, is anthropomorphized as Diana, Goddess of the Hunt, Aphrodite, Goddess of Beauty, and Hecate, Goddess of the Herbs.

As we have seen, the PNW also has stories of White Oak, telling of Acorn Woman. She has a harsh reputation, one does not mess with her. She takes revenge for harms to Sacred Places. When she appears to someone, they are about to die. Settler folks at Sandy Bar Ranch in Orleans, California, tell a version of this story. White Oak savannas were so important to survival that even during drastic climate changes at the end of the last Ice Age, Humans managed to transplant White Oaks and the associated geophytes far north (Turner 2005).

The Japanese Tea Ceremony uses White Oak charcoal. The White Oak there is carefully tended and cut only with permission and etiquette. The making of the charcoal is an apprenticeship craft. The Tea Ceremony is centered around the charcoal fire that boils the water. Several types and sizes of White Oak charcoal are used in prescriptive ways. This is a very old practice.

Undoubtedly, there are other local stories and practices about White Oak. She is central to *Tending the Wild* (Anderson 2005) and *Keeping It Living* (Turner 2005), books that help to orient our work of restoration Horticulture. With a few simple tools and a lot of accumulated knowledge, First Nations women cultivated the rich complexity of the Oak savanna. The digging sticks varied in design and type of wood, by the soils and types of bulbs. In the swales between Oak cloaked hills, they dug Camas and Biscuit Root, Yampa and Rice Root.

Seeds were winnowed in-field with beating sticks into baskets. Kids were sent up into Black Oak granary trees to knock down dead limbs and to beat down acorns with long poles. Acorn storage was carefully organized with rodent-resistant caches and sorted for nuts that would last the longest. Acorns and Buckeye nuts were water-leached in baskets and Maple leaf lined pits, down by streams, to remove bitter tannins in acorns and the **poisonous nerve toxins** in Buckeye nuts (*Aesclepias californica*).

ACORN WOMAN STRIKES BACK

This story was learned at Sandy Bar Ranch, Orleans, California, on the Klamath River. This riverside terrace, the sandbar, has a burial ground in the center of a rare patch of good soil in the rugged river canyon. The Karok tribe, First Nations people, released this village site to a white farmer on the promise that he would not plow the burial ground.

Focused by the landscape on that exact spot were several spirals of influence, including the river in flood, swirling into an eddy and depositing silt and sand in a big event, the daily swirl of fog coming up the river from the ocean to its usual maximum upriver extent, the lines of gullies and waterfalls on the canyon wall, and the movement of the sun skirting the south as it travels toward the west horizon.

The farmer plowed the burial ground and just as he did, a very small woman wearing a basket hat and carrying a burden basket stepped out from behind a White Oak tree and threw acorns at him. He was found dead in bed the next morning. No one has plowed the burial ground since and there is a split rail fence around this portal between this world and other worlds.

THE TRIPLE GODDESS'S REVENGE

It was a hot day, early in the new imperial century. At the mid-slope edge between the riparian forest and the Oak/Pine savanna, where the cabin was being built, the assessment had been made that one Oak must go. The pocket transit angles and compass azimuth estimated the winter sun path: If this one White Oak was cut down, the cabin would solar warm earlier on cold winter mornings.

The tree had been spoken to. Everything explained. All parts would be lovingly processed and used in celebration and appreciation. With much apology the cutting began.

The best pull saw, stainless steel and very sharp, cut nicely and the tree fell south into the open meadow in front of the cabin site. All the limbs were cut neatly from the trunk and the lovely log laid there waiting for the next assignment.

However, all was not going well. The saw broke a tooth in the fibrous tough wood. A sad moment and unusual. The cutting with full attention still failed to be perfect. Something was wrong.

The Sawyer was tired and not entirely happy, even though the job was done and the new opening looked to be full of light. It was mid-afternoon and they headed down to the gulch-bottom shed for a nap. On the way, all hell broke loose; the clatter and whoosh was overwhelming. A Marijuana war chopper had arrived, slipping over the side ridge, and there it was loud and just over the treetops. It appeared to be trying to land on the cabin roof. All these colored lights flashing. Black and awful and studded with guns and ports so it seemed like a monster movie. The chopper moved off after hovering. There was no pot growing in Little Wolf Gulch.

After standing under the big Madrone on the way to a nap, the Sawyer now knew that something was really insistently bad. Umm, they wondered, perhaps that was a group of three Oaks?

The nap went deep. Hot and still buzzing with the helicopter visit and the broken saw and the realization that they were in real trouble now, the Sawyer climbed down from the loft, slipped into the sandals and opened the shed's thin screen door.

A very large Rattlesnake was coiled there, right in front of the door, and yet the Sawyer really had to let some urine out of a very full bladder. They had to pee really bad. They jumped over the Snake and ran down the gulch a few steps and let it all out. The Rattler slithered away towards the creek bed.

This adventure embodied, taught, and imprinted the absolute value of the advice: "Groups of three, leave them be!" The perpetrator of *coupage* had been struck three times. Just to make sure they got the lesson. At least the pitiful Human was left alive. The stories of Acorn Woman, known to the First Nations, do not end so conveniently. One should know the lesson well ahead of time, but then one would have no excuse. Feigning ignorance only got the Sawyer a reprieve, but not much slack. Better pay attention!

A shrine with bright white quartz pieces, a few sprigs of sage and cedar, and a trinket or three, was placed on the stump of the Oak. The shapely trunk is now the center post of the palapa, our favorite open shade structure, where we love to sit and visit. The saw still works fine. The winter sun path was only a bit revealed but the fire safety close to the wooden eaves is much better. The Rattlesnakes have become scarce, Hemp has been legalized, but we do not forget where we are, and the important lessons thereby.

Similar traditional stories involve special boulders in mountain passes, small lakes at mid-elevations, and Oak groves near springs. What we have learned at the forestry camp, from experience, is to never cut too many trunks in a White Oak copse. "Groups of three, leave them be!" Whenever we do cut down a White Oak, we make offerings, explain what we are up to and why, beg forbearance, and use all parts of the harvest ceremonially, for special value.

We do some limb trimming and some dead branch removal so that we can move towards a more fire friendly grove. Every branch we cut is examined to see if it goes to charcoal for the Tea Ceremony, or to the workbench for jewelry, or to the shaving mule for bodging. And we work always, with a lot of "talking back," compliments and praises, gratitude and humility are appreciated. Reciprocation is promised in the form of the return of fire and the reduction of competition from excess stocking (see Chapters 7 and 8).

THE AWESOME WHITE OAK TREES

The leaves are glossy green at midsummer and turn to red, purple, bronze, and brown in fall. The trunks are curvaceous and shapely, covered with lovely shaggy grey bark. The lichens hanging from branches and pasted on the bark are multi-colored and multi-textured. There are surprises hiding in the tree top and in the cavities. They can be inviting to climb. There is almost always some action and drama. The multicolored Oak leaf galls suggest the shape of Apples. Epiphytic life forms are found on all surfaces.

White Oak trees, as we mentioned elsewhere, can survive stand replacement, hot fires. They are capable of re-sprouting from deep roots or thick-barked stumps. They often have fire scars on their uphill sides from burning debris rolling against them. They are tough survivors. Good thing, as so many other entities are entangled with their fate.

But some Oak groves are not doing well. They lack reproduction, the front edge of the cohort of transitional age-classes, that allows replacement as older trees become wildlife snags. This may be due to the loss of regular under-burning, the compaction of soils by Cows and Horses, the reduction of mushroom vitality through drought and water table mining, or perhaps sadness with the loss of tending. Oak trees are often hundreds of years old, so they are still there, but are there any young ones nearby?

Humans love Oak groves and as they are often on poor agricultural soils, they get roaded and divided for housing developments and the development compaction and downcutting reduces their survival probability.

The Oaks miss us and we miss the Oaks, even if not so consciously. Culture can help remind us through mnemonic stories and ceremonies of gratitude and reciprocation.

WHAT WE DO FOR THE WILD

What we have come to understand Social Forestry to be is all about coming back into relationship with Nature; reciprocation through tending is a big idea. Our experience in the woods on the West Coast is that Nature is lonely; it misses the Humans. First Nations humans used to move all over this landscape doing useful things. In fact, the remaining Nature in the Little Applegate valley is a Human-influenced Nature.

There's a difference between conservation, preservation, and restoration; different concepts, with overlapping interests. Preservation is not appropriate; preserving what? People don't understand. They think Nature doesn't appreciate Humans. No! Humans, we are natural beings on this planet and we have been part of these multi-various ecosystems for millions of years. We're ignoring those responsibilities, that marriage, we're not playing ball.

 # 6 Adaptation

Our experience of going into the woods is that they miss us, and here are **the three things** that Humans are really good at:

The **number one** thing Humans are good at is peeing—producing urine. This is really very valuable because our urine is where the minerals are, not our poop which is mostly carbon. Where you pee is actually *Social Forestry*. Are you marking? There are certain places to mark. Are you fertigating? Pee onto some organic material, not on bare ground, unless marking, where it will evaporate fast.

Number two is being absolutely silly and ridiculous. This is a very important Human contribution, and Nature knows this—that we are ridiculous; that we do absolutely crazy stuff. When they hear us doing these things, grunting and giggling and crying and screaming, they actually come around because they're interested in disturbance regimes.

Which is the **number three** thing that Humans do that is very valuable—make messes. They don't encourage this stuff in school. These are the important things: pee, be silly funny idiots, and make messes. These three contributions are strange reciprocations yet appreciated in *Social Forestry*.

When we use our saws to fall, thin or limb up, in the woods, we're knocking down Lichens. The Deer want to come around and eat the Lichens. We're breaking up logs; the birds want to come down and see what we exposed, what we stirred up. We're stirring up soil seed banks and making seed beds by stomping around and dragging wood. We're burning, we're building and feeding burn piles; all kinds of things happen when you're burning. Everyone wants some of the ashes.

29 Loop Law

21 Just Sink Carbon

22 Nutrient Cycles

Read Anthropology books in the *Social Forestry* Bibliography that have to do with what we call Horticulture as opposed to agriculture; find out what these disturbance regimes are about: nutrient cycling and turning over things. Stirring things up and stimulating new growth.

WILD ANIMALS TELL US

*We are surrounded
by dynamic natural dramas.*

When we humans find ourselves outside the hedge, away from the barnyard, with neither Dog nor drove, nor gun, Wild Nature greets us. Small birds come by to take a look. Gatherers, Rangers, Bodgers, poets, and artists, we all have a lot to learn from the less tended edge. Our Human observations are best absorbed with simple curiosity, noticing, as children do. Magical Nature moments help us recognize *the others*, and lead us to respectful relationships.

The denizens of the woods come by to observe us. We present edgy opportunities. What might we knock over or pull down? What is all that racket? Or, what smells are those? Wild beings seek the sounds, and smells, and flash, of Human disturbance regimes! Wildness labors to distribute surpluses from disruptions and abundance.

Here in the Siskiyous, the work Humans can do in the wild-lands is critical to whole landscape integration. Forested ridges provide water brooms and corridors for plant and animal movements that need to be preserved, kept open. The conifer-hardwood forests and savannah foothills are traditional Human forage-lands, and with villages not that far away, there are multiple resilience-supporting opportunities. Most species distribution mosaics and mixed-ecosystem edges co-evolved with Human burning, seed dispersal, digging, and hunting.

Some of the highest densities of biodiversity on this planet may have been associated with Humans who work broadscale Places. On some *fecund* landscapes, possibly the highest biodiversity is achieved—the greatest number of species per local area has been reached recently, just before the dominance of agriculture, through a complex subsistence Horticulture.

If we, an imperfect species like any other, are to re-engage in co-evolution, then careful attention to the traditional knowledge of Indigenous peoples is called for. We can also learn as seekers, practicing open/direct observation, immersion in Nature, *forest bathing* (Plevin 2019), and principled and precautionary experimentation. If our Human motivations are focused on the whole landscape, and remember the Long-Story, we can begin to reciprocate the gifts of the Wild.

THE BEAR, A TRUE STORY

"Good morning to you, Mister Bear," I lilted, as the black-furry-lump stirred and jumped up, and off, from its nap, surrounded by torn packaging and crumpled cans. The tent-cabin still had a door in its frame but the gulch-side west wall was torn open.

Another time that year, the steel food-cache drum was gone. Found it down-gulch about a hundred meters, dented, but with the lid still locked on.

After our seemingly secure stashes were raided, we found ways to better protect caches, with less-evident temptations and invitations. The steel drum food-cache, with the locking lid, became fixed to a tree with chains, and wrapped in barbed wire, and then dusted with the highest B.T.U. rated Cayenne Pepper, on lid and base. All packages of food are well sealed and nothing is left smelly.

The first encounter back in '00 was with a lanky cub, who wanted to knock the privy barrel over. I automatically barked at the young Bear from my upslope tipi, very protective of my own manure! What did I think I was defending?

As the composting systems have become more colonized with local micro-life and fungi, they have become less attractive to larger animals. The composting privy now has *Coprinus* mushrooms, native Dung Beetles, Soldier Flies, and has not brought in the Bear again. I did share the squatting platform with a Hairy Woodpecker one fall morning, as it hunted Spiders.

Then there was my dusk return, with groceries in my pack, a year after the privy incident, to find tipi canvas hanging in sheets. The now young-adult *great-black-one* had gone in one side of the canvas, with a tear of the claw (after hearing mice in the woodpile squeaking?)! Seems he then bit the chain saw gas-can, which sprayed his face, and then he burst through the opposite side. After the canvas tears were pinned up, and the interior scatter got cleaned up some, a fire was made for dinner. Slept fine.

The Bear came back months later, after several meters of new seams had been hand sewed, and he marked the canvas just south of the east door with a perfect claw prick: "I own this". Imagine that the Bear is the landlord, as it collects tribute. Notice the anthropomorphizing, the projection of Human thinking: perhaps fun for story telling but a distraction to careful observation!

BEARS ARE BUILT BETTER THAN HUMANS

Black Bears can reach high speeds in just two moves. They have diffuse, motion sensitive, wide-eyed vision. So move or raise your arms, while you look sideways, on encountering. Do not stare, which is universally impolite, or directly challenge! Sing a silly tune. If charged (say by a mom with cubs), stay cool and get ready to roll into a ball. Usually this rush is a challenge to test you, not hunger.

Black Bears have a much more efficient digestion and metabolism than Humans. They can climb and run uphill faster than we can. Run downhill from a chase. They are also amazing Horticulturalists and we can learn a lot from the *shaggy ones*.

After the resident animals repeatedly invaded our small temporary spaces for water, we put a spring-tub down-gulch in the untended forest edge. Much appreciated as thirsty ones had some privacy. This is a desert forest, and we are the newest and most sedentary campers. After even small messes we had left, invited more messes, we learned to better clean up after ourselves.

One fall day, as I started to handsaw firewood on my porch there was a racket down-gulch. The Bear had been taking a bath in the spring-tub, and I disturbed his ablutions.

WORKING IT OUT WITH THE NEIGHBORS

Learn to follow suggestions.
Accept feedback.

Relations with the Pocket Gopher, Moles, and Ground Squirrels have pushed us faster towards planting more woody perennials, and giving up on annual vegetables. These diggers' appetites and tunneling abilities are prodigious. This adobe mountain-slope is thoroughly burrowed and a long list of species use the well established tunneled infrastructure. Wolf Gulch Farm, over the ridge, laughs at the garden attempts, and gently invites us to just come on over and get some greens.

We then tried to garden in beds lined with one inch mesh galvanized poultry netting, only mildly toxic to plants, as the Zinc releases. That basket is propped against a big log, stone wall, staked boards, or chopped out terrace. Lined with several layers of cardboard, and perhaps a moisture barrier, then filled with compost, it will grow vegetables and herbs while young trees get established. Then the basket rots and the Pocket Gopher busts back in. Eventually, after cracking a spring tub by freezing we repurposed it; container gardening became the best way to limit irrigation, keep compost in one place, and block the tunnelers. Deer and Hare still sometimes nose in through the mesh-tent or brush-mulch, when we are gone overnight.

Similarly, we have learned to compost in a partially buried HDPE trash-can that drains through small holes we drilled, but does not dry out, and that Mole or Gopher tunnel-not. We are considering going to all container-gardening, but irrigation becomes complicated, as the containers dry out quickly, if we have to go away in the heat and smoke season. As water continues to be scarce, we are trying out cloth-wicks from elevated buckets, and other hold-over tactics.

We are glad that there are seldom any motor noises, as such rackets mess up the local Nature orchestra. We only use a chain saw in winter (see chapter 6). Spring and summer use of a chain saw would interrupt the nesting season and early fall is too dangerously dry. One more reason to practice hand tools: we get to pay wider attention and avoid noise pollution.

We learn from the Bear two important principles of behavior in relationships: avoid temptation and know the difference between compassion and sympathy. After all the Quaker non-violent training and peace practice, these two lessons were more deeply learned, from our very local mentor, the Black Bear. "Who owns us," we say honorifically.

CLEAN CULTURE: AVOID TEMPTATION

Clean Culture is a tactic of garden Horticulture where fallen or compromised fruits are not left to rot, on the ground, or on the plant.

Two buckets are carried in the field: one for quality produce, the other for potential processing, such as through kitchen, livestock, and then compost. This way, fewer problem insect larvae become adults and fewer unwelcome scavengers are attracted.

When Banty Chickens are at last let out to forage in the late morning, they concentrate on bugs, slugs, and snails. The young Bantys are first brought into cooperation with the homestead by only tossing forbs and grasses into the deep mulch yard and by confining the kitchen and garden scraps to the well-contained compost pile. No eggshells are tossed into the yard, unless they are oven-roasted to remove odors and then ground in with the grain-scratch, to avoid tempting the Bantys to attack their own fresh eggs, while still recycling the minerals back into their diet (Ward 1988).

Our forestry camp low-energy-use ethic means non-electric passive cooling. A pit in the gulch-bottom stays near 50°F (10°C), with its cold-air trapping shed-roof. Any odors that might tempt, seem to settle down, to puddle. We do have to check for Snakes. The up-slope cabin uses a seasonal-use, screen-box window in summer and we trade out propylene glycol "blue ice" containers (rechargeable artificial ice, freezer packs) into the screen-box overnight. The down gulch cool flow chills them. In the early morning we stash them in an insulated cooler box, quilt wrapped on the NW cool-floor-set. The food stays indoors during the summer days, and only the recharge bricks go out at night. This is discretion in place of temptation. In late fall and early spring, or winter when not freezing, the screen box window stays below 50°F (10°C),.

A window box cooler is best located low on a shady north wall. The outdoor box is accessed from indoors through a window door that also is left open as needed (see cabin discussion in Chapter 10). The earthen tiled floor just inside the window door stays cool year round if managed carefully with perfect timing of draught control and layers of insulation and curtains.

If we leave the curtains open in our cabins and go away for a few days, the Bear will peer in and be tempted by visible food. If we do not wash all the dishes and containers and take away all the recycling, the Bear can smell waste and we are tempting the Bear. Therefore also burn the meat/dairy/fish scraps and rinse and drink the beer dregs. Clean Culture avoids temptation. After almost two decades in the gulch, Deer mowing down gardens are the most common of give-backs and the Bear still comes by and knocks over an empty trash can, or moves something, as if to say "Hi!"

COMPASSION

Empathy is the sense of shared feelings; "we feel with you."

Empathy is not easily controlled in sensitive and psychic Humans. We are moved deeply and want to help. When we can it is best to recognize the impulse and watch the flow with some detachment. This takes practice. Empathy is not bad, but it is sometimes inconvenient; an empathetic sensitive can be valuable in community, with support and appreciation for the channeling.

Compassion is a sort of careful love: observant and allowing.

Compassion is a spiritual goal, a sort of openness to care without interference in apparent trouble that we may not understand.

Sympathy is "identification with": we lose our center and lean over to lend our help.

Sympathy holds families together, it supports a sort of faith that someone has your back. Sympathy is entanglement and one might want to have some sense of the complications.

Good Aikido within Wildness is fully relaxed, all senses open and non-committed to action. Everything to do, nothing to get done. Bears and other wildlife do not necessarily appreciate our attempts to help. If we feed them directly there will be confusing confrontations. Better to plant plenty for all out beyond the hedge and along existing corridors, that we are careful never to block.

Providing water, such as quail guzzlers and open spring tubs, is useful in the accessible edges of Wildness, far enough from our camps to offer discrete privacy for approach and retreat. Predators like to stake out watering holes just in case… Leaving these watering options probably means you should not spend a lot of time there. This is for them.

Wildlife is a non-Human *other* and our Human projection is not useful, but our learning is. Careful and sustained observation, with delicate avoidance of perceived similarities, is strongly advised. The predator/prey dance and subtle landscape-wide bio-fluctuations are mysterious and elaborate. As self-appointed stewards, working and singing through this kaleidoscope, we must practice humility, and learn to think outside our Human box.

If we can accumulate landscape knowledge through generations of local culture with songs, stories, and art, we may eventually manage to rebuild *phrase-languages-of-Place*, such as the new permie-babble and the old mixed-roots trading language Chinook W'ah-wah. A trading language is a creole of colonizer and Indigenous dialects. The Pacific Northwest trading routes were facilitated by the Chinook Nation on the Columbia River. We might learn and reinforce the appropriate taboos, etiquette, and social arrangements that support persistence and resilience.

Animals tell us lots of information we do not understand. Their communications demand our careful attention. Often the presentations are very local; the Screech Owls calling at LWG sound different than those a couple of mountains to the east. As with prolonged observation before building a house, local experience is crucial.

Nomadic horticulturalists and pastoralists keep song-lines in their repertoire, stories that map the traveled and loved landscape. A Culture of Place teaches where to go to hear the news. The Bee-tree, the spring, the big White Oak, and the rocky prominence, proclaim their story via different messengers traveling in various dimensions.

Most four-leggeds move around a lot in the Wild. Humans can become over-attached to shelter and waste a lot of energy on defending camps and villages. Early Human Horticulturalists had very few essential tools and belongings and they mostly lived a semi-nomadic life, with winter-camps and animal-resistant caches. Voluntary simplicity and cultural sharing with elder-counseled options can be a path to enlightenment, or at least it will lighten the load?

MALIGNED AND MISUNDERSTOOD PLANTS

There has been plenty of controversy in seed saving, land restoration, and Permaculture circles about "invasive weeds" and what to do. Some have pointed out the inherent racism in judging species and some have blamed the plant newcomers for the decline of ecosystems and fondly remembered landscapes ("nostalgia guilds").

Often a recently arrived plant will have an eco-spasm. For example: Mediterranean Desert Parsley, *Torrilis arvensis* (we call it Velcro Burr for its affinity for socks), arrived two decades ago, exploding on disturbed soils. There are lots of ways that industrial extraction of bio-life and minerals has caused massive disturbance. Thus if we are disturbed to find our socks full of Velcro Burr, we have some conceptual ecosystem tools to talk about this and get some therapy through our meditation.

Co-evolved, ecological implication implies that native, long-time-resident species have a lot of commensalites: the plants, animals, insects, fungi, etc. that have relations to each other. Oregon White Oak has been around so long that we can pick up pieces of petrified

FIELD HEDGE PARSLEY
Torilis arvensis

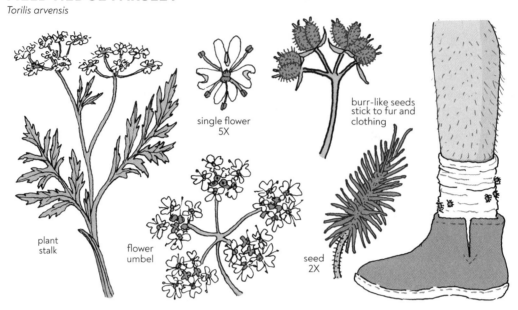

wood millions of years old on parts of the Old Cascade Mountains. This sacred tree is important to both West Coast Indigenous folks and to the Druids of southern Europe. She holds a reservoir of fungal associates, and is very competent at recovery from hot fires by sprouting from persistent roots. The mycorrhizal fungi (who harbor on these deep roots) can help many other perennial species recover by helping them survive catastrophe, or by inoculating carried-in seeds (wind or fur) that then can germinate and thrive. She is also the usual habitat of many birds, butterflies and acorn gatherers. The list of associates goes on and on (see "The Orchestra Assembles", above).

We also have eco-spasms of *global levelers*. These are species of life that have recently been widely distributed by global industrial trade. The Asian Ash Borer in the eastern North American forests is wreaking havoc. We lost the American Chestnut almost a century ago to an imported blight. Sometimes, desperate land managers have compounded the situation by bringing in another exotic species that is supposed to limit the target species but then ends up attacking other natives.

Many plant species considered invasive on the West Coast are from the Mediterranean Basin. Perhaps seventy percent of the now resident species in California are such. Many of these plants came here with stories of their herbal usefulness, and thus are potentially valuable for food and medicine. We can selectively harvest roadsides and clear cuts, where these "weeds" are unsprayed, as cultural products, controlling their spread. Sometimes taking out a *pioneer species* (useful colonizers on disturbed sites) while we engage in regenerative practices using the whole complexity and making useful small adjustments with perfect timing, can result in a set of secondary plant settlers moving in (Ward 2010).

Let's move our thinking and vocabulary away from industrial agriculture and forestry and towards Horticulture and restoration. Meanwhile we can consider rescuing plant reputations. We have all felt misunderstood from time to time in our complicated lives. Haven't we? At least we can ask for reassessment, new better stories.

The best way to make friends with any apparent foe is to get to know them better. That way we gather *compassion* and *connection*. Many inappropriately maligned plants are actually very useful or important in their place-time and evolving relations. Without a cultural practice of nature appreciation and knowledge gathering, Humans can be fearful of the enthusiasm of rampancy. Nature does not, on principle, favor one species over another; she is more interested in relationships. We can learn from her to *not* see problems as insurmountable. It seems, rather, that we are surrounded by insurmountable opportunity.

NATURAL HUMAN MINDS

A caretaker on an island in northern British Columbia once said, "No matter how you try, you will never think like a Bear." Perhaps we can learn to think like a natural Human, holistically. Our ancestral relationship to Nature is deep and if we manage to have a cared-for infancy, followed by a magical childhood in Nature, our minds accumulate the symbolic metaphors that allow mature thinking. If our culture transitions our adolescence with rites of passage, and welcomes our adulthood, we can learn to live with ambiguity and imperfection, to love change and complexity. We come to weather discomfort gracefully, with an expectation of shared experience and eventual community solace.

Our Human thinking and the greater Mind process natural observations symbolically. Words are the last stage of translation, from dream-thoughts to action, and serve as communication with other Humans. Almost all of our information exchange with any energetic entity or Place Spirit is non-verbal and non-conscious or intuitive, through sense-symbols. So much so, that moments in Nature can be palpably magical. How did we come to deeply understand, to *know truth*? Why did that animal come so close? What is it that we seem to be almost getting, and almost comprehending?

Humans are, in their deep essence, ancient natural beings who co-evolved with other Earth beings, for millions of years. Only recently, perhaps in the last ten thousand years, have we become separated and alienated from our birthright. We are Pleistocene creatures surrounded by modern Human artifacts and agriculture and we deeply long for the Wild, just as the Wild longs for us to participate in the *great implicate order* of All Beings. *The All Species Parade*.

Our songs, our silliness, and our sloppiness, have long entertained many other species. We belong here; we have the built-in skills to be useful and through cultural re-membering— coming into our senses—we can dance our glorious part in the multi-dimensional spirals, while we sing the song-lines that keep us oriented to Place.

PART II

In the Forest

The Principles of Forest Ecology include

19.
A forest is an intertwined set of relationships anchored on tall woody perennials and stacked from deep roots to leafy crowns with layers of ecology. Save all the parts: we do not know how to assemble one from scratch. We can help forests get started or repaired.

20.
Forest types are site, aspect, and climate specific. Forests migrate during climate changes. We can only tinker with these dynamics. Pay attention and look for emergent opportunities.

21.
Nutrient cycles are local and elaborate in forests. This net of mutuality supports heavy carbon and water loads, modifies local climate, and allows maximum biodiversity with complex relationships.

CHAPTER 5

Forest Ecologies

This is an ambitious chapter. We are hoping to survey a complicated continent, Turtle Island, also mislabeled as North America (named after a dead white guy). We will be making bold sweeping statements and generalities that may help with local orientation. Where are you? What wants to grow there? How are these forests doing?

INTRODUCTION TO FOREST ECOLOGY THROUGH STORYTELLING

To get the feeling of what is going on in a forest, this next story uses metaphor, reaching between poetry and explanation to characterize subjects as if on a stage. This is the tale of people in action, a peek at what is going on in one forest. This is the children's guide to forest ecology.

CAMP LATGAWA

The May 2019 Three School Gathering at Camp Latgawa, at the Junction of Little Butte Creek and Dead Indian Creek, in the Old Cascades geological complex, draining eventually to the Pacific Ocean.

Latgawa means "upriver people" in the First Nation Dhalgelma languages. The camp is at the end of the road. Tall Douglas Fir trees tower over the cluster of cabins, lodges, paths, and small openings. The soda springs, just up Dead Indian Creek by

(continued)

trail, inspired the first buildings and system developments in the early 20th century. Surely the campground use is way older. The towering trees are not that old. This place used to be more open, probably from regular burning.

This tent cabin with four bunks is labeled number nine. An asphalt shingle retrofit capped with mosses perched on the 1920 long split Sugar Pine shakes, the original roof. The small diameter Douglas Fir rafters sit on full-cut two-by-four rough stud walls. Screens, and some small windows, and the well-used door, show lots of repairs. A typical old bunkhouse (it has been moved at least once across the camp), and for this wet May weekend it is the Nature Lodge.

Groups of fourth grade kids from the three schools got sorted into five clans. First Eagle came to the lodge, then Bear, Coyote, Salmon, and Moon over two days. Surveying the children's personal first names with their meaning-stories leads us to why we call the shiny-three-leaf-one Poison Oak. We can also call it Lacquer Bush because it has the best black shiny sap, when we learn to use it safely. We can call it Guardian Oak when we learn how it is related to forest repair and then learn to plant the next succession.

We—story elder, adult helpers, and more than a dozen ten-winters-past small-folks—at each of the five time-pass sessions, hear about the Two Baby Skunks, the Nature Lodges of past camps elsewhere, and the plan to go out in the rain and look for leaves and bugs.

The first stop outside is at the wet bent but straight arching Guardian Oak flagged on the path. Hello to you, Friend do-not-touch! Then we meet four more plants. First, shady under-big-trees. We learn kind-tending, only taking one leaf per very-leafy clump of Crowfoot ground-hugger three-lobed (but not shiny, and really only one leaf with three lobes!), and also, only-one-leaf from Pathfinder white-under-side arrow-leaf.

Then we walk to the bridge and into west light, and at the edge of the tall-standing-ones, we find big Big Leaf Maple leaves with chewed holes. These leaves snap clean off the twig when the leaf stem is gently bent back. At the rail of the bridge in bright light and on rocky road fill, we find Blackberry leaves on the thorny canes at the tips, still soft enough to tease off a cluster of hairy-three-leaves. "Leaves of three leave them be! If it's shiny, watch your hiney! If its hairy, it's a berry!" We all sing.

(continued)

FOREST ECOLOGIES **111**

Back at the Nature Lodge the young eager eyes use magnifiers and microscopes to search for bug poop, Maple Loopers (inch-worming their way to great delight), insect eggs on stalks and in clusters of barrels, Aphids, tiny Slugs, and molting Lady Bird Beetles. All in relationship to each other and the whole landscape.

After many treasured discoveries and the showing-to-the-elder for names and stories, the leaves are laid out on the floor with the Pathfinder arrows in a curvy row, Crowfoot leaves set on top of each stepping-arrow, walking through a leafy Maple forest with Blackberry brambles around the outside edge. The story of a Crow who eats bugs in the shade and berries in the sun emerges from the children, and then the elder stories the great looping relationship of everything here and now, tying together the clans.

The Moon tugs at us all floating below the stars. To see like an Eagle we use binoculars. The patterns in the stars are the same as the patterns in a leaf. To see the tiny details of the bugs and leaf we use hand lenses. Universal patterns: as above, so below. The Moon monitors the whole earth and the Eagle soars over the forest and canyon.

The Salmon swims up the river of its birth to lay new eggs and die. The Eagle and the Bear and the Coyote (and the Skunks and Humans) all eat the Salmon and go poop up the mountain and over the hill. The mushrooms and soil absorb the Salmon nutrients and the trees, herbs, and berries all grow strong. The many bugs and slugs and tiny-ones eat the rich leaves and some tiny-ones fall in the water, or wash into the stream in a big rain, and the baby Salmon eat them hungrily.

After a season of growing strong, the juvenile Salmon swim down the great river to the ocean and grow mighty on all the river-flushed nutrients that feed the sea-bugs that Salmon eat out in the ocean. The great strong mature Salmon now swims and jumps up the river all the way to Little Butte Creek, finds the gravel to lay the eggs, and then dies as a gift. And the looping of food goes round and round and everyone gets fed. The Moon smiles.

After the crowd leaves, the story elder sits quietly in the Nature Lodge listening to the patter of rain on the old patched roof.

FOREST ECOLOGY TOUR

The broad subjects of forest ecology that we survey are:

- *forest types*, which include landscape **mosaics**, **ecotypes**, and **edges**,
- *flows*, which include water, carbon, nutrients, communication, fire and air, and
- **tended forest stands**, an exploration of anthropomorphic disturbances. We will insert principles and illustrations along the way.

When we walk through a forested valley, or climb up a ridge with views of several *forest types*, we are looking, feeling, smelling, and listening to many clues about what is going on. The forest speaks to us by revealing patterns. Some of these patterns can be seen from far away and many are displayed close by, on the shady floor, or the shaggy trunk.

Foresters used to try to organize what they saw into a definitive list of *forest types*. That was called "discrete ecology" and it allowed a bureaucratic shorthand that "got out the cut" and led to wholesale industrial efficiency, with disastrous consequences. A single-species plantation of cloned genetically-modified timber trees planted on a bulldozed, clear-cut slope, sprayed with chemicals to reduce competition (from ecological complexity) has all the drawbacks of any monocultural conversion, social or ecological.

> *Big fires coming soon!*
> *Biodiversity collapse!*
> *Invasive super-weeds!*
> *Poisoned water and soil!*
> *A Human culture has overspecialized,*
> *and is now obsolete.*

The actual landscape complexity of natural forests is called "continuum ecology" and that speaks to the *edge effect*, where one grove of plant associations (*an eco-guild*) transitions into a different assembly. The edge between *ecotypes* is where the most biodiversity is found and where we can learn the most about the dynamic *flows* of Forest Ecology. And the edge gradates and infiltrates and migrates and thus there is no exact type of forest, only a continuum.

7 Forest Terms

The poster "Forest Terms" is a two-dimensional panoramic sampling of the complexity of one *forest type*. There one finds lists of components and participants and structure and *flows*. What we need to learn about forests in order to do good Horticulture is site-specific, both general and vast. The whole relationship-map of any tree-rich landscape is beyond our Human ability to fully comprehend. We, Social Foresters, have collected clues and names, and seen partial-peeks to patterns, but the whole is vastly elaborated in many dimensions.

That said, the *aspects* (both the direction slopes face and general categories of viewpoint) where Humans can intervene usefully are limited, especially within our ken. We can emphasize the *humisphere*, where the mineral soil mixes with the organic life and litter on the forest floor, because it is fragile and thin and we Humans impact it hard when we burn too hot, or use heavy equipment, or strip the litter for mulch resources, or introduce the wrong worm species who suck down all the surface litter (do not import Night Crawlers, Garden Worms, and Red Wigglers). Learn to look for the special value of the *humisphere*.

The humisphere is an edge where the organic litter lays on the mineral soil.

When we have a closed canopy (*overstory*) of trees, the mineral ground under the carbon litter and under some snow pack stays above and near freezing, with slow-molder composting and soil-life metabolism providing some warmth, and the tree branching structures providing a three dimensional blanket. In northern latitudes and higher elevations, an open field adjacent to a forest grove will freeze several feet deep without the forest to protect the ground. The humisphere thus keeps seeds in the soil-seed-bank viable, keeps fungal-hyphae-nets (mycorrhizal symbiosis) active, and holds the winter root-sleep gently, ready for the spring bloom.

Outputs carry many communications and connections.

We can try to understand the *flow* of pollen, pheromones (perfumes, turpines that act as chemical messengers), and spores in the wind, across the tops of the tree crowns.

Then we can think about trees as windbreaks (see poster 13, Windbreaks), and trees as messengers, and trees in the wind as mixing whips. We can think about atmospheric gas *flows*. The trees breathe in CO2 during active solar-driven photosynthesis, breathing out CO2 overnight and while dormant, over winter. Trees breathe water and pheromones and wind and sunlight and produce complex nutrients and structures (enzymes, sugars, cellulose, lignin, chlorophyll, growth hormones, and much more).

Their roots feed the sugars to a slime-glove around their root-hairs that consists of fungi, bacteria, and microorganisms very similar to those in pond water. This symbiotic soil-life attaches mineral nutrition to carbon and passes the building-block-basics back into the root hairs for the trees to use in their awesome bio-digestion and biosynthesis. We can try to understand if anything is missing or become hard-to-find in the nutrient input and output flows, and try to balance and facilitate forest health.

Lots of biodiversity is perched in and on trees.

The epiphytes (Lichens, Mosses, Orchids, Ferns) and the semi-parasite Mistletoe, elaborate the branching structure of trees and provide important habitat for sleeping insects and small mammals, nesting opportunities, water capture (small cups and pools), pollen and spores, nitrogen fixing fertilization, and those aspects that we do not understand. Birds, bats, snakes, squirrels, voles, and even tree climbing Grey Foxes, as well as Martins, Weasels, Ring Tailed Cats, Bears and Humans can be found up in trees. They all contribute and harvest. We, foresters assessing a forest stand, can look for wildlife structures such as cavities and "witches brooms," and epiphytes to support the tree-crown dwellers.

Biodiversity thrives in tangles of relationships.

Trees and their associates thrive in clumps and eco-guilds (species commonly associated with each other) with the glue of mutual-aid (commensalism). Especially during forest succession, after stand-replacement wildfires and blow-downs, or after Human-caused disasters (clear cutting and old-growth removal), these eco-guilds are dynamic and change as the forest structure re-builds. This succession can be mimicked with Human-assisted replanting, or be encouraged and protected where they are recognized.

Legacy trees are the remnant
of a successional eco-guild
that has been overtopped by conifers.

In a *fire-ecology* mosaic, a large central shade tree might harbor grasses and forbs, along with nitrogen-fixing and berry shrubs, on their south aspect, and harbor vines and conifers in their shade. The whole clump thrives through a synergy that attracts animal cooperators and their nutrient or inoculation gifts.

Natural and patchy disturbances
preserve metastability
and help prevent terminal disasters.

Forests benefit from canopy openings caused by falling old snags or wind-toppled trees through the canopy. A hole is opened, allowing sunlight to the forest floor, encouraging the replacement tree seedlings to establish. The *root-ball* of the fallen tree pulls up a pile of rocks and soil and leaves a pit. When the fallen tree trunk molds nicely into the forest floor and the root ball rots down, the pattern on the forest floor is called *pit and mound topography*. The subtle microclimates and growing opportunities presented in this pattern allow greater forest floor diversity.

The big, long rotting-down log makes its own microclimates of water sponges, animal tunnels, perched moss and berry guilds, beetle and insect habitat, and slow-moldering carbon sequestration.

Fire cycles the most nutrients
when patchy and ground running.
Crown fires are too hot.

When a big tree with lots of limbs falls, Humans can help it get down into soil contact by cutting off the limbs and laying them on the soil (on contour) or carrying them out and converting them to charcoal. This quick fix reduces catastrophic fire-hazard and improves carbon holding on the forest floor. Tall (higher than our knees), three dimensional limbs and twiggy slash is best "*loped and scattered*" as mulch. In Social Forestry we call all tree trunks and large branches that get laid out on the forest floor "*log mulch.*" Logs are most usefully laid across slopes, nearly-on-contour (see below), to catch leaves and water and soil and thus hold carbon and moisture long term. Fires are slowed, not wicked up the slope.

Wildlife perches, foraging, and cavities
make snags critical habitat.
We need a cohort everywhere.

The older trees that die and do not fall become *snags* and are very valuable for nesting cavities, nut storage, bird perches, and forest canopy openings. If they are vertical standing and sit well on their rotting stump/root mass, they can last for decades. The limbs and bark persist for a few years and then fall off, littering the opening around the bark-bare trunk. There is a short initial period of increased fire danger with all that twiggy dry fuel up in the canopy. Humans can hurry unstable dead trees to the ground, get them well bedded, and perhaps high prune some fire-ladder-prone lower limbs on a freshly dead tree, to facilitate the maturation of a well-balanced snag.

Fast fire runs up chimneys (gulches),
ladders (dead lower limbs and dense brushy edges),
and wicks (logs, steep meadows, brush fields).

A freshly fallen tree that lays up and down slope, especially in a position where it can carry fire like a chimney, is a *fuel-ladder*. A tree trunk laying nearly on contour—at the same elevation—slows down a ground fire but can end up burning longer, acting like a crucible, holding a part of a passing fire only to light another one later, when the wind picks up. Pitchy logs can burn for months. Fire can go underground along dead roots, or in dry peaty soils, only to surface months later and start up an unseasonable wildfire.

Nutrients and water
are held in patterns
that divert flows.

At best, a dead fresh fall with ladder-limbs, lying in a chimney, ready to wick upslope and pre-stage a crown fire, can be loped and scattered, and the trunk can be bucked and turned onto contour. The best lay for a forest floor tree trunk is on Keyline pattern (Yoemans 1971), with the off contour slope of the log pointing 1:50 to 1:20 toward a ridge and away from a drainage. This way water, soil, and organic leaves and branches are held upslope for as long as possible.

This sort of work-advice will be repeated in subsequent chapters because telling Humans practical consequences is a way to glamor their attention. What we want to learn here is actually *fire-ecology*. How does a hot fire burn? How do different *forest types* on different *aspects* burn?

A well-watered forest holds weighty stores. Net, sink and loop!

Living trees hold immense quantities of water in their trunks and roots. A well-developed forest topsoil is rich in carbon, well mulched with logs and litter, and the porous soil acts as a quick sponge in big events. Most trees and shrubs in continuous forests have very shallow root systems just below the *humisphere* that spread out very far distances, several times the height of the plant. An eight-inch Maple in a wet temperate forest can have roots more than a kilometer away (SUNY College of Forestry 1967 field study).

TREE PHYSIOLOGY: HOW DO THEY WORK?

There are three ways that trees pump water up from soil reserves and deeper water tables, humidifying the whole forest assembly and supporting landscape scale ecosystem functions. Trees suck up water by the pulling action of *water-transpiration* through leaves and needles. They use very tiny *vascular-tubes* to take advantage of *capillary-water-tension* wall climbing, and they push water up from the soil because they have a sugar and mineral-rich sap that is thirsty for less salty ground water: the famous *osmosis* trick.

11a How Trees Work

Conifers and some broadleaf trees have a central main trunk, radiating branches, and a distinct tip-leader. This form is called excurrent. The drip-line of a conifer runs almost entirely along the outside shed of the most-often evergreen needles or scales. This is good for dumping snow loads. The most active root growth and mycorrhizal entanglement is just outside the shadow of the drip-line. This is where one would add nutrients or relieve compaction, so that the active roots can thrive.

Broad-leaf trees, who usually lose their leaf load in fall and are bare in the winter, classically have a spreading and forking main branch pattern that we called *deliquescent*. Even with a full crown of leaves, there is some stem flow of water down the main trunk to the ground. The drip-line shadow of the crown is still the most active root zone, and one treats that special area the same as with conifers. The less dense shade, wider crown, and some watershed inside the drip-line allows broad-leaf trees to cooperate with a guild of understory plants. Over winter, a lot more snow and rain reaches ground inside of the drip-line than with a conifer. Broadleaf trees, also known as "hardwoods," are more prone to snow and ice load damage and often shed big limbs during events. Evergreen hardwoods, such as Canyon Live Oak, Madrone, and Manzanita are especially prone to snow and ice damage, often to the extreme point of being uprooted and toppled.

*Snow load tip-ups
create important pit and mound patterns
in mature Manzanita groves.*

The cross-section of the stem or trunk of a tree is a concentric ring of tissues, all with practical functions. The outer bark is sometimes thick and fire-resistant, and sometimes thin—although fire may girdle the tree and kill the crown, the root mass re-sprouts or the seeds are co-evolved to germinate after being burned over. Bark has many uses: temporary shelter, tiling pattern roofing, fire resistant mulch, insulation.

*Turgor is sap pressure.
The leaves are limp when it is low.
Beetles bore more easily.*

There is a thin layer just inside the outer bark that is called the phloem. This is the transport tissue for nutrients pumped down to the roots from the photosynthetic productivity of the leaves. When bark is referred to as medicinal, it is this papery thin layer that is called for, not the outer bark. Just inside the phloem is the cambium, a very thin layer of multiplying cells that lays down phloem outside and xylem inside. Xylem consists of spring wood and summer wood, tubular vessels that are added by the cambium (large in the spring, smaller and smaller until the fall) in temperate forest trees. This is how we can count the years of growth in the tree-rings. The newest vessels carry water up to the leaves and older vessels store water, as they lose flow efficiency. Heartwood annular rings often get clogged up with mineral deposits to slow down fungal heart rot, store water, and support the weight of big old trees.

*Watch the leaf characteristics
and the health of the leader
to judge the stress a tree endures.
Colors on leaf edges and spotting are clues
that indicate nutrient deficiencies.*

If one wants to collect sap from a spring Maple flow, the tap goes just barely into the new wood (xylem). The phloem is carrying most of the root sugars headed up to unfurl the canopy leaves. If one wants to kill a tree (!?), one can *girdle* the main trunk a foot or two above the root spread. This means taking off a wide ring of outer bark, phloem, and cambium just after the new leaves have fully unfolded. Then the roots keep pumping water up to the leaves, but the phloem cannot get sugar back to the roots. The roots die first and the top last, and sometimes suddenly, weeks or months later. Girdling is the only way to take out trees that readily sprout from their roots (Populars, Locust, Tree of Heaven). One still may have to chase some diehard root sprouts. This is practical tree physiology; how things work leads to what to do.

Guttation is when the turgor is so high that the leaf tips spit sugar. Look for the shine on rocks below riparian Maples and Alders.

Trees also do things with their roots that we cannot see. Some trees root-splice or join together through roots. In this case, girdling may not work so well. Mother trees direct nutrition and messages through their mycorrhizal associates to their genetic progeny. Family support! Roots, as we discussed above, can go a long, long way. The root mass of a tree can be more than the whole structure above ground. When a deciduous tree drops leaves in the fall, the roots also shed root hairs and fine roots, which are not needed during dormancy. This represents a release of carbon and nutrients into the soil. When we coppice a nitrogen fixing tree or shrub, the roots die back to match the loss of crown, and the bacterial nodules on the fine roots decompose and release nutrients. When one transplants a nursery tree or shrub, the roots are necessarily reduced by more than half. The above ground woody structure should be pruned back so that the re-located plant can grow new roots ahead of growing new twigs and leaves.

Cut Douglas Fir stumps that heal over are evidence of root splicing, keeping the stump alive even though the tree is gone.

WHAT WE CAN DO TO HELP

We can look for water table salting (sea water infiltration and desert water table over-pumping and deforestation) and subsequent tree-stress. We can assess water availability and work to build perched and stored water ahead of dry periods. We can thin forests or return to periodic cool burning to reduce excessive water demand. We can reduce the stress on the critical legacy old-growth trees, and we can work to return Fire, Salmon, and Beaver so that whole drainage basins can re-hydrate, and re-build nutrient and carbon reserves. This is sometimes called climate or drought-proofing. Humans can be useful! They are part of the forest ecology. Now that I have your attention again, let me point out the landscape-wide ecosystem functions and nutrient loops that all this work-advice is in reference to.

INTRODUCTION TO GEOLOGY AND CLIMATE

Social Forestry Principle #20:
Forests types
are site, aspect, and climate specific.
Forests migrate during climate changes.
We can only tinker with these dynamics.
Pay attention and look for emergent opportunities.

Forests exist where there is enough rain, some depth of fertile soil and fractured rock, and local microclimates that allow them to live long enough to reach reproductive maturity and develop *complexity* and *meta-stability*. *Forest types* can migrate, given a slow enough climate change: if it is getting warmer and drier, they can move uphill or around the corner to a shadier *slope-aspect*, if they have mountains to grow on. They can move out onto grasslands when wild or cultural fire is controlled, and/or charismatic mega-fauna (Elk, Goats, Elephants), who browse or crush woody invasions, are removed (by late Pleistocene hunters?).

Forests move north during global warming, when they have a north-south mountain range or basin edge to follow. Thinking about *forest types* as *complex-mixed-herds-in-motion* is a great sit-spot activity (long stationary observation) leaving us humble and awed. This modern Anthropocene time/climate change may be coming on so fast that Humans can perhaps assist in moving the forest types to survivable *microclimates or refugia*, but usually forest migration is slow.

On a much bigger time scale, Turtle Island (the North American continent) is moving very slow, geologically west. Nonetheless, this advancing edge has some of the most dynamic plate tectonics on the planet. A lot of the west leading *edge* of Turtle Island is scraped-up ocean bottom, intruded granite, metamorphosed country rock, and ancient volcanics. The soils that derive from this complex geology are diverse and some are still uplifting and eroding, challenging life to hold on.

Soils decomposed from granitic and basaltic mother rocks tend to have broad-spectrum-mineral-fertility. Add carbon and soil-life and they are considered fertile. If these soils are deep, moist, and well drained, trees would love to thrive there. Sedimentary and metamorphosed mother rocks can contribute unbalanced mineral nutrition and sometimes be alkaline or toxic with metals. Soils can weather-in-place and be easily classified or they can slump (*colluvial*) or water transport and dump (*alluvial*) and confuse their origins. These soils can be studied (profile pits and lab tests) and understood. Sometimes they support very special *forest types*.

All soils can be improved: Just Add Carbon!

21 Just Sink Carbon

24 Soil Water Rock

The ocean waters between the continents are the multi-dimensional braiding heat-machines of climate. The ocean, and the atmosphere with jet streams and wind patterns and solar heating and cooling, drive the weather. The long term *flows* of climate change are difficult to model and predict because of complexity and chaos. The near term weather can be guessed at by using local observations and memory of patterns but as climate changes, so do local patterns.

Snowfall records at Crater Lake National Park (southern Cascade Range) show abrupt snowpack reduction starting in the late 1970s. Thus a historic drought pattern (from Bristlecone Pine tree-rings in the Sierra Range) lasting about 60 years, began. We are about half through the drought cycle except for changes wrought by modern climate-weirding.

The very-long-term climate record shows warming periods that flip into glaciation, over and over for most of modern Human experience. Bog cores show temperate-type forest pollens disappearing at the onset of a glacial period and boreal-type tree pollens replacing them in only a couple of years: Ash, Maple and Beech, turn to Spruce, Birch and Willow. The inter-glacial temperate interludes last ten to twenty thousand years. We are at that approaching *edge*.

One coincidental flow that seems to presage climate flip is the reduction of the Atlantic conveyor-flow, formerly known as the Gulf Stream. Human greenhouse-gas ecosystem-degradation effects could push the atmosphere into a hot-death-spiral (drowned continents and giant reptiles?) or could flip the climate into a big cooling, and ocean drawdown as icecaps replenish. We do not know what happens next. We do know how global industry is tweaking things and what to do to back off, but cannot predict the future.

Climate REFUGIA are geologic areas that preserve remnant *ecosystems* during harsh times. There are famous regions of the world where ancient fossil plants and trees persist because of favorable conditions during harsh eons. The Siskiyous in southwest Oregon, mountains of southeast China, and the southern Appalachians mountains are all repositories of ancient species diversity. The persistent ancient survivors are often the earliest ancestors of whole branches of botany.

Humans can identify and support *refugia* during climate change. We can also mitigate climate with reforestation, carbon-sequestration, water-table-recharge, and revegetation. Reforestation slows thermal-tower desiccation. Water is evaporated and lifted into the stratosphere by these massive updrafts (beloved by hang gliders) where water is a potent heat trapping blanket. Our world-wide destruction of forests has accelerated climate-weirding; there is a lot of water up in the air, instead of held close to ground in carbon sponges: forests.

An aspect
is the direction a slope faces.

On the Pacific slopes of Turtle Island, the interaction between mountain ranges, large valleys, the ocean, and continental weather patterns, paints the landscape-scale mosaic of ecosystems. Forests on slopes and ridges intercept moisture (*fog broom effect*), bounce back cloudy humidity near the ground (cumulostratus clouds), and release pollens and spores that act as rain seed downwind. A mountain range that stands in the path-of-storms captures more rain if it is forested, and releases more rain later to the dry shadow basins.

A bare range or dry-brushy range rakes out less rain and bounces evaporation (*bounce-back*) high into the atmosphere in *solar-thermal-gyres*. The dry area behind a storm blocking mountain range is called a rain shadow. Some rain shadows are giant, think the Great Basin, tucked behind the Cascade and Sierra Ranges— who scrub most of the rain from

Pacific Ocean storms and create great evaporation towers where the mountain slopes (south and southwest *aspects*) are scorched by solar heat as a result of clear-cutting.

26 Water Cycle

Many Siskiyou slopes without forests face south and southwest. Many of these slopes have not supported forests for eons; they may be too rocky, or hold toxic soils, or were burned regularly. The management of solarized slopes is tricky: if we remove a *ridgeline forest canopy* we may never be able to reproduce it. We may be able to manage deep rooted perennial grasses (*Festuca* species) and capture carbon through enhanced photosynthesis by burning regularly (as is done by First Nations) but keeping forests on these exposed slopes is best. Still, the climate-site may be saying *savannah* or open Pines?

24 Soil Water Rock

The common storm-approach from the winter Pacific parade (*rivers-of-rain*) is from the south and southwest and a forest on these south *aspect* slopes will capture a lot of rain. If the soils are deep and the carbon content is high, forests can hold a lot of water. One tree can hold many tons of water in the trunk and branches. After long periods of drought, that moisture content can diminish, leaving trees less resistant to *stand-replacement-fires*.

> *A stand-replacement-fire is a blow-up, often on a steep slope or in a canyon, where the crown fire is so fierce that it cooks out wood gasses, and the air burns.*

The closer a forest is to regular coastal moisture (fogs and light rain), the more carbon can be sequestered in log mulch, even to such an extreme that many huge logs on the ground—in the *understory*—require log walking and bridges to travel-through. These wet coastal forests seldom burn. They recycle essential nutrients through fungal activity. The further we get from humid climatic conditions, the more fire is important in diminishing excess and senescent standing fuel and in recycling phosphorus, potassium, and other essential mineral nutrients.

Figuring out how one can help local forests takes seeing the big picture and looking closely at the local conditions. We recommend that communities collect information on soils, water tables, species distribution, animal population cycle patterns, *solar aspects*, and underlying geology (types of water tables and impervious layers) to start to think about this premise: "*How things work leads to what to do!*" Finding out what previous settlers and First Nations did on the local landscape and overall region will teach many lessons, mostly about the disasters of colonial extraction and the wondrous promise of restoration to appropriate Horticulture and Silviculture.

MAPPING THE COMPLEXITY

*It is complicated
to map complexity.*

Once we have settled in a Place, we have a lot to learn. Ecological sampling, mapping, monitoring, and ongoing observation will be our *quick-fix* while we learn to do good work. These skills benefit from good hard-copy library backup, along with journaling, calendar notes, mapping, and documentation. Posters and maps can be centralized in a drainage basin council lodge, available to sophisticated considerations.

Eventually, the inherited stories from our origin-lands and the acquired local lore will give rise to emergent epics of failures and successes. As we spend more time out on the landscape and as we travel the familiar paths (*transects*) to our restoration tasks, we will pass plants, tracks, scats, signs, and sightings. How things change over seasons, hours, and decades is a most important lesson trove. Very challenging to understand. Cycles of cycles in dynamic, moving *ecosystems* and geologies. Yet patterns can be observed and remembered, given *phrase-mnemonics*, so that we can council with each other and practice precaution, humility, and long-range-visioning.

*A transect
is a line or trail through an ecosystem,
that we walk regularly for observation.*

A staked *transect* can be up and down slope or along slope, from post to post, perhaps one hundred meters long. A sample plot is usually a nest of circles or spheres with a staked center. A series of sample plots can be mapped (located on a map grid) and revisited for observations and notations. A grid of sample plots can make looking at whole drainages more efficient; we get out there and make a bunch of observation bundles, trying to map and think about whole landscapes.

Circle plots surveyed on grids are usually laid out to assess an extensive area. Inventory is taken of herbs in the inner circle, shrubs in the next, and trees in the outer ring, to simplify. This sampling generates lists that estimate species diversity and inventory. The sampling can be revisited years later to see changes. A line transect is often specifically placed to sample an ecotone (edge between vegetation types) or a small area, such as a ridge pad or meadow. The sample notes are taken by walking the line and counting species on both sides, out one meter (a yard). There are short trails that we use every week as if they were staked transits. We can see the bloom season change, animal scat, tracks, and seeds throughout the year. This sort of sampling lends a transect of time to the sampling.

Sampling is a shortcut. We will learn that reality is not so easily represented by data and statistics. The subtleties are infinite. Complications that we deploy are inevitably based in the reduction of complexity. We cannot game complexity, or cop an angle, without consequence. Our Human ken and kin-relations, are based on our best generalizations from ongoing observation and our best *traditional knowledge,* the clues from cultural memory.

The ecological terminology that we use in this survey/review of the broad subject of forests allows for a discussion of complexity and relationships as we float over the landscape in our charcoal-gas-powered dirigible. Only through local experience and community can ecological knowledge yield the best tactics, for the return of Beaver, Salmon, and Fire. May our mountains sparkle with the opportunities from restored old-growth forests.

GEOPHYSICAL GENERALITIES AND CONTINENTAL FOREST TYPES

*Start collecting
field guide books
or find a good library.*

There are whole libraries filled with books about our continent and its forests. We are only going to skim over the complexity, so that the reader can get oriented to their specific local forests in a bigger context. Local foresters and colleges are going to help you if you know what to ask and if you know how to use them best. A land-counselor will bring their own biases but if you get them interested in the drainage basin scale questions, they will inevitably dream and think about your questions, on their own time. If you have the chance to meet them again, they will have cogitated to everyone's benefit and perhaps have thought of other knowledge resources missed in the first encounter.

The entry candy to forest knowledge is tree identification. Trees hold still, you can usually find them again. Some cities have many species of trees along the streets to identify. Bird watching is fun too, but a bit more daunting; binoculars may be necessary, and you have to get to where the birds are just now. Trees change through the season, so twig keys, leaf keys, nut and fruit drawings, and profile sketches are very useful. There is a wide range of identification books available in the bibliography. Look for small books with couplet questions and drawings, such as the Pocket Guide (Watts, 1973) series. An identification key gives you questions in pairs that move you to the next question and eventually the answer—fun! Make a list, go to arboretums and parks to search for exotic species.

TEMPERATE CLIMATE SOCIAL FORESTRY

The mid-latitudes of the northern hemisphere of our one Earth include much landmass on giant continents. The mid-latitude forests, seasonal forests with winter dormant periods, are the focus and experience of this Social Forestry book. Tropical and Boreal forests (along the equator and near the poles) demand a different set of cultural arrangements.

Similarly, Ecotopian Social Forestry is focused on the mid-altitudes. The high-elevation forests are restrained by ice and snow, where we see *tree-lines* at the limits. The forested mountain tops and ridgelines are usually cloaked in conifers (trees with needles instead of leaves) who act as *fog-brooms*. High-elevation conifer forests should be protected as water harvesting and storage forests, holding snowpack as long as possible and keeping the water cycle close to the slopes. The private and heavily logged lands of the Pacific slopes are mostly at mid-elevations, where First Nations communities made their winter village on southeast slopes, between 2,000 and 4,000 feet elevation, above the fog and below the snow. These are now the "hammered lands" and are ready for our residential attention.

Foggy inland valleys are no place to try to live, although this is where modern development has had the greatest impact. The flood-plain and river-corridor (*riparian*), mostly broad-leaf, forests could be part of a complex of regularly burnt prairies, Oak/Pine savannas, and nut and fruit food-forests, visited seasonally for hoop camps[1], used for travel across and along, and not settled. The air quality in these air trapped valleys is seasonally unhealthy.

The Coast Range old-growth forests that remain are carbon and water *sinks* that need protection. On some of these wetter slopes, the Social Forestry opportunities are along south slopes and low elevation ridge lines, where First Nations traditionally kept open trails and foraging grounds. The valleys should be full of giant trees.

---

1   Temporary residence while tending and harvesting on a nomadic landscape scale loop.

There you have the *geophysical* big picture. The Social Forestry and *forest ecology* we are exploring here is mostly on hillsides, mountain slopes, mid-elevation benches, open upslope grass and brush-lands, and along lower elevation ridges.

BELOW THE HORIZON OF THE GROUND

*Forest roots
interpenetrate soil and rocks.
They need nutrients,
water, and carbon.*

The poster Soil Water Rock (7) is a schematic cross-section of geological terms and derived soils. This cross-section starts in the Pacific Ocean floor, the subduction zone deep under the continental plate. We then move up onto the continent with the intrusions and weathering of molten upwellings from deep in the Earth's mantle. The two types of molten intrusion are granitic (crystalized sub-mantle masses that rise slowly and cool beneath the overlaying "country rock") and volcanic (more finely grained broad-spectrum magma that forces itself up through the country rock and builds mountains by layers of deposition, explosions of ash, and flows of lava).

24 Soil Water Rock

Granite is eventually exposed, long after it has cooled, when the country rock above it erodes away; the granite resists weathering more than *metamorphic* or *sedimentary* geologies. Metamorphic rocks are tortured-rocks. These can be created by the pressure and heat of rising granite or by the pressure of lots of layers of burden, and the crush of advancing *continental-plate-tectonics*.

Sedimentary rock is laid down by water (*alluvial*) or landslides (*colluvial*) or wind (ash and dust deposits). This rock starts out as soils broken down from exposed geology and, as layers build up, those soils and deposits are solidified and consolidated into layered rock. When sedimentary or metamorphic rock is uplifted by the growth of mountains or exposed by downcutting rivers, it can re-sort or mix and be re-laid as new, more complex sedimentary geology.

As we promised, this is a fly over. Volcanic rocks (ash beds, lava flows, extruded obsidian, or ejected lava bombs) generally break down (through freeze-thaw, solar-degradation, wind and water erosion, and glacial grinding) into loams and silts. Volcanic soils are generally fertile and have *broad-mineral-fertility*; they develop under forests and grasslands into carbon rich topsoils that hold soil moisture well, supporting vigorous forest growth where they have enough rain.

Some volcanic flows can have concentrated mineral profiles that are special, such as high phosphorus or metals. Volcanic geologies, because they are ejected and layered, are more prone to leaching than uplifted ocean bottoms or granite-pushed country rock. Water moving through deep volcanic layering can deposit silicon and calcium (*agate, jasper and chalcedony*) replacing buried carbon logs with crystalline growths, turning them into *petrified wood*. Some of this mineral replacement can concentrate radioactivity (rock phosphates with uranium).

Near-surface ash layers are often good non-toxic water sponges, but deep pressurized ash beds can crystallize and concentrate toxic metals and minerals. This gets really complicated. The Old Cascades, in the Bear Creek drainage basin in southern Oregon, are overlaid with thick New Cascades lava flows and are sitting on really old sea-bottom sandstones complete with fossils. The Old Cascade ash beds often have petrified wood (White Oak!) and pretty gemstones.

The wedding cake layering of the New Cascades has deep, old water that comes to the surface in giant springs that flow steady all the time, and when tested prove to be fossil water (surfacing tens of thousands of years since infiltrating). Wells drilled into the Old Cascades complex and ancient rocks below that can have high *boron and arsenic* levels, making that water toxic.

> *Soils are composed*
> *of finely ground up rock,*
> *carbon attached to minerals by soil-life,*
> *sequestered carbon such as charcoal,*
> *and—given enough water—*
> *living plant roots*
> *and animals.*

Granite, after it is exposed by the erosion of the overlaying country rock, breaks down by solar exposure ("*D-G, or decomposed-granite*") and is then elaborated by fungal and bacterial digestion into *carbon-rich* soils. When decomposed-granite slides downhill (*colluvial* soils) or is carried away in floods (*alluvial* soils) it sorts out into its component crystalline mineral types: *quartz* travels and tumbles to drop out of fast water flows first as *sand*; *feldspar* breaks down more easily and is carried further to be dropped by slower water as *clay*. Both sand and clay can be ground up finer by flood and soil-life, mixed again by subsequent floods and laid down as fine-grained *silt*.

Both volcanic and granitic soils can form carbon-rich silts (*loams*), through development under forests or grasslands; these *silty-loams* are the favorite agricultural soils as they release mineral fertility easily, hold water well (more slowly drained than *sand*), and tend to accumulate on wide plains at the foot of rivers and along mountain valleys.

Plowing, orchard ripping, blade leveling, and construction earthworks invert and mix natural soil layering. Soils developed-in-place by life have distinct horizons: topsoil, subsoil, and bedrock.

MEANWHILE, SOILS, AND GEOLOGY ARE DEEP: DO SOME MORE RESEARCH

Where agricultural fields get abandoned, they may not be able to support trees. Grasses and shrubs might persist, but many trees, such as Oaks and Pines (early invaders of feral fields), send down taproots from germinating seeds to secure deep soil moisture. A *plow-pan* is a plowing-caused barrier that builds up just below plowing-depth, as the mixing and inverting of the soil allows fine clays and carbonates to settle into a hard impenetrable layer (the plow-pan) that resists tap root growth. Agricultural soils can also be missing fungal inoculations that are necessary for tree seed germination and survival. Deep ripping may be required to break a plow-pan, but extreme measures can invert redeveloping natural soil layering or just trigger a deeper plow-pan.

Soil patterning is a remedial tactic for compacted, crusted-over, and over-grazed soils to help rain sink in and seeds germinate, using a tool to create micro-pit-and-mound divots, or shallow rips that divert erosive sheet-flow.

Another soil damage is compaction. Big heavy machinery, used over and over on soft fertile soils can consolidate the topsoils and subsoil to such an extent that the air and water capacity, as well as the penetrability is compromised. Bringing in even heavier rippers may be heroic and not regenerative. Repairing deeply damaged soils in order to get trees to grow is beyond our level of detail (above our pay grade), and may be inpatient industrialism rather than restorative, slow-succession.

Some soils and rocks can absorb carbon and form *carbonates*. These can glue soils so tightly that roots and water cannot penetrate (*limestone caliches*). Calcium and magnesium leached from soils and rocks can deposit lime-coating (carbonates) on downstream rocks (this white-wash coating fizzes when vinegar is added), and cementing sands and gravels into *conglomerates*, such as *quartzite conglomerate* (with river rounded cobbles cemented in *sandstone*). These cemented gravels can then be buried deep by geological deposition (*over-burden*), transformed into rock and re-exposed through erosion to become good rock climbing cliffs (the Schwagunks in the Catskills, New York, and Payne Cliffs at the edge of the Old Cascades in southern Oregon). The books to read for big geology time scales are by John McFee (1993).

The best vertical-sampling access to the layers of local geology is in publicly filed well drilling records. Although the drillers are not always good geologists, with the local context in mind (from reading McFee), we can usually interpret the notes (for example: 6 ft. red-brown clay, 30 feet fractured greenstone, 30 ft. 20 gpm, 30 to 60 ft. granite). This drilling core record likely shows the edge of a *pluton* where it meets the fractured metamorphosed ocean bottom *basalt* (now *greenstone*), which is permeable, and the water is found at the contact with the impermeable granite. The clay at the top is likely iron and magnesium rich, alkaline, and possibly difficult to grow trees on without lots of organic topsoil, perched on the lime-clay subsoil. This soil might also support rare and endangered plants, so do due-diligence and learn everything before disturbing what might be a *refugia*. This geo-story can be found in the Applegate-series of tortured rock types exposed across Jackson and Josephine Counties, southern Oregon.

There are many stories to tell
from all drainage basins
and these stories give us clues
to soils, water tables, and geo-dynamics.

Some intersections between soils-vegetation-climate and social arrangements can be seen in the poster: "Geo-General Places, 23." And geo-political and geo-historical intersections can be found reviewed by Robert Kaplan in *The Revenge of Geography* (2012).

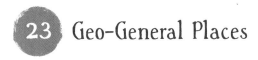 Geo-General Places

THE HOMES OF FOREST TYPES

Silviculture manuals and textbooks explore the utilitarian values of trees (USDA Forest Service, 1965). What are they good for? How can we monetize them? How can we get them to grow to industrial specifications? These references are useful for some tree-lore and alternative uses, but do not tend to consider the whole-systems context and *continuum-ecology* of forests. We can find bio-regional maps of generalized *forest types* and those are interesting. You may find the big context of forests for your region.

There are Silviculture (the science of growing trees) textbooks that lay out the *bio-regions* of North America, as part of explaining the ranges of timber species. These maps can help us orient our landscape-scale restoration strategies. *The Face of North America* (Farb, 1963) is an excellent natural-history of Turtle Island's geographical-zones, with good examples of *bio-regions* and *vegetation-types*.

The broad-strokes list of geographical-types includes mountains, plains, valleys, basins, deserts, lakes, marshes, and swamps. Every regional *ecosystem-type* is actually a mosaic of several *plant-community-types* finely adjusted to their very local geophysical footprints. This *complexity* gets really fuzzy, as the edge between *vegetation-types* (the *ecotone*) often contains the greatest *bio-diversity* and *microclimate-diversity* within the *eco-mosaic*.

These mosaic-tiles support guild-clumps of species-that-work-together to enhance persistence and fecundity. The perfect place to plant a tree and the perfect species to plant, remove, coppice, or thin are both very site and association-specific. Here below, we dabble in clues and stories in case the reader is inspired to observe the big pictures carefully before investing effort in the regimes of restoration.

TYPES OF GEOGRAPHICAL FOREST FEATURES

Meadows are openings in forests where grasses and wildflowers are not shaded by trees. Usually these are wet, at least seasonally, and sometimes they are sitting on clay or ash-pans that force shallow root-zones. Thus, trees cannot colonize meadows that are too wet or poorly drained.

A *lake* is a natural body of water and a *pond* is made by Humans or Beavers. Only lakes and ponds that freeze overwinter and experience spring and fall turnover (this is part of the science called "Limnology") can avoid *eutrophication*: the filling up of bodies of water by *anaerobic* accumulations of carbon (muck, bogs, peat, fens). Bodies of water below *snow-lines* that do not freeze-over become marshes and then meadows. Then, as shrubs move into the meadow from the edges, trees may eventually survive and shade out the previous meadow *plant-community*.

Beavers spread water with their dam building and tree and shrub *coppicing* (for browse). They create *ecotones* numbering up to a dozen, where before there were maybe three. Beaver work holds carbon and water in sponges, in upper drainage basins (*net, sink and loop!*). The first place to intervene and slow down erosion and excessive drainage is high in a drainage basin system; there we re-introduce Beaver (they were most likely there before us), after we plant Dogwood, Willow and other native shrubs so that we *hold nutrients and water as high as possible on the landscape.*

Or we imitate Beaver with gabions, brush dams, and coppicing shrubs. Because we want, in principle, to reforest ridges and ranges to rake in rain, establishing sponges high in primary valleys and basins should allow forests to be planted adjacent to the improved water retention.

> *Beaver colonize mountain basins,*
> *the Salmon swim up whole drainages;*
> *cool burning prevents catastrophe,*
> *and protects soil moisture,*
> *by thinning competition.*

Closer to the ridge itself, perhaps only on north and northeast aspects, Humans may be able to build net and pan structures (*microtopography* imitating fallen tree root pits), capable of supporting tree seedlings. *The Siskiyou Crest* (Ruediger 2013) tells the story of a famous example of inappropriate ridgeline-logging during the early 1900s along forest route 20—"the Loop Road"—up from Ashland, Oregon. This created a high-elevation desert that is still resisting reforestation a century later.

Many near-desert-climate ridgelines and mountain ranges in the Great Basins and inner coast ranges of western Turtle Island would benefit from reforestation to better hold water and biodiversity during climate-change. Humans do this most efficiently and sensitively with *local ecological knowledge* and cultural follow-through. We need water *detention* earthworks[2] for water table recharge and appropriate Fire or carbon sequestration to improve *soil-sponges.*

Ridgelines with *conifer forests* rake moisture out of passing clouds, which is why these trees are called *fog-brooms*. The shaded north-aspect slopes under these conifers hold snowpack late into the spring. These fog-broom-forests are critical to drainage basin water tables, stream flow downstream, and weather downwind, out across dry basins where enhanced water content clouds are swept by the next forested range (as in the Great Basin). Forests are elevation-limited by geology and climate.

---

2 Ponds and dams that catch and sink water, perhaps leaking slowly, building water tables.

Usually, nine thousand feet elevation or less is where we find the *tree-line* along the mid-latitudes. Above the tree-line we have alpine meadows and shrubs and dwarf trees, struggling to hang on. Many Pacific slopes, near the coast, or along interior ranges, are not high enough to have alpine ridges. These lower-elevation ridges have been hammered for log extraction, as the easiest place to build a road is often along the ridgeline. This is commonly disastrous.

*Understanding
how trees and woodlands work
will lead to what to do.*

Forest-groves are dynamic multi-layered carbon-capture systems. There are specialized species at all layers: from the canopy, down through the sub-canopy all the way to the understory plants of the forest floor. Forest groves need some disturbances, such as holes in the canopy left from wind throws or lightning-killed snags for wildlife, to keep pioneer plants and replacement tree seedlings in the mix. This temporal complexity is called *meta-stability* (see above).

Grasslands and *Sagebrush Flats* can be complex in their own way. As in the Great Plains of the central continent, woody species are moving out into new territory, as fires (wild and lit) and Buffalo no longer limit tree reproduction, leaving us with a challenge. Grasslands, when dynamic and regularly renewed, can be great photosynthetic carbon fixers (*soil carbon sequestration*). Properly stimulated grasslands can build abundant carbon rich topsoil, and with regular burning they can hold significant biochar carbon sequestration as well. Invading resinous evergreen trees, such as Juniper or Pines, can actually reduce biodiversity and fuel hotter and more destructive fires.

If a drainage basin council can figure out how to build a multiple-species woodland that allows periodic underburns and has meadows and grassland, persisting in a grand mosaic, reforestation of open grasslands might be justified, when appropriate to local climate-change challenges. Bringing back the whole, open, *grassland-ecosystem*, with great herds of grazers moving through in clumps and with a mix of seasonal burning, might be the best strategy for restoration that still sequesters significant carbon. There are many factors, cultural and ecological, to juggle.

Grasslands co-evolved with herds of dinosaurs, and learned to put stored energy into root growth after disturbances. With their growth tips (*terminal-meristems*) protected in the ground-level clumps of perennial and fungal-associated bunch grasses, they can respond quickly and regrow new leaves after being grazed or burned. Grasses respond to fire, grazing, and mowing with strong regrowth, and if the timing and amount of pressure is perfect, they are great carbon capture plants.

Marshes, lake edges, ponds, and seasonal pools (vernal-pools) are the powerhouses of carbon capture.

The most effective (net photosynthetic capture per area) carbon-fixation ecosystem is shallow water (Limnology), where sunlight can penetrate to the bottom, and the whole column of warm water is alive with Algae and pond-life; this is why *eutrophication* lays down so much muck and peat.

Forest-groves, with multiple stacked layers of leaves, capture and store massive amounts of carbon, but most degraded, over-cut, and compacted woodlands breathe out as much carbon as they capture in photosynthesis. These are carbon-neutral woodlands and Humans can move them towards net positive carbon capture with regular cool under-burning, canopy closure (with snags and openings), water sinks, and nutrient recycling. If leaf area can be increased, interspersed grasslands can be kept vibrant, and logs and branches that have fallen can molder slowly instead of burning hot in wildfires, with a tended woodland moving toward *net-positive carbon-capture*.

A *savannah* is a wooded grassland, usually with Oaks and Pines, and a few patches of shrubs. They can be fecund and bio-diverse with lots of edges and open enough *glades* that browsers and grazers have some cover and sight-lines (predator spotting and escape corridors). This is the White Oak woodland mentioned in detail in Chapter 4. All across the temperate regions of the world, savannah's bioregional collection of ecosystems is a Human favorite. Fire and grazing are key to keeping savannah open and productive. Brush-clogged Oak woodlands, overgrown with fuel, after the long absence of controlled burning, are ripe for *stand-replacement-fires*, which are followed by more brushfields. Oaks and Madrones are capable of re-sprouting from deep roots and on their root systems (surviving the maelstrom), thus preserving fungal associates which assist in the establishment of trees from seeds that float or are carried in. The care of savannah is not for timber production and has "pined" for appropriate stewardship. Savannah none-the-less holds multiple Social Forestry opportunities. All the skills of stewardship culture and many of the benefits of ecologically appropriate skimming are found in savannah.

Savannah is the comfy landscape for Humans. Check out our parks and campgrounds.

Swamps are forested bodies of water. *Marshes* are open and grassy. Both are usually braided with water ways. *Riparian-zones* are the strips of trees and shrubs that are adjacent to running water. Swamps have the advantage of being handy for floating logs; most Cypress swamps have been logged, over and over. When a swamp has a closed canopy of *tree-crowns*, the water in the shade stays cooler and better oxygenated. Where shallow water is in sunlight, the nutrient-laden soup thickens with captured carbon and loses its oxygen. These factors dictate the wildlife, fish and reptiles, and Human utility that swamps could have with good stewardship.

There are huge complicating flows through big swamplands; seasonal flooding delivers silt and flushing. Many swamps on giant river deltas are subsiding as flooding is controlled and canals are punched through (increasing drainage, dropping water tables, and decreasing silt deposits). Decisions regarding these giant wet drainage-basins rests with multitudes of upstream players. The benefits from whole (continental) systems-coordination and common-values, such as erosion control, river plain (flood plain) restoration, levee stability and utility, silt-dressing (natural fertilization), forested riparian-zones (cooler water flows), and carbon-capture maximization, accumulate with Human stewardship.

Bogs are *topographical* depressions where there is no outflow of water. Usually they are found in glaciated areas and are sink-holes left by melting ice masses. They are basins with no outlets. Over thousands of years, bogs tend to become acidic *refugia* for cultural-use plants such as Cranberries, Blueberries, and insectivorous plants. A floating plant-mat can evolve with enough integrity that clever Humans can walk/float/bounce with "bog-slats" tied to their feet.

A bog can be quite large and have a series of shrubby *ecotones* around the basin below the surrounding forests. Beavers can work incoming stream flows before they get to the bog pool. Bogs lose water to evaporation and to bottom drainage; the underlying soils and geology are key to their persistence. Bogs can be stable over very long time-frames, accumulating peat (and mummies). Sample cores drilled into ancient bogs show a record of pollen captured year by year in undisturbed layers.

THE LIMITING FACTORS IN FOREST HEALTH

Yes, Dorothy,
there are limits.
The laws of physics still hold.

Forests grow within a set of limits. The evolution of life on this planet is within the conditions found. Nice optimization, considering the long-time-scale of disasters and asteroid impacts. Great extinctions have happened at least five times before this *Anthropocene* geological era, when Human industrial and agricultural activities have accelerated a natural global process of repeated glaciation.

As we have said elsewhere, biodiversity re-evolves after extinction-disasters but it takes a very long time, as long as the whole cultural story of Humanoids: millions of years. As we Social Foresters consider how to pump down carbon and how to learn to live within limits, we can map and consider the types of forest that may be possible. Or type of grassland, brushland, or wetland stewardship that may help move the whole world towards restoration and stewardship goals.

We have already discussed the limits of altitude. These limits are changing with climate shifts but forests growing near tree lines grow slowly and our intervention needs be delicate and humble. It is probably a good idea to leave these alpine forests intact. The similar limit is latitude. Boreal forests grow far north and are limited by short growing seasons and extreme cold, as well as soil types (*acid bogs* and *permafrost*) that limit nitrogen and phosphorus availability.

So far with climate warming, tree species have been moving up in altitude and north in latitude. When we consider *aspect*, the limits of shallow soils, and extreme heat and solarization, climate change is pushing forests in the Siskiyous onto north and northeast slopes, where there is longer lasting snowpack and deeper soils, developed over millennia by forest roots. Rocky areas such as bluffs and cliffs, lava flows and craters, landslides and scree slopes, all limit soil depth and forests. Sometimes Humans can terrace severe slopes or develop benches to establish some trees, within limits.

Some generalized geographical sets of limits are familiar to Humans. The *savannah* is not a closed canopy old-growth forest because of limits of water (Mediterranean climate: wet winters, dry summers) and periodic disturbances such as fire, browsers/grazers, and Horticultural interventions. When a previous regime of disturbances is interrupted by neglect or disastrous change, such as the loss of *keystone-species*, the cascading results may move the system toward simpler and less fecund states; the loss of *disturbance-regimes* can be a limit, such as through extinction or migration, or the suppression of regular burning.

> *Senescent vegetation is not thriving:*
> *it is holding nutrients out of loop,*
> *outgassing more carbon*
> *than is being fixed by photosynthesis,*
> *and is prone to catastrophic fire.*

The accumulation of standing vegetation from the loss of grazing or burning regimes can lead to an available-nutrient reduction; this is called *senescence*. The ecosystem has slowed down and is "napping" because so many nutrients are tied up in dead or non-vital vegetation. Thus forest over-stocking also limits water availability through too much competition. A senescent grassland, brushland, or forest is not sequestering carbon, as much as it is losing carbon to outgassing from solarization and decomposition. The dead and dying vegetation is ready for a stand-replacement conflagration. Better that Humans tend ecosystems to maximize carbon capture while we can.

The nutrients available in different *soil-types* are well studied and can be referenced through the government agriculture and forestry departments. All soils benefit from carbon content, which is why industrial agriculture has been mining and exporting carbon inappropriately for centuries. Carbon (humus, charcoal and biological) content in soils allows complex soil ecosystems to thrive by holding water and fostering fungal and bacterial attachment of mineral fertility. Micro-life breaks down rock and mineral soil components and parks the minerals on carbon, allowing plants to take nutrients up into the work of leaf holding structures and sugar factories. It then takes carbon dioxide out of the air, with solar assist, and water and minerals from the soil, with fungal assist, to assemble biological complexity and fix atmospheric carbon.

Salmon bring back ocean nutrients to fertilize naturally poor, or overly leached soils.

Many of the soils and geologies on the Pacific slopes of North America are mineral-nutrient poor, as they are uplifted sea-bottom or metamorphic rocks. The Salmon has been crucial in closing a big nutrient loop from erosion and nutrient-leaching off the continent and into the ocean. These land losses are converted into *crustaceans* in the river-mouth nutrient-soups, then scooped up by Salmonids, until they cannot wait but run back up the wild rivers to spawn, die, and be spread by grateful Bears, Cats, and Humans, all over the headwaters' landscapes.

Restoration of stream corridors (*riparian-zones*) includes making sure there are *large-woody-debris* (LWD: logs and stumps) in the stream bed to slow down the flood flows, deposit gravel and sand bars, dig out deep cool plunge-pools, and nurse riparian shrubs and tree seedlings. Alders and Cottonwoods will only germinate on fresh sand and gravel and one can date the last big scouring floods that came through a stream or river basin by the age of tree lines, of Alder and Cottonwood, along the edge of the flood gravel deposits.

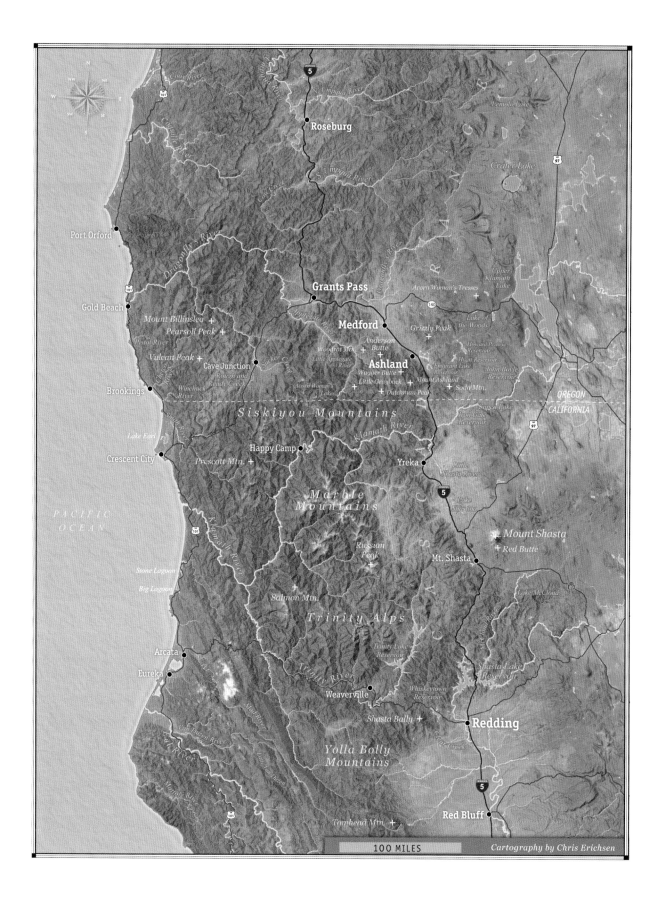

1 Observe*

2 Indigenate

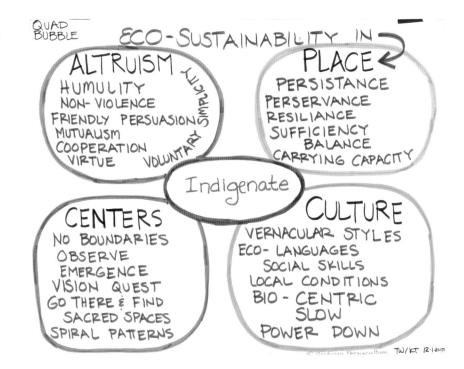

*Editor's note: the community-generated posters in *Social Forestry* retain their original spelling to keep the integrity of the original hand-drawn artwork.

3 Sensible Drama Diet

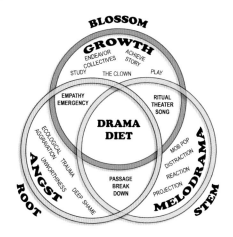

THE SENSIBLE DRAMA DIET

4 Ecological Analogs

DEFINITIONS

Ecological Analogs

Functional Systems must have a minimum number of parts (elements) in right relation (principles).

Principles are derived from observation of **Complex** natural systems.

Whole systems tending connects parts with principles to allow complexity (elaboration) to unfold

Avoid flatland by noticing dimensions

Work will judge value

Search for Lacunae

5 Animals in Ecosystems

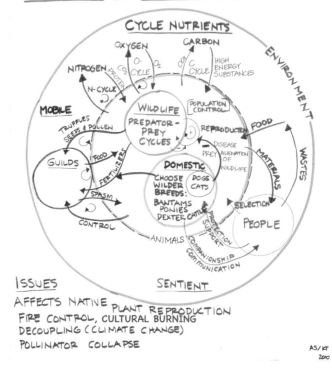

6 Adaptation

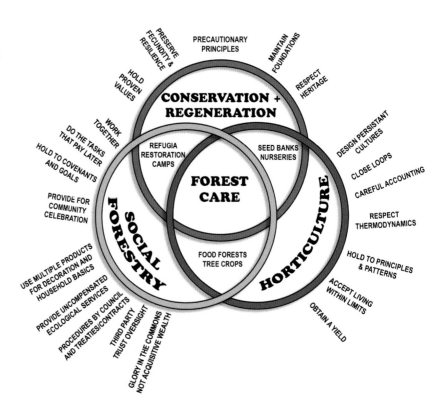

7 Forest Terms

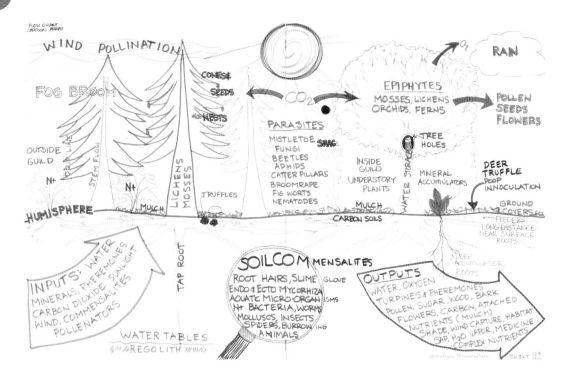

8 Prescription

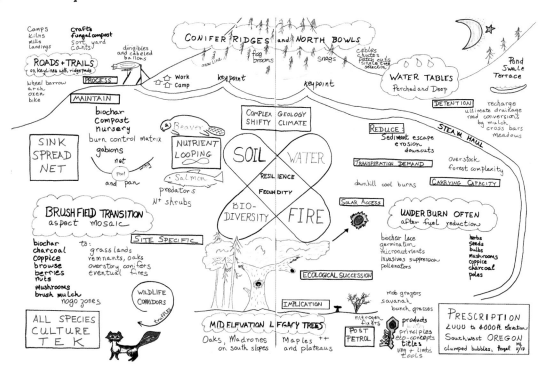

9 Social Forestry Transition

10 Social Forestry Terms

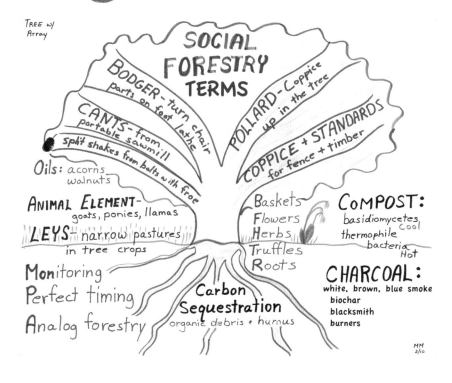

11 Tree Shapes

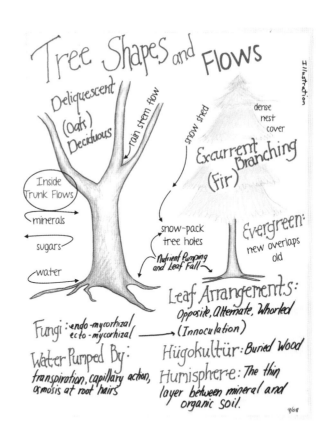

11a How Trees Work

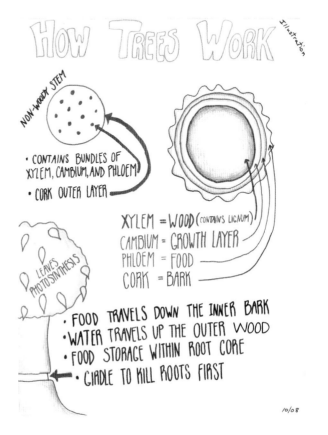

12 Hedgerows

13 Windbreaks

> *Complex geologies*
> *can concentrate natural toxicity.*
> *Local knowledge is vital.*
> *Water testing for Mercury, Lead, Arsenic, Boron,*
> *nitrates and nitrites is advised.*

There are soils that derive from parent *metamorphic* rocks that release metallic toxicity, raise the *pH* to extreme alkalinity, and limit the types of plants and soil-biota that can survive. Sulfur from volcanic vents and springs is extremely acidic. Ultramafic rocks, such as serpentine, break down into magnesium rich alkaline soils with pH 12 or 13 (caustic!). This alkalinity limits plant uptake of *nitrogen* and *phosphorus* from soils. Just as the bog supports insectivorous plants (where the water is acidic) so do alkaline seeps and flats.

Some plants facing challenges figure out a way around: Either plants evolved long long ago in a *reductive* (no oxygen) atmosphere or ecosystems co-evolve to allow life even under strict limits. The *Kalmiopsis leachiana* plant in Southwest Oregon is an ancient relic of the early evolution of the Heather family that includes Manzanita, Madrone, Azaleas, and Huckleberries. It was thought extinct and only found in the fossil record, until discovered by Ms. Leach in the rugged alkaline soils of the Chetco River drainage.

> *Industrial poisons*
> *are often new to life-on-earth,*
> *especially in the quantities*
> *concentrated by the empire.*
> *Plants and soil-life have not yet*
> *co-evolved to deal with this.*

Toxicity from toxic chemicals is much harder to talk about. The brown-fields (industrial waste-lands) left by agriculture or chemical forestry need mitigation. The most promising approaches use mushrooms, especially *Pleurotus* (Stamets 1993) and *carbon-farming* to tie up and trans-mutate dangerous chemicals. This promise would require a lot of laboratory support for testing to affirm progress. We should all hope that significant cooperation is aimed at clean ups. There are some Mustards (*Arabis sps.*) that have been shown to concentrate heavy metals in the above ground plant which can then be processed to extract the toxic metals and sink them in safe perpetual deposits or useful artifacts (machines?). These clues are hypotheses that demand testing and trials but they are encouraging pointers to find options for biological mitigation.

It is the Human condition to live within limits; the laws of the universe demand this of us. We can do this gracefully and modestly, and perhaps even be useful to the greater natural context of our home Places. The sensitivity and memory necessary to perfect timing, and appropriate skimming, can only be held in community. For our behavior to be pertinent, we hold etiquettes and taboos in the circle of drainage basin councils, with mutual-aid arrangements, within whole bioregions. All Beings and cultures should have delegated spokes in drainage-basin councils.

*Our Human wisdom
is limited.
We need
All Species Counsel.*

Trees hold council as well. The gathered presence of a sacred grove is palpable to all visitors. Individual trees have character and spirit, and some radiate in abundance. We can identify the *legacy-trees* that have persisted through regime changes and who speak of past *ecosystem* balances. A legacy-tree is most often a broad-leaved tree that has been overtopped by conifers (especially Douglas Fir and Ponderosa Pine, on the Pacific coast inner ranges and valleys). The widely branched structure of legacy-trees bespeaks a time when the landscape was more open, and this tree was able to spread out for lack of competition and shading. Perhaps because of regular burning, or having stood in pastures now overgrown to woodlots.

Biodiversity can be limited by a lack of forest complexity. A Douglas Fir dominated stand (perhaps the grow-back from a catastrophic fire?) can lack species diversity. Legacy-trees can be saved, where they are being overgrown by Douglas Fir (perhaps from the genocide of First Nations people, and the end of regular burning?), by opening up the *crown-closure* on their south and southeast *aspects*. They can still harbor nurseries of tall thin Fir poles in their north shade sector, as long as that overstocking is thinned regularly and competition is controlled. Or the whole, new forest structure that still contains legacy trees can be worked up so that under-burning can be re-introduced.

Ecotopian Social Forestry is about visioning an *ecological-succession* from overstocked senescent woodlots and monocultural plantations to complex mosaics with clumping, groves, glens, prairies, coves, and mounts. Our Human tending-work is driven by the threat of catastrophic fires and declines in ecological vibrancy. This tending also offers the gift of a period of carbon-skimming that can support the forestry-work necessary before we can return to the easier work of regular burning and Horticulture. We in the Pacific Northwest have a surplus of Fir poles and might switch from milled stick-frame houses to pole buildings.

> *Ecosystems are limited*
> *in their fecundity and resilience*
> *by the actions of Humans,*
> *because of degeneration*
> *through excessive taking.*
> *Restoration can bring regeneration*
> *through tending.*

Our overarching Social Forestry work-goals (see Chapter 6) derive from forest ecology. We make decisions based on forest structure that support biodiversity (cavities, Mistletoe-brooms, nesting-tangles, snags, acorn granary trees), opening or closing the canopy based on species-mixes and ground-covers to reduce catastrophic stand-replacement fire danger (via landscape-scale strategies) and sequestering as much carbon as possible, while recharging water tables. We read the vegetation and the soils for the site-story and see tactics emerge from the intersection of *ecological knowledge*, cultural interests, and site-spirit clues and feedback. We learn to listen, and we learn what has worked, looking for the surplus yields that we can divert, which we know we have to reciprocate on the *gifting-hoop-loop*.

As we learned in Chapter 4 ("Relationships"), the whole set of reciprocations-in-ecosystems has a vocabulary to help us appreciate and talk about complexity. The *bioregion* we walk in, while considering forests, has its iconic set of *keystone-species* (such as Salmon or Buffalo) and implicate-species (such as Oaks and mushrooms), as well as producer-species like grasses, and tree leaves, and Cattails. Every landscape has an optimum evolutionary potential that reaches that Boolean-logic edge of co-evolved, interconnected diversity, where the interactivity is more important than the number of species. The whole complexity at this theoretical *climax* (a mature, dynamic ecological state) is not a system that can be freely exploited by catastrophic harvesting, without collapse.

> *Carrying-capacity*
> *is the functional ability and measure*
> *of a landscape*
> *to feed and shelter*
> *a maximum number, and weight,*
> *of species.*

Overshoot is the ecological term for going too far, and taking too much, without giving back. *Bottleneck* is the term for the constriction of species-diversity, forced through resource-declines, after overshoot (Catton 1992, 2009). Ecological-collapse is often accompanied by *eco-spasms*, where an opportunistic species (such as Humans) over-populates and overshoots the *carrying-capacity* of a landscape. Usually a keystone predator has been removed (such as Wolves or Sabertooth Tigers), causing a previously controlled consumer-species to explode in a spasm of resource-depletion, almost going extinct. Voles in the Klamath Basin in the 1950s twice went into eco-spasm. All the predators had been removed after the marshy lakes were drained for agriculture, and the exposed, fertile, muck-soils produced plenty of vegetation to feed an exponential-growth plague of small boxy rodents (Ashworth 1979).

The promise of Social Forestry is that we can find, once again, the appropriate diet, clothing and material-culture for the drainage-basin landscapes that we live on. Within the limits of the Place and with the modesty of "just enough," the simplicity of *right-livelihood* still leaves room for appreciation and celebration. It is a glory itself that Humans still have a part in this ecological situation after all the modern history of planetary exhaustion. Hopefully, the mistakes we make with our best intentions are learnings, and Humans come back to the land.

A SHORT SURVEY OF A BIG REGION, THE PACIFIC NORTHWEST

Climate change has already left a lot of standing dead trees in northern California. The fire hazard is extreme and getting worse. The resource-extraction footprint of a huge, hungry growth-fixated population of Humans has reduced the *carrying-capacity* and increased the likelihood of disasters. Holding onto the headwater-forests during collapse of empire is always a task for Peoples of Place. Every investment is risky but the outcome of making no investments is dire. The state of the forests in the PNW is not as bad as many other mid-latitude forests. We still have some primordial groves. It is the industrially hammered forest-lands that beg for our help.

Most Pacific Northwest vegetational-provinces can hold several forest-types.

Northern California has a complex set of natural *vegetational-provinces* (Hickman 1993). The west *aspects* of the coastal range hold fog-forest remnants, yet even in these seasonally-wet forests, competition from Douglas Fir and True Firs is reducing water holding capacity. Before the US Forest Forest Service was formed, cultural burning outlawed, and fires suppressed, fires in the late dry season were common in the Redwood giants. Some understory tree removals or falling for log mulch is justified, especially when moving towards re-introducing fire as an option. Even the coastal rainforest is periodically under-burned to good effect.

The east and south aspect slopes of the inner coastal ranges are often much drier and Oak/Pine savannah is a common component. The ridgelines in the inner coastal ranges were regularly burned for passage, geophytes (open meadow bulbs), and basketry material. *Riparian* old-growth forests sometimes thrive in the canyons below. The northeast and north aspect slopes of the coastal ranges often hold an Oregonian forest of mixed broad-leaf and conifer trees.

The Oregonian vegetational province is the classic mid-elevation forest-types collection that cloaks the industrially hammered-slopes where we now live and work.

The Oregonian vegetational-province is a forest-type that is found all the way north into British Columbia. This forest-type exists on north aspects in California and south-aspects at the northern latitude limit. As we float north in our visionary dirigible, we see below us some of the highest biodiverse vegetational-provinces on some of the most complex geology in the world. The Klamath Knot (Wallace 1983) holds conifer *species-densities* unmatched anywhere. The Klamath and Trinity Rivers cut through the jumble of pasted terrains and captured archipelagos (McPhee 1993).

The Klamath River starts far out in eastern Oregon, on the Gerhardt Rim in the Great Basin, 100 miles east of the Southern Cascades Range. This east-west river-transect takes us over open Ponderosa Pine forest, holding wet shrubby basins (Beaver!), and Aspen groves to the Pacific Coast Redwoods. The Gerhardt ridgelines hold northeast-aspect conifer forests of True Fir and Lodgepole Pine, and other Pines, with perhaps some Hemlock.

The federal homesteading legislations encouraged ranching in brittle-desert-basins. The carrying-capacity was quickly broken with imported farming strategies. Hard country to thrive in and there is lots to learn.

Often, even these desert-range forests (at the edge of climate-limits) have ended up overstocked as a result of fire suppression and old-growth removals; with climate change and extensive road building, these pockets of Great Basin forest-types are overdue for wildfires. Industrial logging of the Ponderosa Pine forests on the west slopes, above the basin marshes, has overroaded and severely compacted the fertile, but dry, volcanic soils.

Our Social Forestry work means moving fire back in, saving some roads for fire lines, and putting most to bed (turning forestry roads into water infiltration earthworks[3]). The ultimate restoration opportunities are upland water infiltration, thinning for carrying-capacity (reducing competition for remaining fire resistant legacy old-growth), and restoring Beaver meadows wherever possible.

Many exotic grass and forb species have spread throughout the settler homestead ranches and adjacent federal grazing lands; a lot can be done by *mob-grazing* with close supervision (portable dwellings, Dogs and Ponies?) and the re-seeding (and fungal-inoculation) of native Bunch Grasses (*Festuca, Avena, Elymus, Bromus sps.*), and tending of bulb-grounds (*Camassia, Lomatium, Perideridia*). The return of Wolves is wonderful (Elk and Deer herds bunch up and stay away from riparian zones[4]—mob-grazing with Wolf shepherds!), but we may want to wait a bit before we welcome Grizzly Bear back.

Bears
are excellent Horticulturalists.
They transplant, thin, fertilize,
inoculate and till/pattern
many types of plant-ground.

Berries are a highly valued vegetation-type. Regular burning is necessary to renew bountiful harvests. The shrubs commonly valued (*Vaccinium, Arctostaphlos, Sambucus, Rubus, Ribes sps.*) are pioneer-species that follow stand-replacement burns after birds and Bears deposit berry-seeds. Slopes and flats that hold berry-fields are found on all but north and northeast aspects in the southern Cascade and basin-ranges, and then move to the south and southwest *aspects*, further north. Clearcuts (where not sprayed with defoliants) carry good berry fields but these are usually overtopped and shaded out with returning conifer overstories.

Good berry grounds are shared with Bears and traditionally burned in the fall after harvests. Berries can persist along meadow-edges and on rocky ground too harsh to support forests. Once berries get established they serve as a nurse cover for tree seedlings. Keeping berries in the mosaic mix all through Ecotopia (traditionally by burning) is a forest succession set-back strategy. It has to be done carefully to keep fertility and productivity high, and it may be necessary to move berry patches around, allowing forests to grow back. Ah, the challenges of disturbance-regimes; do we know what we are doing? Are we sharing the berries?

---

3   Ditches and swales that quickly divert and soak in runoff.

4   Stream and riverside forests and brush strips.

*Tended eco-mosaics
provide lots of edge,
berry crops move around
as forests migrate
and transform dynamically.
Controlled burning,
with seeding and transplanting,
preserves sustenance opportunities
while supporting meta-stability.*

One of the remarkable observations of forest-type tending is the unexpected extreme location of White Oak savannah ecosystems, far north in BC. The early colonial botanists (almost exclusively men) thought these outliers were remnants from an interglacial warming period but it is most likely that they were moved north and established on hospitable slopes and benches by Horticulture; the savannah so usefully provides medicinal herbs, geophytes, basketry materials, mushrooms, and acorns. The First Nations women moved whole ecosystem-types around, as part of tending vast landscapes (Turner et al 2005).

The wet west-aspect and cooler-slope north-aspect conifer forests, found throughout the Pacific Northwest, loft such dense and tall closed-canopies, that there are very few understory plants. There is usually thick, carbon-rich litter and log mulch, along with some mushrooms, but the awesome value of these fire-resistant cathedrals is water-harvesting and carbon-sequestration. Not a lot of food or materials there, except where big Cedars might be selected for single-tree removal. Pit-house boards (often up to one by three meters broad and twenty centimeters thick) were hand-split from carefully fallen Cedar and Sugar Pine snags. This meant camping on-site for a while, as the giant slabs were teased out with wedges from the fire-felled logs.

*Old-growth deep shade,
with tall green crowns,
and closed canopies,
presents few food resources to Humans,
but has immense value
to the whole-drainage-basin
and All Sentient Beings.*

We can learn from the First Nations the rare tasks that take place in these old-growth temples. Most of the ecosystems, where Humans can find useful strategies for restoration, are nearer to the mid-elevation winter-camp and permanent-village sites. Most of Ecotopian restoration work is likewise on these slopes and flats throughout the PNW. The west-aspect Pacific rivers, especially the interior valleys, have *flood-plain* forest-types that consist of Cedar, Ash, Maple, Cottonwood, and Alder. Better drained rises hold White Oak and Douglas Fir. Isolated hills often carry a savannah forest-type and some large meadows in grasses, with riparian corridors, that hold many understory fruits and berries. Much of these *original-vegetation* types (there before the settlers) were burned regularly, especially the grasslands and savannah.

Farming has altered the natural patterns, straightening drainages, removing complexity, and suppressing fire (except grass seed field burning), turning over and mixing soil profiles and making giant earthworks (roads and settlements). These bottomland forests are on fertile soils, with high water-tables, and can capture silt during flooding. The exotic trees, shrubs, and vegetation introduced with intensive agriculture are a new set of forest-types, now co-evolving.

The native Oregon Crab Apple is cross-pollinated with exotic European Apples and no longer bears true offspring; all the seedlings are hybrids. This offers new varieties that could be selected and encouraged. The Black Locust from the eastern continent is unstoppable as its roots sprouts so readily that removal is very difficult so we might want to manage it for its values. Exotic Poplars, Willows, Broom and Butterfly Bush can be coppiced, burned, and browsed to keep them in immature woody structures, and avoid maturation to seed setting.

*Humans
have lots of restoration work
to do in big-valley-basins,
but they need not
live there in the winter.
Nice Place to visit
in the harvest and burning seasons.*

Because these inland valleys have persistent fog and high spore and pollen-counts, as well as long periods of stagnant weather with Pacific high-pressure atmospheric domes that park over the region, the First Nations did not have winter villages in the valley bottoms. The adjacent hills and mountain ranges provided better perches above the valley air-soups.

The restoration options in these inland valley basins include replanting food-forests, bringing back native grasslands and Fire, and giving up room for Beaver and Salmon.

Vast bottom-land forests, with seasonal trails and waterways, are visited during harvests and ceremony, Salmon runs and burning season. These rich and thick bottom-land forests, marshes, and adjacent savannah, should be liberated from the pall of development: move upslope into the side valleys, locate rail and road on keyline pattern and reduce the impact of traffic. Our regional carrying-capacity for All Sentient Beings would be massively improved by restoring and tending the valley-forests.

*Urban forests
can be a saving grace
in the midst of dense development.
They clean and cool the air
and help absorb storm waters.*

Urban and suburban sprawl in the PNW and elsewhere has largely replaced floodplain original forest-types and wetlands. The remnants of settler civilization might be retro-fitted to support intensive food growing and natural drainage pattern restoration (daylighting buried waterways). Ridge corridors and river banks, returned to forests, will support lively wild-interpenetration, with smaller decentralized-clumps of salvaged-housing and processing materials.

The street and yard-tree groves and strips, on neighborhood-scales, can be seen as natural structures providing ecosystem functions (wind breaks, air conditioners, bird corridors). Colonial parks, originally set aside for recreation, represent potential nodes of ecosystem restoration, with native food-plant nurseries and seed-crops, allowing the industrial/commercial mess to be re-greened and brought back to productivity, supporting the Human restoration efforts and the celebration of the returning *keystone-species* and *implication*.

*Tree crops, nuts and fruits,
are ultimate Horticulture.
Their care and propagation are essential
to support Humans during hard times.
Tree crops provide multiple values.
Turtle Island was thick with tended tree crops
before European invasion.
These food-forests were under-appreciated;
white folks thought they were just Nature's bounty
and put their pigs out to scoop up the "surpluses."*

Scattered across the Great Basin ranges, in the rain shadow of the Cascades and Sierras, are tree crops that were regularly tended not that long ago (before 1850, imperial calendar). Pinyon Pines grow slowly, and are slow to mature to cone bearing, but the seeds bear such important nutrition that preserving and protecting these remnants is crucial. Some groves are being tended again or still. We need to learn how to move these groves north and onto amenable slopes and flats, with climate-change. Black Oak (the favored traditional storage acorn) can be found on the dry side, such as the upper Pit River drainage, where they probably were tended, not long ago. Tree crops are awesome in many ways but they take generations of care and maintenance (reciprocation) to preserve.

HOW TREES LIVE

Woody perennials (shrubs, vines, and trees) find a living through strategies that allow them to move more slowly, to become very, very local, and to feed other Sentient Beings in a way that benefits the larger context of nutrients, water, and Fire that allows and preserves forests. The intelligence of woody plants is so different from our fleshy high-speed animal running-about that we do not appreciate their wisdom, without slowing down and sitting a bit. We look like blurs to the time-consciousness of a tree.

*Trees
teach Humans
about Time.*

The oldest trees are thousands of years old and have witnessed modern civilization from the start. Bristlecone Pines east of the Sierras have reached 7000 years and have taught us much about drought cycles and climate. Legacy trees are often hundreds of years old. They tell us about the more recent colonization of the Pacific Northwest. Old-growth forests contain trees half a century old and older. The age of a specific forest stand is more distant. There are Chaparral brush-clones in deserts that are thousands of years old and Aspen grove clones high in the mountains the same. Although the grove may have young branches and trunks, they are all root spread from the original seed planted ancestor, and thus one old being. Some trees propagate more by root sprouts than by seeds.

Trees develop character. We Humans use their shapes to hang stories on. Children love to climb trees and hang-on-tight in the highest branches, while the wind swirls the canopy. Trees sing in the wind, and sometimes one branch rubs another and strange music is heard. In ice-storms or heavy snow, and/or high winds, old trees, especially hardwood legacy-trees, come apart: large limbs are dropped, sometimes ripped off explosively, with great noise.

FOREST ECOLOGIES 149

> *Trees know*
> *how to survive wind*
> *and snow-load pressure*
> *by letting go of some parts.*
> *They shed their limbs to live.*

When you go into a tall multi-species conifer old-growth grove during a windstorm, and lie down or look up leaning on a cane, you might have the balance to see the trees dancing with each other, keeping a certain spacing and moving in circles and figure eights. Any branches that touch, tear off twigs which are floating down wind. This is dangerous because of spear-like dead limbs falling far and piercing the soft forest floor. Old-growth groves in big wind are not good places to camp.

Trees communicate with each other in more ways than we can imagine: chemical messengers (pheromones and turpines) through the air and through the soil fungal net, electrical signals both ways as well. A forest reacts to disturbance (caterpillar eco-spasm, or aggressive logging) by sending out warnings to make leaves more bitter, or move nutrients to root storage, or send out psychic bombs, even howling in low frequencies.

Social Forestry Terms

Nutrient gifts can be delivered through root splices. Seed trees favor their own progeny. Browse favorites, such as Deerbrush, vary in sweetness; you can see which ones are shaped round by nibbling. Some Southern Arizona First Nations children know which Mesquite bean crop is sweet. Certain Southern Oregon Pines are "scraping trees" where a patch of bark has been removed and the good pitch of this Pine (who is named and celebrated) is harvested as it seeps from the edges.

Ecology is about relationships. Humans have multi-dimensional connections with trees: Forest Bathing (Plevin 2019), prolonged observation (sit-spot meditation, vision-quest), and non-invasive/non-toxic tree houses and platforms (primate nests) are good ways to get acquainted. Without trees, we would not be Human.

The Social Forestry Work Goals from Functional Principles include

---

22.
Restore the wildlands: regenerate nutrient cycles and foster resilience and complexity.

23.
Sequester carbon: hold carbon on-site in soils and vegetation. Live well within the solar carbon budget (net photosynthetic capture).

24.
Bring back broadscale burning: use underburning (after clearing excess fuel) to recycle nutrients, stimulate carbon capture, and preserve biodiversity. Also called "cool burning."

25.
Move local forests toward Old Growth: facilitate fire-tolerant stability and complexity by managing the understory.

26.
Close the loops: nutrients, water, carbon, wildlife and soil biota must not be exported or lost to erosion; loop, net, and sink. Beaver, Salmon, and Fire are our icons.

27.
Hold water as high as possible in the drainage basin: keep it cool by shading the riparian corridors and building water tables upslope.

28.
Bring back the Beaver: especially in the headwaters. Learn to live with their engineering; they create several layers of riparian sponges and sinks; they build for Salmon.

29.
Perfect timing: become tuned to the land. The shape of Place holds the stories, and we can learn to come into alignment with potential magnification and elaboration.

30.
Use the right hand tool for the job: sharp, well-tended tools are convivial. We work together safely and expediently while attuned to our culture and surroundings.

These nine principles above are our work goals, now we will continue with further social and cultural principles

31.
Restoration to what? The regenerated ecosystem functions will not be like any previous regime. Tending changes the Wild dynamics. Humans are implicated in co-evolution. Pay attention and stay nimble. Observe and interact.

32.
Human cultural work-arrangements are reflective of the drainage-basin ecodynamics. We mimic a dance beyond our bipedal grace. If we cherish all the ways Humans can be useful, we can honor all skills with non-exclusive—but skill-focused—work-guilds that cooperate with the whole-regenerative-effort through drainage-basin spokescouncils.

33.
Ceremony and ritual, guided by taboos and etiquette, keep local People of Place oriented. How we go about our tasks speaks to the sensitivity of our attention. Keeping an open-view of complexity benefits from layers-of-reminders, ornamenting our demeanor. Let us, at least, entertain All Sentient Beings.

34.
Work skills are learned from example and honed through practice. Re-skilling is the work of manifesting the basic-hand-skills common to Human cultures but forgotten in modern-industrial-consumerism. We need to get back to muscle and attention knowledge (poise). Our bodies and our culture benefit from movement and gathering skills. Ergonomic muscle dance with sharp objects. Work together, but do not get distracted; avoid dangers.

CHAPTER 6
Forestry Work

CONTEXT FOR THE RESTORATION OF SOCIAL FORESTRY WORK

One question Dennis Martinez always asks is "restoration to what?" He means perhaps that we do not know what we are doing? Certainly true! Just to get started, it would be great to have some idea of *original-vegetation*, and sometimes one can find speculative maps, but even the term "original" is questionable. We might better think about "where-to-from-here." What are the opportunities for useful intervention and *disturbance-regimes* in the ecosystems where we are? How do we ride climate-change, or learn to flow with cyclic-wildlife-population-patterns? What is *perfect timing*?

The practice of speaking gratitudes for all the ecosystem functions that Humans benefit from, that precedes a centered and deep council session, helps a Culture of Place stay humble, to take time to observe, and report, and consider. As Human participants-in-forests we do have useful tasks to do, and helpful attitudes to practice. All our efforts need to be held in the light of the principles of precaution and conservation (*right-livelihood* and *simple-living*). Resilience, fecundity, and persistence are our banners-of-intention.

 Observe

OBSERVE AND INTERACT

There are Advices for Observation,
to help us learn the skills.
Beyond direct observation,
we learn to Interact.
First we talk back;
say something nice, but engage.
Then observe some more.

Learning to mimic bird calls and animal noises is very useful for subtle signaling between Humans. Mimic too well the wrong call, in the wrong context, at the wrong time, and we get serious feedback. Don't be rude. Don't imitate a Great Horn Owl near a Screech Owl nest cavity. Especially if you are hiding in a tent or shack. Big no-no! Nature is not fooled if we remain amateurs.

As we start to make disturbances in preparation for Horticulture or Fire, start small. Use small scale experiments in handy-to-observe (trailside?) pockets. See what happens. Try again. Plant just a few seeds, transplant only three trees (and follow up with water at first). Find the least intrusive purchase: the corner where if something goes wrong, we can get to it.

Any installation leads to maintenance. Our test systems degrade. Some call buildings from natural materials "melt-down-houses." We ask, "Who is going to move in, Pack Rats?" Plan for maintenance. As we learn how chores may accumulate, we learn to build better and simpler. A basket that works, is a basket that lasts.

Feedback up-front from our experiments, followed by thought-out investments, leads to less maintenance chores. Household chores are purposefully tied in with the homestead eco-nook, so that we avoid pollution and temptation. We still need to be as simple as possible, and everyone knows how-to-do. If only one or a few folks can fix complicated systems, we have a brittle situation, both ecologically and inter-personally.

Work with the flows, start upstream. If we try to stabilize a riverbank before we stabilize and sponge-up the headwaters, we will waste our effort; the next flood will tear out our retaining tricks. The upper drainage-basin priority is probably Beaver. Go talk to them. Move already companionable pairs to the high-wet-meadows.

Ecosystem functions
dictate principles which,
deployed through patterns,
lead to details.

As workers, we must *obtain-a-yield* in order to sustain our efforts. The opportunities we have to accept gifts from natural systems are balanced by our reciprocations, both in attention and stewardship, and in our ceremonies of recognition and mutuality. To be social, both with All Sentient Beings and with each other, the tools we use (the very extensions of our bodies), are shared, special and subtle. The art of using appropriate-hand-tools is in the dance of attention and co-creation, which proper-cultural-tools of Place facilitate. Hand tools, properly maintained, express our carefulness, while they allow a gracefulness; we work quietly, while we communicate and cooperate. The pace of our lives and the fullness of our simple house-holding keep us happy-at-home, with its great-wide-context, and engaged in great and small cycles.

*We dance
our way to
fecundity and regeneration.*

Social Forestry welcomes reciprocation, no-go-zones, offerings, feedback, monitoring/remembering, and the re-enchantment of whole-drainage-basins, with stories illustrated, and indeed taught to us, by the actual shape of the landscape-geology. We learn our stories from where-we-are, and we remember the *ecological-knowledge* that we have accumulated in Place, by constantly referring to the shapes-of-the-land that remind us. Phrases in our language repeat the lessons and wishes of our culture of *forest-tending*. We sing, dance, and apply our knowledge in all our mumbles, motions, and movements, all-together-now with rhythm-in-harmony.

PLAN

The understanding of *context* and the holding of *Ultimate-vision* are necessary to our orientation, but the work of planning what-to-do is first about *Quick Fix* and *Retrofit*. Neither of these phases of plan and focus should compromise the potential of the *Ultimate Intention*. Even Quick Fixes need some context assessment, reinforced by practical experience, so that they are immediately useful and leave no long-term harm.

*Quick fixes
are right at hand and obvious.
Retrofits
need plans and budgets.
Ultimate Intentions
are visionary and long term.*

When we first look at a forest-type complex, there are priorities in our scoping (the holistic-inventory). Where are we in a local drainage-basin, that is connected to a larger drainage-basin, or a big body of water? Take walks, or let them take you. Find the access trails and old roads. Look at the water courses. Erosion? Logging debris? Thick Brush? Lots of standing and leaning dead snags? Limb filled and vine-twined under-story?

There are rules-of-thumb to consider. Where is a fire most likely to start and where will it go? When you have a likely ignition corridor (such as a road or a power line) and you have a heavy fuel load to the northwest of a potential starting place, any fire in the northern hemisphere of the planet has almost always gone northwest first. If you then have buildings and escape routes that are in a fuel-loaded "fire sector" (the slope or valley uphill and NW of an ignition hazard), then you have a priority of reducing fuel load on those slopes first.

Is there a core-area of infrastructure, cabins, barns, and sheds, that need protection or stabilization? How are the roads—do they provide escape routes? Are there too many roads or too few? Are they in awkward shape, leading to erosion and stream sedimentation? Assessment comes first, before we know what opportunities and necessities are being presented, and before we even imagine a *plan*.

HOMEWORK FOR FORESTERS

Maps, and lists, and documents, are very useful. Here is a homework list that can serve as an information wish-list in advance of a forest-planning session.

Here is a typical wish list for homework:

1. Get the seven-and-one-half-minute USGS topographic sheet(s) for the area that includes the larger drainage basin, so that we can see the collection of smaller drainages, ridges, basins, flats, mines, vegetation patterns, and standout rocky features. The print-out maps that you will find at some camping stores are very poor quality. You may have to order good ones. We've also found topographic maps through the US Forest Service.

2. Get the soil-survey-sheets for your farm. The Natural Resources Conservation Service (NRCS) gives out this information for free. You can find their office by searching for your local County Soil Conservation District. You may also find this information and print it out, perhaps at your local library. It will be lovely to see the soil-types map and the text descriptions for every *soil-type* you have.

3. Get the well drilling record for nearby wells. You may be able to find neighbor well records if there is none for the well you are using. Try the State Water Master's office? You will need a property description, with tax lot number, and range and tract (from the USGS topographic-map.)

(continued)

4. A large format aerial-photo of the area is extremely helpful. You can try printing one from searches at your local library. Some properties had aerial-photos taken when they were last owner-transferred or by the County, Forest Service, or BLM. You can find maps of inferior quality from your County Property Data. Ideally this photo will have the tax-lot-lines superimposed on it in color. Even better, are topographic lines superimposed on this photo, sometimes 5 feet apart or at one-meter intervals. We can then use a sheet of vellum or tracing paper as an overlay of the same size, to use as a sketching see-through for planning (tape a base map with the planning overlay on a bright window, to see through the paper layers).

5. If you haven't done this yet, it is often helpful to have the County Soil and Water Conservation District visit your site. They are non-regulatory so you don't need to worry about whether things on your site are permitted or not. The worksheet planning guide and free site visit they offer will better prepare you with ready-to-use walk-through advice for connecting the elements and flows on the site.

6. What are the goals and scope of the drainage basin super-project? Perhaps you can make a cartoon sketch or a list of elements of what you may be thinking so far? This can be any size but it could be a useful way for you to convey big dreams.

7. We would love to see any old homestead site plans, septic-field maps, the neighborhood tax lot map, property descriptions (from the title insurance packets), any historical maps (Metzgers, for example) and historical documents, any weather maps (rainfall patterns—USFS?), and any utility-location maps (underground lines and pipes). Are there traditional water rights or irrigation and fire district documents?

8. If you have the billing documents, a utilities cost summary for past years would help us understand energy use patterns.

Assessment Elements

Typical elements (pieces on the plan game-board) and features to be considered in assessing a homestead in a drainage-basin include:

Solar access
Gardens, orchards, and animal systems
Drainage, infiltration, and pooling
Ponds and aquaculture
Water catchments, rain barrels, and cisterns
Soil regeneration and fertility
Wind and storm protection and influence
Natural home heating, cooling, and ventilation
Energy efficiency and household economic budgeting
Privacy and sound barriers
Personal security (crime, predators, etc.)
Native and edible landscaping
Beautiful and aromatic embellishments and ornamentation
Best use of local materials
Natural building materials
Fire protection and fuel reduction potential
Alternative forest products

The *relative-location* of elements is based on functional-relationships and position-in-flows.

This Homework-for-Foresters is voiced as if it were for homesteaders in forested drainage-basins. In times of *Transition*, existing homes and structures will need Retrofit planning to reduce energy needs, and conserve valuable assets. The plan that the whole community imagines includes the very-local homesteads and looks at the whole-drainage-basin with a willingness to change the-way-things-were-done-just-before-now. Let's bump out the perspective to the royal We, the whole-community-in-council. Perhaps the area under consideration has long missed a loving communal tending?

Where are the remaining commons?

After assembling all the documents that can be found and after listening to the testimonials of observers and Rangers, a rough map of the ecosystem-types can be laid over the topography maps, so that slope-aspect, soils, and underlying geology is assembled to feed our three-dimensional imagination. The drainage-basin-council can begin to consider where to start and in what directions to proceed.

The Ultimate Intention
needs a sequence-of-priorities.
Thus the mime:
Sequence Stage Sort Stash.

Once your drainage-basin has good maps, you can see through layers of information by printing them on plastic sheets (or drawing on thin paper), all maps at the same scale. This allows a semblance of three-dimensional seeing (*palimpsest*). Remember that you can also use a big, bright, glass window as a light-box and tape maps over each other to see through the layers.

A sand-box-model is way better; one can pretend to be Eagle and hover above the model, seeing the big-picture. But learning to use maps and drawings will help keep us oriented; maps are like lists. They hold information on a two-dimensional sheet, which we can learn to project 3D by imagination, and enlarge our concepts of Place. We still need to get out there and walk about, perhaps taking notes on maps while we explore?

The map
is not the territory.

When as much information as immediately possible is assembled, it can be brought to council. The People of Place will hold a holistic discussion of what the priorities, sequence, and Ultimate Goals might be. This is an ongoing process—as we learn about Place and Culture and Work, many of our assumptions will need revisiting. And perhaps outside opinions can be solicited? If the drainage-basin-community is ready to listen to visiting wisdom, it is best that such "experts" have time to be there for a while and get oriented themselves, or that they visit several times, perhaps in a range of seasons.

Whenever a Ranger or Forester from some other landscape comes to visit, they will pay attention, and probably be unable to stop thinking about the questions being asked. Give them time to digest and consider, perhaps send them away, or give them some leisurely hospitality and local culture time. Let them dream, just like the whole-community and landscape needs dream-time to see *Deep*.

The landscape itself, and All Sentient Being there upon, certainly have much to teach any new settlers and will continue to offer input to any culture of re-localization. Ceremonies of gratitude and propitiation help us listen carefully to what-the-land-has-to-say. Then, ritually-spoken intentions-of-reciprocation can be a way to open up to wider communication and co-evolution. We dance the circles, travel the *hoops* of our seasonal work, and spend time being present with the subtle ways that Nature talks to us.

There are many methodologies of planning that help organize the layers and layers of information that inform our community and surround our deliberations. In the *deep-past*, Cultures of Place made many mistakes, and if they persisted, they accumulated very important protocols and stories to keep them oriented to the delicate-dance of resilience and enhancement. When faced with fast climate-change and emergent-cultural-start-ups, we will need to use our best tools of planning, just in case we make it through the *transition-times*.

Cooperative agreements and covenants are the Human treaty-way to keep track of promises and ways-to-proceed (the protocols, such as etiquettes and taboos). Stewardship contracts can be a transition document that arranges whole-drainage-basin oversight while regional powers sort out their long-range priorities. All empires now need to devolve power to drainage-basin-councils, just to ensure that ecosystem functions are restored and maintained. Local decisions that direct local repair are key to effective and regenerative strategies.

Eventually, after drainage-basin cultures-of-regeneration are persistent and progressive, treaties of mutual aid between whole-river-system basin-councils will replace any overweening continental-hierarchies. A good part of planning has to do with *invisible-structures* (our social arrangements, the organizational templates). Mapping cultural-sectors and flows can be aided by systems-posters, such as bubble-diagrams, quadrants, and categories of relationship (network maps) with spiral overlays showing evolution through sectors of experience, personal and collective (Wilbur 1995). See "Rural Report" in Chapter 3.

18 Inventory

27 Water Care

17 Now Next Forever

Sectors, as used in Permaculture Design (Mollison 1988), refer to flows through a landscape, in any dimension. As we mentioned above, a fire sector is the area that is most likely to burn; planning will expose these critical sectors. After we have realistically understood the work-zones of our intended tending, the sectors will cross—or intersect—our work-zones and show us where to plan our sequence-of-priorities. Sectors might be water on the surface or the slope where predictable winds move or the corridors-along-ridges and across basins where predators and prey funnel, which may coincide with Human travel routes (daily and seasonal). There are many other flows we can characterize and perhaps put in stories, such as the *Song Lines* of Indigenous Australians (Chatwin 2016). We can learn to sing our way to work, across a familiar (as we are all in relationship), multi-dimensional-complexity.

The most intensive zones of our intentions and inhabitations map our refuge and nurturance so that we can persist through the work on the land. We may have a semi-nomadic series of camps and still use a winter-village-site. Our work may center on small hamlets, with craft sheds and material-caches. Near our camps, farms, and villages, the work is focused on protecting assets, preventing unseemly impacts on the surrounding landscape, and building intensified Horticulture with storage and processing facilities.

*Keep the common chore-routes
close to the kitchen door.*

These areas, close to our hearths, are where we focus the planning for most-used-work-zones. Just as the whole-drainage-basin has a mosaic of ecosystems connected by flows, the Human enhancement close to home (right around dwellings) and just outside the hedge (supporting those dwellings) is mapped for an intensive-use landscape (with more details), where there is plenty of complexity to warrant careful planning. We place our intensifying-activities (*biological-magnification*) at nodes of interaction, where our efforts are to facilitate a yield while advancing fecundity and resilience.

A general pattern appropriate to the PNW is finding a site for a seasonally occupied winter-cabin, above the valley fogs and on a southeast slope, where it gets early morning light from a low-angle winter-sun-path. The cabin is also tucked into a small grove of fruit and nut trees that shade the cabin in the hot summer. Sheds, cool-pits, firewood piles, outdoor-cooking, and gathering-spaces are all set in useful nooks where, for example, an evening breeze sweeps out the mosquitos in early summer, or a winter sun rising over an east ridge lights a bench, where the day is greeted with a *hot-cup-of-tea*. Patterns of natural but intense use try to limit compaction, control runoff and nutrient flows (pee and poop), allow wildlife flows, create some special habitats (bat-boxes), and allow very-low-energy-use footprints with comfy shelter when needed.

THE WORK OUT THERE, AWAY FROM HOME

Similarly, as we plan out into the wood-lots—where perhaps we have already reduced fuel-loads, established trail-systems, returned periodic-burning, and managed seed/herb/root crops, tree crops, seasonal-pasture-meadows, water-capture earthworks, and a few sleeping shelters and lookouts—their patterns, revealed by overlaid-mapping, will show us the potentials, and we can map those *conceptual-plans*. The process of implementing plans—actually doing the work on a specific site—requires standing on the spot where the map suggested thinning and seeing the complexity in real Place and Time. *The map is not the territory.* By the time we are out in the headwaters at the forestry and hoop-camps, when we pass through wild-lands (seldom visited, still occasionally burned), our collective-attention is to Wildness and the eco-complexities where our interventions are broadscale, diffuse, and rare.

The work sequence (from the plan) is done in waves, by a series of craft-guild crews moving through with seasonal timing and careful observation. As crews visit different glens and slopes in a big drainage, they take notes and bring their reports and testimonies back to drainage-basin-council, informing the next steps. A lot of very specific decisions (skim one tenth of this seed crop?) are made on-site and in-flow.

Observe and interact.
Start small and learn from mistakes.

The closer we are to our intense habitation and work-camp nodes, in the most active work-zones, which we examined during planning, the more specific we can be with sequence, estimates, and investments. Near home, we will be focused close, and we will receive feedback, as we sit/work/sing there, and watch. We can adjust our tactics as we learn from our mistakes. Further out into the Wild-lands of the upper-drainage-basins (the headwaters), the most remote work-zones, we need wider attention over longer time by several craft guilds and Rangers, to even begin to plan for how we can do the *least work to the greatest benefit*, as we are so far from base-camps.

The community plans its restoration and regeneration strategies through the inventory and mapping process, which then informs the worker-guilds' deliberations and discussions. The tricked-out council-hall holds us in the wrap of maps, art, crafts, icons, and memories. The drainage-basin holds us in its arms, bosom, and lap. All dream together through the return of Salmon, Beaver, and Fire. All Sentient Beings chitter and squawk around our common Human impertinence, that we dare to live this way, but at least we Humans know we are silly and awkward; we apologize as we remember to practice reciprocation and gratitude.

INTRODUCTION TO FORESTRY WORK BY STORY TELLING

To get the feeling of what is going on in the forest, this next story uses metaphor, reaching between poetry and explanation to characterize subjects as if on a stage. These are the tales of people in action, the peek at what is going on in the forest. The broad brush picture sets the scene in general terms. This is the consumer guide to forest-ecology and forestry-work.

THE STORY BEHIND THE PRESCRIPTION POSTER

After giving a talk near Central Point, OR, to the Small Woodlot Owners Association, the audience questions made clear that we did not have the poster they wanted. The six posters clothespinned up across the stage (on the rope I brought) all had to do with the nature of how-things-work. The audience was not interested in thinking about design; they wanted to know what they ought to do. Once again, as we often find in teaching, some folks do not want to learn how to think, they just want to have the recipes.

The next day was supposed to be a day off, during a very busy fall, but instead the previous evening's notes and the demanded-task got organized: outline lists and ideograms of the proper forestry tactics for the complicated Siskiyou Mountains. These elements ended up clumped by subtopics: landscapes, principles, tasks, products, and features.

The new poster has a mountain scene layout, with detail lists hung on the icons of the eco-type clumps, so we call this a systems-poster of "clumped bubbles" (clumps of clumps). The ecological concepts are boxed and colored red; the vocabulary of the work is lettered in green; the products are listed in purple; and the water-terms and topic-clouds are in blue. This is almost a board game, where we can visit different topics and concepts as a story telling sequence. After the poster emerged and got colored in, we used it for the annual December story to stage a visionary tour of "Winter Forestry in the Oak Pine Savannah." Several other posters were hung and fresh boughs, winter-berries, and wreaths were brought into the story lodge.

When we taught the Social Forestry course a month later, the linear story of recipes and prescriptions needed some elaboration. This poster is vulnerable to exploitation. A greedy or desperate gatherer might use the lists to search for high value, rare, or in-demand commodities, and ignore the larger-picture and all the covenants and principles, taboos, and etiquettes in order to "cop-an-angle" and "seize-standing-assets-for-liquidation."

This is colonial-thinking and we have had too much of it already, but in order to talk about how to do "Forestry Work" we need to look at many aspects. The stewardship tactics and the allowable-by-principle skimming of "ecologically-harvestable-multiple-and-alternative-forest-products" must come from drainage-basin-council deliberations and be implemented by cultures of forest tenders, not distracted, independent actors.

So sure, this prescription is just what land owners (who assume powers-of-dominion instead of cooperation-and-reciprocation) would love to have seen (and photographed) that evening. How can we talk about all this without looking at all the "way-things-work" posters and without knowing a lot about the specific slopes and groves that are in our home-Place? So please be considerate, if these opportunities and necessities do not find their proper perspective, we will make the same mistakes all over again, and the whole drainage basin and local Human culture of Place will suffer further.

*Do not exploit
short-term advantage
for private gain.*

Commodity extraction always looks to bundle single values and reach for "economies-of-scale," which works out to be greedy extraction that does not support the workers, the local culture, or the forest ecosystems. Managing multiple products requires caching, and sorting, and complex crafts, that add value, on-site, to complex-materials-flows. Sorting yards for single-species commodity bundling do not work economically, as the broker always takes too big a cut for the logistical challenge. The overvaluing of information-processing by merchant-elites has always stripped the land.

Cultures of Place that use forestry products to support simple house-holding economies are subsistence-cultures. Their cooperative work forwards ecological restoration, when it is within All-local-interests to persist-in-common. The gathered-materials and value-added craft-flows for local culture do not produce a surplus that might allow for taxation or skimming by imperial-demands. The margins are too close to break-even. Natural growth and repair rates are way lower than the cost of financial services on debts and investment-driven limitless-growth.

*Get a grip;
dress up
and stay home.*

This poster shows how we might support a culture of Forest Work, that is, doing regenerative Horticulture and restoring ecosystem functions. The forest will present us with gifts, and we should understand that the beaded-belt-treaty offer is to come into relationship with Place, and to be humble and ethically selective in our tactics and strategies. We cannot remove more than can be replaced through natural processes, and we cannot remove *keystone-species* or reduce *biodiversity* and *implication*.

8 Prescription

Here is a discussion of the poster,
pasted with phrases,
words, icons, and colors:
rules of thumb
for an imaginary composite forest-scape
on mostly north aspect basins.

MID-ELEVATION LEGACY TREES

On any southeast, south, and southwest-aspect slopes, where there is history of logging and road-building, and where regular under-burning ended long ago, we often find remnant hardwoods, being overtopped with Pines and Douglas Firs. These can sometimes be big (2-3 feet diameter-breast-high) trunks with big branches. This spread-out tree-shape shows that they were once growing in open savannas that have filled in with conifers, after fire was suppressed a hundred or more years ago. They have often dropped big limbs that died from lack of sunlight and have a remnant leaf-crown that is competing with the straight-up evergreens.

As Madrones, Black and White Oaks, and Maples all have extensive root systems with many fungal associates (*highly-implicate* trees), they hold a reservoir of mycorrhizal relationships, that will prove important to regeneration, following hot "stand replacement" wildfires. The Maples and Black Oaks will probably have to start over from seed blown into the ashes, if the fire was too hot, but the Madrone and White Oaks can re-sprout from surviving roots and stumps. It is impressive to see that the first green rising through the ashes is often Madrone sprouts.

This portends successful germination of the blown-in—or bird and animal dispersed—new seeds, as some mycorrhizal fungi have survived on the deep tree roots and can facilitate the regrowth. The legacy tree may be killed to the stump by a hot fire, but the roots that are co-evolved with fire will survive. Often we find big burls at the base of Madrones, formed by repeated cooler burns and repeated sprouting. Fire-coppice-burls. White Oaks can be seen growing in circular clumps of trunks, where the central stump long ago burned or rotted out, but a ring of new trees has risen.

Many of the mycorrhizal fungi produce especially delectable mushrooms.

The forestry work opportunity here is to preserve the scattered legacy trees, by opening up the south aspect of these overgrown and crowded out old-timers and by cutting the conifers for building poles and log mulch to allow in more sun. Any conifers growing up through the crown of the hardwood can be carefully taken (avoid falling against big hardwood limbs). Any young conifer poles that grow straight and tall, with few lower limbs, in the north shade of the hardwood, can be thinned.

This pole clump can then be managed for ongoing production in the shade sector, or north aspect, of these *legacy-trees*. Conifer poles can be "farmed" in the shade of these great remnants, but new seedlings should be controlled with cool-understory-fire or thinning on the south aspects so that the legacy-tree can grow a healthy crown of leaves and thus *persist*.

Once a mixed conifer and hardwood forest has been thinned and prepared for under-burning (see Under-burn Often after Fuel Reduction, below), cool understory burning on a three-to-five-year cycle will reduce new conifer reproduction and increase mycorrhizal-associated bunch-grasses, along with other savannah herbs, flowers, and bulbs. The new forest-structure will not be the previous (pre-colonization) Oak/Pine savannah, that was maintained by regular burning for hundreds, and thousands, of years. It will be a mosaic of conifer-clumps in the shade sector of legacy-hardwoods, with savannah in the new, sunny openings.

Careful observation will notice tangles, mistletoe clumps, and cavities that are important for wildlife. Thus, the legacy trees are also "wildlife-trees," and should be marked (with a big red W) and/or known by name, and celebrated. Some tangles are inhabited with nests, and if we remove the lower-fuel-ladders during our stand-conversion, we can perhaps still burn underneath soon? Or, to move fire in, we make temporary fire lines to exclude ground-fire from some wildlife-clumps (small no-go-zones) of the mosaic.

Under-burn Often,
After Fuel Reduction

After we have taken some forest products during the clean-up phase, we can then use fire as a tool for ecological management. Cool-burning stays on the ground and converts fine-fuels (grasses and dry twigs or leaves) and some limbs (slash from high-pruning) to biochar. The downhill-burning style is still hot enough to kill small conifers (bark scald), but not too-hot so as to bake the soil and the savannah root systems. Many perennial bunch-grass seeds require light fire to break dormancy and germinate. Read Chapter 7, "Fire."

Neglected mid-elevation lands with excess roads, compacted soils, and tattered post-logging remnants of mixed hardwood, conifer, and brush, do not usually thrive enough to have a net-positive yearly-carbon-fixation. Restoring light-fire, after some thinning and fuel-reduction, puts some nutrients back into the *humisphere*. The standing *senescent* and standing-dead woody materials, clogging the understory, can be laid down (slash) to return some available soil-fertility nutrients, by being converted into charcoal and ash during moderate and cool-burning.

These first returns of fire might germinate some remnant bunchgrass seeds. The opening of the forest-crowns allows sunlight to reach these understory grasses and herbs. Deep rooted bunchgrasses respond to light-fire and mob-grazing with vigorous regrowth and net-carbon-capture. The annual-carbon-budget for well-tended mosaics can become net-positive.

> *Perennial grasses are stimulated by light-fire, quick mob-grazing, and high-mowing.*

Mob-grazing is a natural-pattern where sufficient predators, especially Wolves, push grazers and browsers (Elk, Deer, Antelope, Caribou, Buffalo) to clump up and move often for group cohesion and escape. To copy this ancient co-evolved pattern, we can use small versions of Cows, Sheep, and Goats, moved by Dogs, or confined briefly with mobile fencing, to stimulate photosynthesis.

Mowing with scythes and sickles can also be a bio-stimulation, with perfect timing and species-specific focus, removing less appropriate species by cutting them before viable seed set. Mow, leaving the mulch and favoring perennials; rake the cuttings and favor annuals by exposing mineral soil and scratching a seed bed. The *soil-seed-bank* can hold both rare endemics and invasive exotics: change the conditions and germinate exposed or scarified seeds. There is only so much to say in general, as so much depends on the local site conditions. All these rules-of-thumb are hints. We might be searching for meaning and thinking, "Could we be looking at this or that? Seems to remind us of . . . "

RIDGES, VALLEYS, AND NORTH ASPECT MOUNTAIN SLOPES

Riparian zones are critical to the return of Salmon and Beaver. Fish-carrying streams need shade and insects to nurture the juvenile *Salmonids*, which return as they mature, bringing such nutrient-gifts as nitrogen and phosphorus, back up-river from the Pacific Ocean. Salmon eggs can be transplanted to re-opened spawning-basins after the dams have been removed and the culverts pulled out so that there is clear-passage.

Sometimes this means first removing old roads and bringing back Beaver. Starting at the headwaters in any drainage basin and building water *detention-structures* will hold the winter-storm-moisture back and insure cool, late-summer flows, with shaded resting-pools for the smolt to shelter in on the way out to the ocean and the-return.

Decommissioned roads can act as infiltration earthworks.

Roads, trails, and landings for forestry and ecosystem restoration can support fog-broom effect and snowpack *retention* below ridgelines, at mid-elevations, where the ridges can be forested. Using Keyline-patterns (Yoemans 1971) and concepts, we can rescue old, but appropriately located, forest roads and put the great majority of old forestry roads to bed.

When we find the Keyline mid-elevation ridge pads (landings) and saddle-flats (between hills) we can connect roads and trails to our work. Survey out from the flats (which could be seasonal-water-infiltration sites) and to our restoration work-zones. Assessing the upper slopes of any drainage-basin for opportunities to improve forests looks at many factors, as discussed everywhere in this book. On the Pacific coast-ranges, no more clearcutting and ridgeline clearing should be allowed.

The conifer forests above 4,000 feet (1,400 meters) elevation on the interior ranges are critical to the whole region including the Great Basin. Conifer forests can be managed to rake clouds, hold water, maintain biodiversity, connect wildlife corridors, and *sequester carbon*, by thinning (fuel-hazard-reduction) to avoid catastrophic-fire and by periodic under-burning (*cool-burning*, see Chapter 7) to recycle nutrients and build up *bio-char* (see Chapter 8) soils. Logs and limbs can be left on the forest floor, if in ground contact, and laid just off contour (on Keyline patterns).

A few big trees can be harvested, with eco-sensitivity, and only for special uses.

Some single-tree removals, for high-quality-timber and wood-products, can be justified, as long as soil-compaction and leave-tree debarking (from sloppy falling and skidding) is avoided. This can be done with logging balloons (to lift the tree as it is cut off the stump). Or if the log is being skidded out on the ground, use break-away pulleys on crooked winch cable paths (to snake a log over to a road), rollers (to prevent skid ruts), and slash-cushioned-chutes and skid rows. By rigging the skidding cable on slings and pulleys that pull away from anchor trees as the log is pulled past, a log can be wiggled out of the woods without debarking remaining live trees.

Animal drayage (Mules and Draft Horses) on Keyline skid-trails can reduce compaction and disturbance and perhaps fit into a local work-way. We can forget heavy machinery. There is even the possibility of using portable mills and Human-powered pit-saws to render big logs into *cants* (squared off logs) that are more easily carried out of the woods, down to re-sawing at drainage basin sawmills.

11a How Trees Work

The reproduction of complex-forest-groves is species and site-specific. Firs (*Abies sps.*) and a few other conifer species are capable of persisting as small understory seedlings for long periods, while waiting for an opening in the canopy that will allow enough photosynthesis (solar access) to grow tall and join the *forest-canopy* top-crowd.

Closed-canopy forests hold soil moisture longer and allow periodic fires (with perfect timing) to stay on the ground without harming the overstory, as long as *fire-ladders* are removed. Re-introducing fire depends on whole-drainage-basin strategies that reduce overall excess-fuel-hazards, so that wildfires tend to stay on the ground. This means reducing fire-ladders (brush-ramps) at the edge of closed-canopy groves and coordinating regional controlled-burning to reduce the risk of catastrophic stand-replacement wildfires.

Working with an intact old-growth conifer forest grove is very different from re-building old growth characteristics, whether on logged-over mid-elevation slopes or clearcuts, or in over-stocked riparian-zones and woodlots. The vast majority of compacted, over-cut, over-roaded lands are found at mid-elevations, on lower mountain range slopes and in valleys and basins. These are usually appropriated lands, stolen from First Nations, stripped and hammered, over and over. They are easily accessible, below persistent snow-packs, and have been left to the vagaries of market forces and settler desperation.

These are the forests that need us the most.

Mixed conifer and hardwood (broad-leafed) woodlands have a complex story to unravel. The appropriate restoration tactics are dependent on a complete assessment and a wide-species-cooperation with Human Cultures-of-Place, who are making decisions through drainage-basin councils. As these interior-valley forest-lands are above the valley fog-banks and below the snow-packs, they can be accessed year round. The best season for forestry work is winter, with less wildfire danger, most birds and animals dormant or out-migrated and the cool season weather more convenient for strenuous work in protective clothes.

PERFECT TIMING

The secret to effective farming and forestry is *perfect timing*. Part-time remote-stewardship by absentee-landowners often misses important windows-of-opportunity; the workers are not there at the right moment and ongoing observation is weak. Willow and other hardwood coppicing is best done December through February. Stumps sprout in the spring, from nutrients stored over winter in the root mass. The right tools and the proper thinning choices will make for better and better wand quality and longer lasting *stools* (the stem-bristled, still-living stump).

Wildflower-emergence and song-bird-nesting happen in spring, when we should only minimally disturb the woods. No chain saw-work. Leave plenty of wildlife-trees and no-go-thickets in any woodlot.

Burning season timing is critical in the Shasta bioregion. Broadscale-burning is best done in mid-winter after sufficient rain and before early-flowering in February. The art of proper burn-timing is complex; it has always been the province of Indigenous women's societies.

The secret to effective house-holding, Horticulture, and Social Forestry is also *perfect timing*. Getting ready, ahead of the opportunities, to practice good sequence and timing is an art. This is when "having too many irons in the fire" pays off. We can sketch a plan and bundle or clump the materials and tools necessary, so that when the right guests arrive, with perfect timing, they can get rolling quick and have rewarding results. Keeping tools sharp, clean, and organized, means we are ready for *quick-fixes* with *exquisite-timing* and *elegant-results*.

Lists and plans are good practice, but mindfulness and attention to principles and patterns facilitate flexible-visions.

As we are overwhelmed by modern-juggling and too-much-information, managing to notice or remember the contingencies (as challenges present themselves) means we need to take breaks, open our senses, and enter *observation* (take a walk!) on-site several times per day. What with climate weirding, our ability to respond and pivot with unexpected events depends on our sensitive sequence and timing-adjustments, a sort of meditative but steady-dance.

A good example of the benefit of perfect timing is the challenge of the narrow "entry window" for working most clay-soils. If the clay is too wet one can trigger clumping, or collapse, of crumble-structure by mixing. If the clay is too dry, one cannot "enter-it": the implement bounces. Crumble-texture is the clay-loam goal, and if you are deep-forking, or double-digging (French Intensive Method, Alan Chadwick, John Jeavons, 1979), you may have only one or two mornings a year to aerate or work compost into such clay-loams. Clays can be the foundation of very fertile soils.

YEARLY WORK CYCLE

Many traditional peasant (*paysan*: people of the land) cultures based the sequence of seasonal-work-and-celebrations on the pattern-dance of the sun, moon, and stars across the *ecliptic-sweep* of the firmament. The rate-of-change of the rate-of-change, of day-lengths and sun-height along the northern-temperate-latitude-band of the Earth is palpable to the people in touch with Nature.

The European Gaels used solstices, equinoxes, and cross-quarter days to schedule cultural events and work seasons. The "quickening" of *Imbolc* (February 2nd) moves us out of long nights with little change (deep-winter) to catch the early lambs. The rush of the spring-equinox re-awakens us to the promise of fresh greens. The full-buzz-slipping into the long days of summer at Mayday (*Beltane*, May 1st), settles us as we prepare to plant the field crops.

The languor of heat and short nights at summer solstice keeps us busy (with naps). The intimation of fall at August 2nd (*Lamas*) starts the harvest; the great cooling and shortening of days cascades past the fall-equinox and hurries in the winter stores. We are "whistling past the graveyard" at *Samhaim* (Hallowe'en, October 31st). As we enter the feasting-season (eat everything that won't last), we indulge the crafts, theater, singing, and decorations to cheer us up through the winter-solstice. This is the template wheel-of-the-year and every locale in the Global North elaborates on this, according to the highly variable specifics and specialties of *Place*.

*The seasons
go round and round.*

Community gatherings at the quarter and cross-quarter days provide distraction from the dizziness and help us focus on the season and its changes. These are the good times for planning, comparing notes and observations, and councils and clan visits before guild workers scatter. To better orient ourselves, careful observation and a sense of being embedded in Nature inspires ultimate design.

Keeping the homestead calendar and diary with weather, blooms, pollinators, and surprises leaves us records where we can find patterns. The practice of daily documentation keeps our observation skills attuned. Perfect timing is the goal. Although our modern lives may have us scheduled weeks ahead, the best morning to plant or harvest comes at a more immediate moment. This is where extended family or cooperative living arrangements can allow elegant, flexible, and effective timing. Elders, for their wisdom and experience; children, for their original minds; and folks with limited mobility, who watch, can all be honored as key councilors as they report and witness.

SISKIYOU MOUNTAIN STRATEGIES

Here is what we have learned in the Siskiyous: a set of examples of accumulated-principles for seasonal-sequences of work in Social Forestry. As we map our opportunities-by-season we find something to do in all times and conditions. Everyone can help.

Small Organic Farms most often have woodlots and forested lands nearby. Spring, summer, and fall on a farm are taken up with planting, tending, and harvest. Winter offers timely work in the woods and craft-sheds. We can cut and thin fuel loads and convert the wood to fuels: seasoned firewood, charcoal, and brown gas (wood gas). Ethically, we cannot replace petroleum with charcoal: Many epic eras of Human "civilization" (e.g. the Bronze Age) have failed for too much wood cutting and charcoal making. The atmosphere now has enough carbon already; we can grow some atmospheric carbon back into forests for long-term-sequestration.

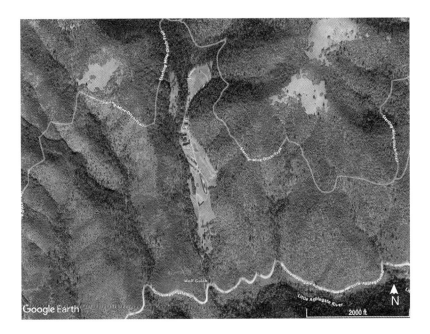

The most cherished harvests from local forests are of ceremonial foods, cultural materials, craft goods, and biochar (used to treat the soils in farm-fields). Ash and ramial tissues (small twigs and branches) should remain in the woods or be returned later as compost. Only the surplus carbon-fixation materials that do not hold complex-nutrients can be exported. This is usually wood without bark. Some flowers, saps, and essences can be exported but the mash from processing (the "mark") is best returned (perhaps via composting and then nursery trees) to the forest.

Willow, Hazel, Dogwood, and other hardwood-coppicing is best done December through February. The winter felling of the selected tree leaves a stump that sprouts in the spring from nutrients stored over-winter in the root-mass. Stumps that have been worked for some seasons become "stools" that are bristled with small cut stems. The right tools and the proper thinning choices will make for better and better wand, withie, and rod quality and longer lasting stools.

Wildflower-emergence and song-bird-nesting happens in springtime; be careful of noise and brush-cutting; do not disturb the nesting territories. No chain saw work in the springtime! In our winter-work, we leave plenty wildlife-trees and no-go thickets (clumping pattern!) in any woodlot. We learned to use forestry hand-tools, such as Japanese pull-saws for felling trees, to allow more social work-time, and less cumbersome safety-gear.

Burn season timing is critical in the Shasta Bioregion. Broad-scale burns are most effectively done in mid-winter, after sufficient rain, to preclude catastrophic runaways, and before early flowering in February. The goal is a biochar-lace-net, suspended on the stem-stubs of bunch-grass clumps, with still-green emergent seed-leaves (of wild-flowers) under the net—unscorched.

The art of perfect-burn-timing is complex and is almost always the province of Indigenous women's societies. During mid-winter dry periods, the fine grasses snap when bent on propitious days. The very-best-time to burn is at night, if the fine fuels are dry enough; the cold air moving down-slope carries the fire gently.

Compaction of delicate forest-soils is also critical to avoid. The use of heavy machinery in the forest reduces reproduction success and recovery time. Pond and road building is best done in the fall, after fire hazard is lowered by early rains and before the ground becomes too soft, becoming vulnerable to compaction. Woodlots and whole-drainage-basins need the permanent road and skid-trail systems laid out on *Keyline-principles*, with log-landings doubling as detention ponds in late winter, to build *perched-water-tables*. Single-tree-selection logs can be snaked out of the woods, with long cables and break-away-pulley-blocks.

Bundled-materials can be yarded to roads, with temporary chutes or cables down-slope, without dragging soils and slash. Animal and Human-powered *drayage* is good use of biological-intelligence. Bicycles with carts, as well as small walking tractors (steam, brown-gas, or electric), are small-scale-technologies that reduce machine compaction on roadsides and along fire-break-trails. All construction and earthworks in the forest needs to be justified and *stacked-with-multiple-functions*.

The timing and sequence of Social Forestry processes follows the seasonal-sensitive-condition of the soils, water tables, and ecologies. The most compacting efforts (log-skidding, felling, and earthworks) are done first, after fire-danger passes and before soils soften. Our work on a snow-pack, logging on frozen-ground and packed-snow can dodge compaction. The roadside, and ridge-pad, materials-yards and decks, are filled with perfect timing, for processing or loading later in the winter.

As the early-winter snow-line in the Siskiyous drops onto the mid-elevation slopes, after the heavy logging is done, the slash (which the Bodgers and Charcoaliers have gleaned), is piled for biochar-burns in the spring. The charcoal-logs are clumped at kiln-sites. Working on a snow-pack is slippery and cold, but materials are easily slid to caches and stashed for later processing.

GUILDS, CREWS, AND SEQUENCE

Social Forestry craft guilds are worker-societies that specialize in a forestry work-skill bundle. Guilds work with other guilds through a spokescouncil process we call drainage-basin-councils. Crews form temporary work-arrangements to get a job done and combine workers from different guilds. Sequence is the order in which guilds approach a site-task. Chapters that follow tell stories about the work of cooperative societies organized on Clan, Guild, Village, Ceremonial Society, and Elders concepts. These formats appear to fit our work arrangements.

Craft-guilds sport costumes appropriate to the work that celebrate skill and natural-elegance. Hats or vests are common as markers. Sets of tools in baskets are a ready sign. Guilds hold traditions and initiation rituals, with stories-of-origins, or miracles (and disasters). Guilds cooperate with each other through the drainage-basin-councils and at markets, craft-sheds, materials-yards, and seasonal-festivals. Guilds come to be conservative in their resource-use and protective of their members and traditions. They tend to introduce concepts of *precaution*, insurance-warehousing, and planning-for-persistence.

The Guild titles we have found to be functional and popular are (not in any hierarchy):

Rangers—The long-legged adventurous sort—these are the scouts that run ahead of the stand and basin assessors and do preliminary observations. They keep remote camps functional; lay out trail and road systems; find eco-work-opportunities; communicate with Travelers, Traders, and Rangers from near-by drainages; and report to and connect Guilds as they work and move on.

Mothers—Also called "roots, seeds, flowers, barks, bulbs, saps, and mushrooms." These folks follow the suggestions of Rangers to visit a proposed work-site, ahead of other guilds. The ground-cover and overall productivity of the site is assessed and flagged. Rare species are clearly marked and seeds, delicate medicines, and lichens are skimmed, while future harvest priorities are flagged with the Mothers Guild tag. Mothers are in charge of all broad-scale-burning: They direct the other guilds on burn days and facilitate Fire. They also support the hearths with small-wood and brush-wood.

Elders—The wise facilitators call drainage-basin council to meet at all the propitious seasonal turns. The Rangers report the scene, the Mothers add details and concerns. Then all the guilds discuss a sequence of work and the likely materials flows. Plans are made for harvest, eco-restoration, caches, materials yards, craft opportunities, and warehousing. The council lodge holds the wisdom of the ancestors and is held with love by the Elders. All guild-workers know other guild skills. Often, tasks such as burning will need all hands present.

Hunters—This is a very social guild. They work together without competition to have complex relationships with animal societies in the wild. This takes much skill of observation and cooperation. Usually, a crew has about eight men, one of whom is color blind and sees through color to pattern. Humans have evolved with special hunting abilities that can be socially integrated through council and interactions with the wild. Like Rangers, they can cover a lot of ground or sit very still for long times. They use cultural fire for hunting purposes after consultation with the Mothers.

Sawyers—This is the engineer guild. These folks know how to get things done and move heavy objects. They do more than just saw. Trail-building, bridges, lodges, cabins, barns, and ceremonial pole-structures are all built from logs and poles, harvested with care, moved magically long-distance, and raised with community effort. Earthworks, roads, ropes and pulleys, leverage, levitation, and dangerous-tree-removals are manifested. Tree falling, pole thinning, high pruning, and big single-tree-extraction are in their ballywick. Fire-lines, wildfire control, post-fire mop-up, and monitoring smoldering underground fires get their attention.

Bodgers—This is a crafts guild. They can make anything work. They know the materials and the tricks. Bodgers are always looking for the right piece of crafts material. The word refers to furniture made with small round wood. Bodger has come to mean "quick-fix" without elegance, but that is unfair, a slur on real goods made locally from appropriate materials. In order to sell industrial-consumer junk, first they demote art. Crafty fixes and objects are elegant in their own right.

Bodgers follow Sawyers to select out craft wood from the limbing and log off-cuts. Bodgers might get into a stand after the Rangers have marked the trees and before the Sawyers extract the logs and poles. The Bodger tag marks an art piece to be saved and put aside. Some Bodger cuts can be made before the Sawyers, to skim burls or stools. **Basketry** can be its own guild (and include **Fiber Crafts**) or cooperate with Bodgers so that the finest materials are procured. Bodgers work in craft-sheds seasonally where tools are bundled and materials are cached.

Charcoaliers—The makers of charcoal and artisans of carbon-sequestration. These folks like it hot. With their kilns, burn piles, quenching-drums, and mushroom-logs, they distribute and deploy carbon so that the site can thrive after forestry treatment. The materials left from falling, limbing, and brush-cutting can be treated four ways for carbon capture:

- lop and scatter (logs and limbs off-contour—ground contact)
- shredded and moldered with mushrooms,
- fed to browsers as cut-fodder
- converted to bio-char

Charcoaliers walk the site with the Ranger and the Sawyers to suggest ways to lay cut wood so that it can be dragged in chutes to burn piles or be left for log-mulch. They want to survey the condition of carbon-capture on-site. They are the systems thinkers, monitoring net-carbon and suggesting tweaks to the work of other guilds.

Some wood (especially Manzanita and Oak) can be pyrolyzed to make artisan-craft charcoal for smithing, cooking, tea ceremony, and art. Sometimes a **Shepherds' Guild** is appropriate on a landscape, or Shepherds can work with Charcoaliers, managing carbon.

Herders—Often nomadic, Herders and their Dogs move goats and other small browsers through the seasonal Hoop Camps, following restoration thinning with nutrient recycling and worker supported yields of fiber and food. This is a colorful guild with bells and music, song, and epic stories. We welcome them as they pass through. Other guilds provide corrals and camps to help this interspecies animal/human guild.

Farmers—Although perhaps a bit more sedentary, Farmers know how to do intensive management. They enjoy the poles, crafts, bio-char, fuel, and animal support from the forest. We need them at council.

Traders—Passing through with news, special crafts, invitations, and stories, these folks might be a social group or family. The local drainage-basin council maintains and supplies a campground at the down-river eco-village where up-river local folks come to market.

Gate Keepers and Guides—Eventually, a culture matures in Place to support pilgrimages, eco-tourism, visiting delegations, and inter-basin social events. The complexity of the Guild Roster depends on the carrying-capacity of the drainage-basins and the maturity of the local culture. It is also limited by population dynamics. A village of over 200 folks will probably need careful facilitation since some folks will not know others well. A council meeting of less than 11 is short on perspectives. The work of eco-restoration will dictate the guild array and the Humans will catch on with attention and time.

Small villages and hamlets (extended-family farms) probably get things done with crews, mixing up the on-site guild talents and working together to get some big chores rolling.

HOW TO FALL A TREE AND WHY

The theme of Pacific slope forests is overstocking. There is an increasing fire hazard and a lot of nutrients are not being cycled for landscape health. The way to the return of Fire and Salmon is through tending. The big work here is carbon management. Ecotopian Social Forestry has the materials to rebuild Cultures of Place, through the work of thinning and conversion.

Douglas Fir (genus *Psuedotsuga*, actually more like a Spruce), White Fir (and the genetic swarm of Silver, Noble, Shasta Red, and Subalpine Firs, genus *Abies*), and the brush species eco-type called Chaparral (after *Larrea*, genus of the *keystone-species*) are the gifts that our challenge promises. The removal or conversion of these select woody plants has the restoration potential of bringing back Fire and recycling potassium, carbon, phosphorus, and important micronutrients.

We will visit the brush-field conversion strategies elsewhere. Here we want to think about taking out small trees in overgrown pole-stands (grown up after hot wildfire or logging) and tall, thin trees overtopping legacy-hardwoods or filling in open (fire-formed) stands of old-growth overstory. Even the mighty Redwoods are overcrowded with Firs in their understory since the suppression of Fire.

Some of these Firs would be old-growth size on the drier interior slopes. Falling these is a logistical and skills challenge and needs to be done by experienced Sawyers. Big trees under huge trees should be laid down as log mulch in preparation for the return of Fire. We will talk here about much smaller conifer poles that we can take out with hand-tools and social cooperation.

In most of the Siskiyou Mountains, this means taking Doug Fir and White Fir

poles, fewer than six inches (15 cm) diameter-breast-high (dbh). They may be as tall as forty feet (12 m). These firs are dying back on many overstocked and over-roaded slopes. We watched the White Firs die out from 2,000 ft (600 m) up to 5,000 ft (1600 m). Suddenly, one year they turned red-brown and left a giant fire hazard. Climate-stress die-off. Doug Firs are now following suit. As Doug Firs move north and up-slope, they are failing to thrive where they once were invasive and competitive.

The eco-mandated vernacular building style is therefore pole-building. Doug-fir poles that have grown too thick or under a canopy tend to be tall and slim and dense; they grow slowly. Some two inch (5 cm), thirty foot (9 m) poles are over fifty years old. These are very strong poles for carrying loads and are very flexible (tensile strength). Great for pole-frame buildings. See Chapter 10, "Forest Shelters."

When we decide to take out smaller understory trees to move old-growth characteristics forwards, we are going to consider multiple angles and aspects. We need to stay safe and we need to preserve the forest structure while we pull out a tree. Brush-cutting is a different set of skills and uses many of the same tools (see Chapters 7 and 8).

Even a small diameter cut can go wrong and injure the Sawyer. Working in crews allows for spotters (who watch everyone for safety and coordination), Sawyer-limbers (who can work on a downed tree while the Sawyer-fallers move on), and Charcoalier-slashers (who drag and organize the cut-off side limbs and tops).

The Rangers have previously scoped the pole-stand and marked the cut with blue. The Rangers marked the leave-trees with orange; the wildlife trees with red; and the chutes, skid trails and fire-lines with green. The Mothers have cruised through and used orange and red to leave their mark. The drainage-basin council has discussed the priorities and organized the cooperation of the guilds.

The first work-entry into the pole-stand is by the Charcoaliers, who pull down the dead standing wood, both lower limbs and poles that have died but are hung up in other trees. They look for deadwood that is hanging dangerously and figure out how to get it down before the Sawyers try to fall live poles. This low-nutrient fuel can be lopped and scattered or bundled for firewood or charcoal.

The Sawyers arrive at the site and talk with the crew that has assembled. Safety, sequence, spacing, and yarding are all reviewed. Fallers need good boots, hard-hats (bicycle helmets might do), heavy cloth pants, gloves, goggles, axes, and sharp pull-saws. Binoculars might be handy; it may be hard to see the individual tops of crowded small poles or hidden nests.

The Sawyers have checked this side-box list (from memory?) and agreed with the marks of the Rangers, Bodgers, and Mothers. The fallers have priorities of safety, avoiding hung-up-trees (that then need a crew and tools to extract), laying trees down where they can be limbed and prepped for removal, leaving a low-cut stump, and placing log-mulch nearly-on-contour (Charcoaliers will move carbon around later). Some pole-saw work may be needed to do high-pruning before falling.

Often the tall poles are held up by the close neighbors and will need to be pulled out of the canopy. These stump cuts can trap a saw or jump off the stump sideways but still vertical and break a saw or threaten injury. Falling small-timber has its own challenges. If the fallers cut the poles in sequence from up-slope, laying down poles sideways, they may be able to leave a deck of poles on-the-ground and ready for the rest of the crew.

The following list of considerations is the checklist we provide in the Social Forestry Winter Camp Course. This is a crib sheet, carried in the field and used to take notes and ask questions. This list is specific to the desert alkaline eco-mosaic of the Little Applegate Canyon. Any other drainage would need its own crib sheet, and we all need an experienced Ranger along to explain and point out what looking at all these considerations really means on the ground.

The Rangers walk through forest stands that the drainage basin council has designated for thinning and tending, thinking in multiple dimensions while they scan the site. The Rangers then flag trees with different colors to show different treatments. An experienced Ranger seems to make quick decisions, but the process is actually complex. A Sawyer following the Ranger's flagging also may seem to be moving along swiftly but is actually considering many aspects of the work and the forest.

60-ODD THINGS TO CONSIDER BEFORE WE FELL A TREE

*Checklist for Stand Exams
at Wolf Gulch Ranch*

This annotated checklist is shorthand for complex decision making in field, falling saw in glove and hard hat on. The actual decision is made faster than going down the list. Experienced Fallers are looking at many clues and focusing on priorities. Previous craft guild workers may have already done some gleaning ahead of the Fallers. The drainage basin council may have already sent the Rangers through to flag and mark the suggested cut.

Red flags for keep. Blue for cut. Green for trail. Local customs and dyes will evolve. As the Fallers may be preceded by Bodgers, Herbalists, Charcoaliers, and Rangers, the Social Forestry community will have already had lots of observation ahead of the Fallers. The guilds' and settlers' drainage basin spokescouncil offers consideration before the dramatic opening of tree falling.

During the work of transition to fire friendly forests, we have a period of opportunity. The excess—perhaps senescent or dead, standing woody stems—needs to be processed wisely while we move back to regular underburning. In the PNW, our most common species for thinning and falling are Douglas Firs, True Firs (White, Noble, etc.), Ponderosa (and other) Pines, and some Cedars. As the list suggests, we are looking for overstocked, suppressed (slow growing), and dead standing candidates. We want to leave any viable old growth conifers and we want to preserve any remnant legacy hardwoods. Lots of things to consider!

The ecological imperatives yield a temporary surplus of tall skinny poles that drive our vernacular architecture: We must learn to build pole structures that last. We will be producing more dead wood than needed for log mulch, so we must learn to make charcoal and cultural tools and devices for convivial persistence. We will learn new/old forestry in the next phase after we have done the basic restoration work.

*We are mostly talking about
conifer trees six inches and less
of diameter-breast-high (dbh).*

(continued)

Analyze Vigor. We thin the young to leave better stand resilience and harvest the slow growing (suppressed) to recycle the nutrients.

1. Tree Age. Old, big, fire-resistant trees are protected. Some small diameter trees are actually 50 years old! What type of material will this be if cut and where do we yard?
2. Percentage of crown on stem. 30% or less may not be enough photosynthesis for release.
3. Signs of new interior bark. On Pines and Firs, red bark in deep fissures shows vigor.
4. Signs of disease/infestation/stress. Lots to learn here, but again, is the tree vigorous?
5. Proximity to other trees/canopy closure/canopy competition (crowding, sun sector, etc.).
6. Coppice regrowth? Will this stump re-sprout (hardwoods)? Are we thinning a clump?
7. Adaptability to new climactic regime? Firs may be targets where Pines are held.
8. Tip dieback, multiple leaders. Seeing the tips takes perspective but is telling.
9. Leader vigor. Compare with other tips? Fewer than 6 inches is slow and fewer than 3 inches is poor.

Analyze Wildlife Value

10. Signs of nesting. Whip out the binoculars? Look especially where trees intertwine.
11. Signs of feeding/browsing. Some trees and shrubs are especially tasty and should be kept.
12. Seed/nut/fruit/flower/browse production. Some conifer overstory may support other plants.
13. Seed cache. Hollow limbs, old nests, mistletoe clumps?
14. Squirrel runways. May have to be local to see these but are there limited skyways?
15. Crown Shape/Decadence (lots of dead branches and cavities?). Wildlife trees!
16. Wolf Trees/bedding areas. Remnant legacies? Protected slopes with obvious beds?
17. Dog Hair/Wildlife screens. Clumping pattern can support cover and safety escapes.

18. Proximity to wildlife trails/corridors. Do not block passage areas, even with log mulch.
19. Clumpy mosaic. Sometimes Humans think parks where they need some messes.

Analyze Fire Resistance

20. Historical fire regime. Burn scars—perhaps not on bark but can they be seen in tree rings after falling?
21. Logging history. Bark skinned, slash—recovery may require pile burning.
22. Proximity to structures. Shade yes, fuel ladders to roofs edges no. SE and NW sectors.
23. Tree Species. Some resist fire, some will be fuel for underburns when young.
24. Proximity to other fuels/fire ladder. Stand structure shows how fire can spread, break ladders and ramps.
25. Fire Zone Location. Access trails as firebreaks and area definitions.
26. Removal Impacts on wind speed. Take out too many small windbreak trees and lose big ones to winter gusts.
27. Check for flagging, be careful at edges upwind. Check tips and tops for evidence.

Analyze Hydrological Regime for Erosion and Compaction

28. Slope/Aspect. Balance fire prep with steep slope holding, especially on North aspects.
29. Soil Type/Root Shapes. Try to see below the surface and avoid compaction and erosion. Root zones can be stacked and the right mix will help water retention, fecundity.
30. Depth of debris coverage/carbon soils/old log orientation. Some slopes may need isolation from burning while log and litter mulch molders down.
31. Water capacity. Die back of upslope conifers indicates drought. Thin, mulch and soak in.
32. Yarding feasibility. Keyline skid trail. Fire breaks and access also move materials.
33. Ground vegetation impacts. Prepare soft cradles to fall into; on contour bio-swale?
34. Proximity to riparian zones. Keep seed trees, avoid impact, keep shade, large woody debris in flows? Leave buffer zone but perhaps burn? Canyons can be chimneys!

(continued)

Analyze Botanical Value

35. Prevalence of species in local area. Know the maps and look for the Herbalists' flags.
36. Prevalence of species in bioregion. Climate change means some species are coming in and others moving out.
37. Prevalence of forest/ecosystem type locally, regionally? Clumping mosaics? We can preserve some types on appropriate aspects and soils in Refugia.
38. Novel Ecosystem function (nurse plant, mycorrhizal associations etc.). Site specific!
39. Seed trees. Can be flagged and fitted with Squirrel guards. Usually isolated from limb skyways (see above).
40. Trees impact on understory/overstory. Positive? Negative? Shade tolerance? Some trees are shade tolerant and will persist waiting for an opening in the overstory, while others (such as Douglas Fir) do not persist and can be harvested where overtopped.

Analyze Economic Value

41. Felling feasibility. Good drop zone and skid potential? Will it have to be left for log mulch?
42. Basketry, poles, saw logs, firewood, incense/smudge etc. How does this tree fit?
43. Edibility. May want to preserve famine foods and high-density nutrition, such as nuts.
44. Use for check dams, swale wood, etc. Close uses may subvert hauling out for building.
45. Labor Access. Is the site going to have lots of entries in the future?
46. Tool Access .What saws and limbing tools will fit on the slope and in the stand?
47. Yarding to decks and mills, *mise en place*, process. Quality materials need the right approach.

Energetic and Aesthetic Analysis

48. Clumping/Single Stem. Clusters of poles or clumps of species may harbor nests and skyways. Does this inspire? Is this where we would love a future big, old growth tree? Can this skinny pole mature?

49. How will the area look and feel? A fire-ready landscape is different; can we look forward?
50. Proximity to energetic gates. Special portals will be found by the guilds and stories told! Do not block or fall trees into these veils.
51. Unusual or unique features/growths/patterns. They will speak to us and let us know!
52. Ask permission. Explain what you think you are doing. Be polite and humble.
53. Receive permission. Take time to listen and feel. Do not be surprised by responses.
54. Oak. Groups of three, leave be. May be fire coppice ring, or special clump. Give thanks and respect to the Triple Goddess and Acorn Woman.

Special Considerations

55. Portage markers. Special tall trees visible from a distance are very useful. Plant new ones?
56. Princess and Camp Trees. These are story trees with lineage and Place marking value.
57. Leave gifts. Our willingness to gift and sacrifice keeps us remembering our blessings.
58. Come Back and Check In. Bring the children and tell the story of the grove.
59. Sing and Dance. The woods love it. We get to be Human.
60. Reciprocity. Closing the loop is true of nutrients, blessings, and ritual. We act in Place.
61. **Direct Communication. Talk back. Listen carefully. Use our special senses.**

WE FALL THE TREE

And then, after all the preparation, the Sawyer-faller steps up to the tree. Says hello, touches for a bit, explains the overall goals. Asks for forbearance and looks up. Gets focused and oriented. With the ax, the faller reaches around in a circle to be sure of an open work zone. Using the ax again, as a plumb-bob, the tree is assessed for lean. Which way does it want to fall? Can we get it to fall any other way? How will it get to the ground? Will we have to push or pull it out? We want the lookout person to stand where they can see the tip move when the cut is almost through the stump.

Back at the tree trunk, the faller makes a face cut. A horizontal cut, one quarter through the diameter, is made near the ground. The second cut is a diagonal slice from above that intersects the first flat-cut. The wedge of wood is then knocked out with the ax, and this "face-cut" is smiling in the direction of a desired fall. The ax-head can be fit in the face-cut so that the handle points in the direction of the fall. Look up again, check in the the look-out, glance at the escape route (away from the fall path and behind a protective nearby tree—in two steps!)

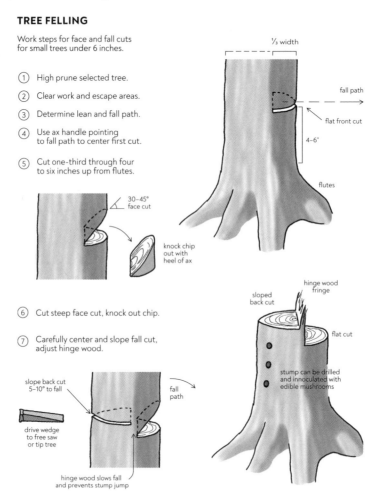

Now the falling cut is made from behind the face-cut, slanting slightly down into the middle of the wedge-cut. If we want the tree to fall directly, we try to have the falling cut meet the wedge-cut all-at-once, with no hanging wood. A thin "hinge" can be left to keep the trunk on the stump and let the tree down smoothly. A corner-hinge can be left to pull the tree out of the face-cut direction to one side or the other. This is how the faller controls the fall: good clean cuts, careful assessment, checking progress during the falling cut (by assessing the remaining hinge wood, and (with small poles) giving a bit of a push.

Beware recipes!
Think with the crew!
Cover the bases!
Pay attention!

The dangers are many, some are mentioned above. A Sawyer-faller needs guidance on-site and an apprenticeship with the Sawyers Guild to learn the craft and crew cooperation.

After the falling is done, the crew has many opportunities to deploy and move materials. To season poles and small logs, the bark should be removed with draw-knives (to slow down the beetles and hurry the drying time), the side limb stubs cut flush, and the poles set up into big trees to dry in the shade. They will need to be turned once in a while to keep them curing straight. After the poles dry and season, they are much lighter to carry or transport. The bark can be mushroom-composted or used in many other ways (medicine, incense, graywater matrix, resin extraction, insulation).

SEQUENCE REVIEW

SEQUENCE STAGE SORT STASH

The Four S's

All guilds are gifted with observations and experience. The community grows with Place.

They gather together in council and map out sequences of cooperation to align with perfect timing.

All guilds stage their sequence, setting up bundles and tool kits, camp gear, and provisions.

When the sequence is triggered by natural signs, the guilds follow each other, sorting the surplus.

The sorted materials are bundled and stashed so that all guilds recognize other guild-bundling and stashing.

The material culture of the people and the return of Salmon, Beaver, and Fire is elaborated.

Before logging or brush-clearing, the Mothers, seed-collectors, mushroom and lichen-gatherers, and craft-wood-gleaners, "high grade" the treatment area, selecting special out-puts and perhaps doing some seeding and mulching ahead of the heavier disturbances. The Bodgers move through tagging special tree parts. After the logs have been drawn out and decked, the Bodgers sort and stash the chosen craft pieces, and Charcoaliers cut up the slash into graded limb sizes to bundle, carry, and yard them at the kilns.

The remaining *ramial-tissues* (twigs, leaves, bark, and small-branches) are either fed to goats; shredded for mushroom-moldering-piles; cut-up further for "lop and scatter" to provide forest-floor-mulch or continuous under-burn fuel; or they are piled for the biochar burn-and-quench crews.

After all this harvesting, sorting, and stashing, the broad-scale fires, supervised by the experienced Fire-guild (the Mothers) members and tended by the whole-community (all-guilds-present), convert the fine fuels to mineral fertilizer (ash) and biochar. These are cool-burns and do not damage the soils or excessively pollute the air-shed. Seeds are spread (Mothers Guild) in the hot-ashes. Perhaps the next time the treatment area is worked-with, it will only need to be cool-burned again.

Regular underburns
are minimum-maintenance,
following appropriate stand preparation.

The charcoal kilns at the ridge-pads and sorting-yards can be burned, quenched, unloaded, and reloaded, with the sorted-by-size wood, over and over, long into the winter. The brown-gas from the kilns can be filtered and used to run small portable sawmills and drum-shredders (Jean Pain).

The lean-to shelters near the kilns host Basket-makers and Bodgers with their spring-pole-lathes. The mushroom growers (Charcoaliers?) tend their big piles of spaw, and distribute the compost to plant nurseries and restoration efforts, while also producing hot water for worker comforts. Keep water and nutrients as high on the landscape as possible! Return fertility to the forest!

One Social Forestry alternative-scenario looks like this. First, the artisans go through the forest, and they look for the best pieces of small wood that they can use for furniture, in molding, in flooring, in the turning of bowls. That's called high-grading (different from the industrial timber old-growth high-grading). Let in the craftspeople first and they tag and harvest the artisanal materials, leaving behind some slash.

Then Charcoaliers come in and look for opportunities for mushroom growing—take the materials that are left over, or still-standing hardwoods, that need to be coppiced; move them into the shadiest, coolest places in the bottom of gulches; and inoculate them for mushroom-culture. There is still a lot of material such as poles and small-timber for other uses, such as building housing-infrastructure for the workers—and that's where natural-building fits in with Social Forestry: temporary-shelters for itinerant or seasonal-workers.

The sequence of harvest attention might begin with the mushroom and lichen gleaners from the Herb, Root, and Bark Guild (the Mothers). The most delicate of bounty gets skimmed first. The sequence is always site and season appropriate. Perhaps the Mothers Guild has flagged some rare plants, or special discoveries, that are not ready for gathering. The Bodgers then cruise the area and mark the special pieces of tree and branch or sprout that should be sorted-out and not burned. The Mothers Guild works ahead of the Sawyers to sweep up herbs, seeds, berries, barks, and roots that would be lost or crushed in the cutting process.

Have the Basketry Craftspeople checked in? Do they want a certain type of cut for the species that may make good *withies*? After the various guilds case the job, it is time for a huddle, a last-minute-council on-site to review and evaluate. What's the ultimate plan for this slope? Will we be re-introducing broadscale-burning soon? Are there Pines to plant here? What are the balances of overstory and understory? Do we want *crown-closure* and shaded-ground that we seldom burn? Are we opening up a savannah here? How can we choose to work, so that we are increasing stability, avoiding catastrophe, and building resilience?

The principles from working with Fire include

35.

Fire is spirit and light. Fire is magical. Fire is powerful. Our Place and Culture are built on Fire and cooperate with Fire. Plan for catastrophe.

36.

Ignition in dry-fine-fuels that climbs fuel-ladders to windy-dry forest-crowns can quickly explode. Usually, after ignition, the wildfire moves northwest on the ground. Blow-up is a massive burn where whole slopes or valleys are vaporized and the air burns. Escape is difficult.

37.

Cool Fire can be seen at the end of a wildfire, as the weather cools and the burn drops to the ground, moving downhill. We can mimic its benefits and learn to work with moderate broadscale under-burns and meadow-burns.

CHAPTER 7
Fire

As storytellers are wont,
first a story
to stage the set,
to be ready
for the details
of the practice.

SCOUTMASTER BAILEY'S WOODS

We were young Scouts, the Beaver Patrol, and this was our official campsite. We wanted to spruce things up. Didn't they give Watermelons at camp to the neatest Troop, the ones who painted the stones white, along the woodland trail? So we decided to burn. Perhaps we were too eager and green. The fire ate the grass and weeds well enough but then the wind picked up and a wildfire blossomed before our helpless little boy eyes. Whoops!

Away it went, first twenty feet to the northwest and then with the wind, south. The village volunteer fire department was on its way, sirens blaring. We watched it eat into the Sweet Fern, Bracken Fern, New Jersey Tea, Pitch Pines, Red Maples, and Aspen. Right up to the edge of the nearest industrial sand-pit, where it pooped out. No more fuel, wind at back. Saved! I do not remember what the punishment was, perhaps Scoutmaster Bailey understood the lesson learned. Be careful with Fire.

NATURAL AND INDIGENOUS BURNING CYCLES

To generalize from the literature, stories heard, and local clues, First Nations burning was yearly on prairies and grasslands, three to five years on eco-mosaics, five to ten years in various other eco-types, and on Horticultural time-tables for basketry, medicinal-herbs, and bulb-meadows. Natural wildfire cycles, without Indigenous broadscale-burning, seems to be about twenty years in the Siskiyous.

We have a story from up on Griffin Lane of an old Incense Cedar stump that got cross-slabbed during a construction project and showed three to five year burn cycles in our drainage-basins. The poles cut at Wolf Gulch Ranch show twenty year burn-scars in the tree-rings. The Kalmiopsis Wilderness has been on that wildfire schedule since the late last century. The literature seems to agree.

If a cycle gets skipped or there is a wet climate period, fuel can accumulate that feeds big fires when dry periods come. It only takes one hour of sunlight, at almost any time of year, to dry out the standing dead grasses on mountain meadows (*one-hour-fuels*). After the First Nations cultures were removed and destroyed, a different fire regime was initiated by colonists to clear forest land for pasture, to open up fields for plowing, and to clean up after logging. These ill-planned patterns have left the PNW with explosive-fuel-loads and degraded ecosystem functions. The First Nations were invested in a long-standing set of disturbance-regimes using broad-scale and site-specific Fire. They knew what the contract was. Place, Fire, and People were dancing a Sacred-Way.

Aboriginal burning in the Pacific Northwest has a long history (Boyd 1999). There are charcoal traces in soil layers, as well as the tree ring records. By staying on schedule with observed cycles of fuel build up, First Nations in the PNW were able to disaster-proof whole-drainage-basins and valleys. The regular burning ate up catastrophic-fuel-loads, fire-girdled young conifers (keeping meadows open and preventing thick pole-stands), maintained an open old-growth-tree dominated forest/grassland landscape, and stimulated cultural-use plants. Kept the Beaver and Salmon happy too.

The European colonizers were not impressed. They wanted to move Cattle and Horses long distances, and their wagons were pulled by hungry Oxen. When early trappers wanted to head south from Fort Vancouver, through the Willamette Valley, and the grass was all gone up in smoke, they complained in their journals.

As soon as farms and permanent settlements were established, they felt vulnerable to these vigorous range-fires. The traditional burning practices were outlawed. As late as the early 20th century, "free-burning" was still practiced and fought. The remaining Indigenous people learned to set delay fuses with old wooden boxes, candles, and kindling, so that they could be gone when the fire took off.

Much of the extensive burning of grasses was apparently "forced-burning." The ignition would be upwind and roar across grasslands as sheets of flame, only to slow down and become patchy when entering any forest lands. Round prairies and meadows were burned from the perimeter inwards, with the flames converging in the middle and dying out. The burn was done efficiently, just by ignition on the right edges at the right time. Least work for best effect!

They could do this because the greater landscape was burned often enough to prevent catastrophic-forest-replacement-fires. As this pattern of Humans and Fire is so old, everywhere on the planet, we know once a fire-regime is established it is risky to stop burning. The fertility and response is so good following moderate burns that accumulated-fuels would quickly become dangerous.

The village was safe on the river, until they stopped the burning.

Apparently, the uses of controlled Fire are legion. We have only clues from stories and field studies to intrigue us. Big trees were felled with fire, to burn into lengths and split with wedges into big, fat house-boards. Fire ate out logs to make boats. Fire treated crafts materials, such as Jasper (for chipped-edge-stone-tools), hardened digging-sticks, or straightened arrows.

Baskets were sealed with pitch and clay to make boiling-baskets where hot stones (ones that did not explode, Greenstone) were rinsed off and dropped in Acorn-mush. Wait for the right burbling sound, sing the *pluk'a-pluk song*, and the mush is ready to eat. Many food, medicinal, and crafts plants have co-evolved with aboriginal-burning-practices and are in decline without the reciprocation of tending.

Human use of Fire is especially associated with savannas everywhere. Where there are still mega-fauna, such as Elephants, the big browsers knock trees down for fodder and preserve grasslands, preventing forestation and hotter fires. In the northern hemisphere, the Pleistocene mega-fauna disappeared with the glacial sheets. This was a period of Human adjustment only ten to fourteen thousand years ago. Tree crops that needed big browsers to process their large seeds came into Human management and Horticulture (Barlow 2000), which developed new fruits and nuts. And it would seem the whole forest-ecosystem complexity was moving north as the ice-sheets retreated and moving around as the Great Basin dried out. Great lakes appeared for a few thousand years, then the Cascades rose up higher and the climate went through big long cycles and folks had to move around and learn new ways.

Now, we new settlers and remaining First Nations groups are learning what to do. Restoration to what? There is too much accumulated fuel-hazard to get back right away to landscape-scale burning. First we have to do small and strategic burns.

We have the opportunity to harvest or compost a massive over-supply of senescent woody fuels. Some pile burning is called for, and that needs to be done carefully at a distributed small-scale. As we manage to secure small parcels of woodland and meadow, with fire-trails and fuel reduction, we can *mob-burn*. Plenty social dancing with cool-burning.

> *Get all the Guilds together*
> *and we can mob*
> *a small burn safely,*
> *a social-fire-blitz.*

THE NATURE OF MODERN FIRE

The priority in Transition Times is survival. Our salvaged infrastructure is worthy of some effort toward protection. We won't get going on ecosystem functions support until we secure our tools and buildings. This requires intensive fuel management near infrastructure-clusters. This also means knowing how to suppress, block, or deflect wildfire. We also need to know how to dodge escaped wildfire at our practice burn-sites by keeping good containment and having the skills, tools, and persons in position to round up escapes and quench smoldering-pockets in mop-up after a broadscale-burn.

Fire fighting has been monetized just like everything else in late capitalism. Big wildfires that destroy taxable property are profit opportunities. The organization, communications, support lines, and heavy equipment are all war-like. Even losses are "creative destruction" and then the economy has "growth" cleaning up and re-building. This is becoming an economic-growth-sector as climate-weirding tweaks expectations. Lots of power and excitement, news and drama.

Backcountry, up-river, cultures are under stress during Transition times. In the Siskiyous, the equivalent of hurricanes, tornados, flooding, or other magnified weather is wildfire. Staying Home and trying to tend the headwaters is a risky endeavor. The promise is that eventually we can modify the risk to the benefit of the whole-drainage-basin. Meanwhile, we need to respect the threat.

WILDFIRE HAND TOOLS AND TACTICS

Institutional training in wildfire control started in federal agencies and has moved to universities, non-profits, and First Nations programs. A lot has been learned. The personal gear used by professional fire control includes fire-resistant clothing (industrial petrol-fabrics); hardhats and face masks; emergency pack with fire-shelter tent; heavy leather gloves and boots; and special padding for chain saw work. The hand tools are for wood-cutting, ground-scraping, digging out ground-fire, knocking down fuel, lighting back-fires, and local water spraying. Many of these tools are appropriate for cool-burning and mop-up.

Pulaski Ax. This invention of a combination ax-blade/adze-hoe on an az handle has proven versatile. One can chop through roots and limbs, scrape hot coals off of trunks and logs, fell small trees, and cut into soil to rough out a fire line (ahead of the McLeod). Mr. Pulaski famously saved a fire-crew by getting out into a grassy meadow in a blow-up and lighting the grass to burn in every direction away from the worker-huddle. It worked.

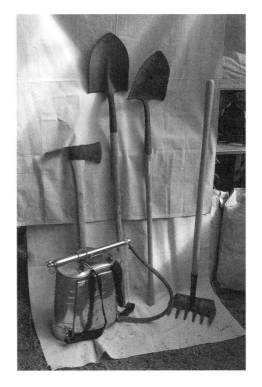

Shovel. The favorite fire-shovel has a triangular blade that is deep shaped and small. One can move a lot of soil fast with smaller moves than with big scoops. A shovel can dig out ground-fire, clear a fire-break down to mineral-soil (*sub-soil*), throw mineral-soil at the base of flames (as good as water), and move fuel around. The essential tool. There is a lot of skill in using a shovel safely and continuously. Learn from an experienced teacher.

McLeod. This is a combination of a wide-blade hoe and a thick-tined rake. Invented to work with fire-breaks. The long-handled-hoe can move lots of material with drawing strokes instead of lifting and throwing (as a shovel), and the rake-side can move a lot of fine-fuels off of mineral-soil without taking too much soil. A McLeod can clear a fire-break fast.

Backpack Water Pump. The pump is in the handle, which is pulled and pushed in and out to throw water in a focused squirt or a wider spray. The backpack tank is HEAVY, five gallons of water weighs forty pounds (< 20 kilos), and is made of metal to avoid melting or ignition. The straps used to be leather to resist flames (petrol-fabrics in Transition). Controlled burns on federal property, sovereign territories (reservations), and conservancies are most often backed up with water tankers and pre-laid flexible water-lines.

Drip Torch. The burn specialists brag that they use bio-fuel instead of petro-fuel. The metal fuel can has a hose that drips flame slowly. The worker can walk along, leaving a line of Fire to start a back-fire or to start a controlled-burn. We use bundles of grass, torn from outside the fire-breaks, folded in half and twisted. These *"Fire-dollies"* are used to start and spread the line of Fire in cool-burning.

Hardhat. The old tin hat is classic and does the trick. We want to protect our head and shoulders from falling debris and burning embers. A lot is going on during a wildfire control event. A version can be made from hardened-leather, wire-sewn bark-helmets, or thick-pounded wool-felt (kept wet). During Transition, old heavy felt hats are better than nothing, or some found salvage items (metal pots that leak anyways?).

Logging Saws. The classic two-person whip-saw (also called a "misery-whip") is actually a wonderful tool for cutting down burning snags (wear a hardhat!), bucking up downed logs, and cutting firewood. Takes some skill and cooperation; sing while we saw. The one-person 16–24-inch pull-saw for falling can get through a log fast, fall a tree, or reach up and high prune limbs in tight spaces. Once again, let us repeat: "use the right tool for the job and get some training."

The wildfire-control-tactics are stop, starve, deflect, suppress, and contain.

To **stop** a ground-running wildfire, either knock down the flames, break continuous fuels, or let it burn up to a pre-built fire-break. If you can keep up with the spread, it might be best to break fuels on the sides (two crews?), rather than try to cut off a flame-front with heroics. Lots of contingencies here; get an education and experience from a fire-school. We are reviewing these concepts so that we can help folks learn cool-burning.

Building a fuel-break—or fire control line—alongside a wildfire is a crew task. To use our worker-guild titles, the Ranger and Sawyers move up to the front of the line and clear the way by removing dead standing fuels and logs lying across the fire-break path, while arching limbs above the path with saws and axes. Then these two guilds work together to **throw fuel *into* the fire, *not away.*** Takes practice.

We want to remove fuel from the other side of the fire-break and we want to avoid throwing burning debris outside the fire-break by mistake. Throw the fuel onto the ground if possible and reach up to take down *fuel-ladders* and dry limbs on both sides, throwing the fuel towards the wildfire.

Wicks are dead logs, or bundles of dry limbs that convey wildfire sideways and uphill. If they can be broken by cutting through, the wildfire spreads more slowly and stays on the ground.

Knock down the flames by throwing mineral-soil at the base of the flames (where the fuel is) or spraying water at the base of the flames. This is usually mop-up. Build fire breaks first. Knock down flames only when an escape jumps the fire-break. The back of a shovel blade can be used to knock down fuel and flames to the ground, where they burn more slowly.

We are experimenting with *Fire-brooms*, made from green Hazel boughs and branches. The broom can be used to knock down flames in fine-dry-fuels or to fan flames in cool-burning for better ignition.

Contain a wildfire by surrounding it with fire-breaks and letting it burn out the fuel inside the circle-of-containment. On a big wildfire this might take months and heavy equipment. The conversation here is about containing a start-up fire near Home. The only wildfire local folks can suppress is a new ignition, caught right away. We still need all these control skills but hopefully the containment is fast and pre-existing fire-breaks hold.

A Fire-sector preparation strategy (see below) is pre-containment (with any luck). When we do controlled-burns and cool-burns, a great deal of preparation is done ahead of time including fuel treatments and a continuous line of fire-breaks around the burn-area. There also should be back-up fire-breaks and crews, just to be safe.

WILDFIRE BEHAVIOR

We need to learn a lot from wildfire-behavior before we try to tend broadscale-burns. Fire is a vigorous element in the universe. Fire plays with us.

Many ecosystems are co-evolved with wildfire and many plants have seeds that will not germinate without Fire-scarification. Part of the Gaea hypothesis is that life on our planet has modified oxygen levels so that wildfire is mollified. If the oxygen levels were much higher, everything would burn. If the oxygen levels were much lower, life processes such as photosynthesis and animal-respiration would be seriously constrained.

Wildfire is seasonal in the PNW. The hot, dry summer and fall, with thunderstorms and Lightning, set up **Fire-Season**. The late-summer-ignitions (stupid behavior, T-storms, and random events) quickly run into dry fuel. If the winter rains come late and the snow pack disappears early, even if there are late spring rains, the grasses are sparse.

If the winter is wet, the snow pack lasts, and the spring is wet, fine-fuels can grow tall and lush, and then dry out fast in July. August storms bring Fire, but little rain, and extensive wildfires can burn until late November (difficult summers for smoke-filled valleys). Thus the right dry-winter reduces wildfire danger and a wet-cold-winter, followed by a hot-dry-summer, encourages wildfires. Especially in the Siskiyou-south and Great Basin-east, edges of Ecotopia.

Long-term accumulations of dry-dead or dying vegetation and continuous-patterns of fuel-build-up and distribution; overdue periodicity (the time between) of wildfires; low moisture-content of fuels (time of day and season); dry-hot- weather; and ignition-sources (such as Lightning and Human behaviors) all set up the potential for wildfires to behave wildly.

Wildfires can be slow (downhill at night), patchy (jumping-around), and even go underground, **or** they can *blow-up* and march across forests and grasslands, very fast and furious. Some parts of our planetary ecosystems resist fire and seldom burn (too wet and too cold), while some ecosystems are dependent on periodic Fire.

Climate-weirding has brought wildfire to where it seldom previously visited (jungles and boreal-forests) and can encourage unexpected behaviors (*fire-tornados*). Humans have been living with wildfire and changing landscapes, with adaptive practices, as long as we have been co-evolving with our cohorts and landscapes.

There is a long-story of the study of *fire-behavior*. When Humans have tried to modify landscapes and wildfire-patterns for industrial purposes, big mistakes have been made. Here in the Pacific Northwest, fire suppression by government agencies has been a priority of colonization (see above). The imperial goals of resource-extraction and control-of-Nature, led to the interruption of traditional First Nation practices and the subsequent build-up of catastrophic fuel-loads. Inappropriate extraction-processes, such as clearcutting and conversion of complex-old-forests to even-aged-plantations, has encouraged blow ups and severe, hot, fast burns that outgas nutrients more than a patchy or under-canopy ground fire.

The land management priorities of commodity extraction have accumulated a body of fire science that calculates wildfire-potential and expected-behavior. As climate-weirding has thickened, wildfire is jumping the expectations: previously ambitious development of the landscape, conversions of wild-ecosystems to agriculture, and industrial-forestry are becoming more risky. The wisdom of previous wildfire-adapted Human cultures has been forgotten.

The *ultimate-goals* of coming back into balance with natural systems, along with the critical needs of restoration of nutrient-looping and biodiversity amelioration, can be effected by re-learning the arts of living-within-limits and coming-into-balance with natural-processes. The return of Salmon to the Pacific slopes of the Cascade Mountains and the Coast Range will depend on a parallel return of regular-burning to our complex-drainage-basin-topographies.

Modern civilization requires a miracle-of-cooperation to come back into balance. Many disasters of modern-settlement, and development, and industry, are being magnified by Natural-Chaos, with accelerating declines of safety-expectations and carrying-capacity (Catton 1981).

Meanwhile, while we watch the fallout of our mistakes collapse our convenient and inappropriate expectations, those folks who are re-investing in a positive future for the *greater average health of all beings* (Kauffman 1997) can start to experiment with changing our diets, living practices and lifestyles that move forward to a renewed fecundity and resilience of the natural processes that all species depend on.

This is a daunting vision but it is full of potential. Re-learning about broadscale-burning, a process-of-renewal, fascinates Humans. We are not actually so changed by our modern experience that we cannot reconnect with ongoing natural co-evolution with Fire and All-Sentient-Beings.

WILDFIRE BEHAVIOR WE CAN LEARN FROM

Archeological evidence, tree-ring records, and TEK shows that Humans have a deep learning with wildfire. All the so-called controlled broadscale-landscape-burning that we have deployed in the past was learned first hand from wildfire and our mistakes that ballooned into loss-of-control. Some loss-of-control seems to be inevitable considering the dynamic Nature of Fire.

The most ecologically-beneficial-wildfires in the Siskiyous are patchy and leave a mosaic of different burn-intensities. The highest biodiversity, and therefore resilience and fecundity, is found on landscapes with a complex of mixed-age-stands of trees and lots of edges between brush-fields, riparian-strips, grasslands, old-growth-forests, and patches of disturbance (*pioneer-ecosystems*). Many wildfires, especially in wilderness or roadless-areas, are ecologically beneficial.

The worst wildfires occur on industrial-forestry-lands with heavy logging slash-loads and even-aged plantations. These Human-modified landscapes, with lots of roads, encourage inappropriate behaviors that can blow up and burn disastrously, leaving behind erosion, continuous-soil-sterilization, *soil-seed-bank* destruction, and severely damaged settler infrastructure.

We can do better than we have done so far, to prepare for these inevitable consequences of landscape-scale industrial-leveling. Our buildings could be better located and made more fire-resistant. The forestry-practices could change to foster a more fire-resilient mosaic of treatments. The logging-practices could focus on leaving less-dangerous fuel-loads and on moving previously-logged-forests back towards the re-introduction of regular under-burning.

Our settlement-practices could discourage McMansions and isolated Manor-houses and scattered summer-cabins. Our eco-villages could be down river, surrounded by tended fire-breaks (pastures, fields, orchards, and work-yards). Our hamlets of sheds and cabins on farms could be fire-proofed with tile roofs and adobe walls. Our up-basin camps could be kept simple and safe with regular local burning. We could encourage a culture of land-tending that allows cottagers and hoop-workers to do the ongoing Social Forestry that this book is visioning.

THE FIRE SECTOR AT HOME

As we have discussed elsewhere, *Social Forestry* is best hosted on hammered, mid-elevation, private and public land grids. Used-up forests have been over logged and over grazed, as well as cut repeatedly for firewood. These lands are hammered by heavy equipment on too many roads and end up with compacted forest soils from all these abuses. Conifer forested ridge-lines and ranges are best managed for *fog brooms* and upland-water and snow-pack retention. This is where we assess the fuel and Fire flows that help us understand potential local fire-behavior. As the priority on private-lands is most often the protection of homes and buildings, those would be our core area, our inner work-zones, hear Home (sedentary-base hamlet?).

Radiating out from these assets to the southeast is the *fire-sector*. This is the area that will carry fire to the buildings, if a fire starts in that sector and if there is a fuel-load burden that could accelerate a fire towards Home. Looking at the topographic maps, we should see if there is an ignition-potential and if there is a continuous-fuel-load from that hazard to the buildings. If the slope in the Fire-sector is uphill to the NW, and/or there is a gully or valley that runs up to the NW, there is an initial *chimney-effect* that will focus and accelerate the wildfire momentum towards Home.

Assess the Fire Sector for ignition, fuel, and slope.

Reducing the fuel load in that sector is the highest priority. Say there is a road to the southeast, where neighbors or passersby might cause a Fire to start by throwing a smoldering cigarette out the window, or stopping to take a break and dropping a match, or throwing sparks from a dragging trailer-chain, or pulling over in dry grass with a hot, low, muffler and driving away without realizing that they just started a fire.

Fire sectors

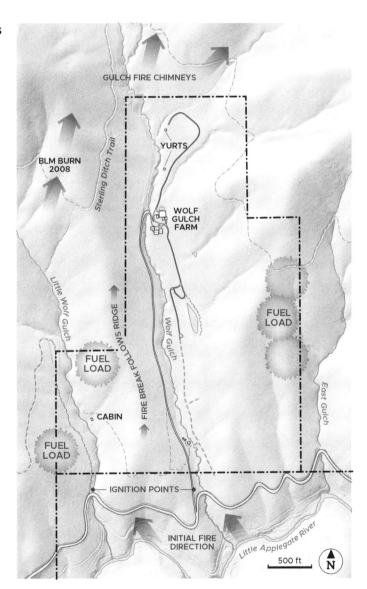

Primary fuel reduction along roads

Strategic clearing on ridges and fire break trails

Now, let's tell a story of what could happen. The window-of-opportunity to stop this forest fire start-up is small. In the first hour or less, wildfire start-ups usually move to the northwest in the northern hemisphere. This is a rule-of-thumb that proves to be reliable. Once the wildfire starts to move and climbs up into tree crowns, throwing Fire ahead, she gets strong enough to create her own fire-weather—the wildfire goes with the prevailing-winds, available-fuels and Fire heat-updraft. A wildfire that has already gotten rolling cannot be easily stopped.

If your Home is being threatened by a large wildfire, you will want to evacuate, not try to fight it.

Any fire creates an updraft of smoke and water vapor. This *fire-tower* sucks in oxygen and wind as the fire grows. Thus a wildfire starts to create its own weather. If the initial start-up has an uphill slope to its northwest, the slope can act as an accelerator. If the slope is already warmed up by the sun, the slope itself is throwing up a thermal-tower of sun-warmed air.

A canyon, such as the Little Applegate, has a predictable daily, diurnal, wind pattern (when there is not a storm coming in or a strong weather-change wind). At night, around the Little Applegate River drainage-basin, cold air falling off the 5,000 to 7,000 feet watershed-ridge sinks down the canyon as a cool-heavy steady-flow (*laminar*).

One can cool off a canyon or gulch cabin quickly after sundown, on clear summer nights.

At sunup, on a clear day, the east and southeast mountain and ridge *aspects*, especially those that are meadows or rock, heat up fast. A set of thermal-towers rise-up and reverse the night-time down-canyon flow, as the towers pull air up from the slopes and gulches below. As the day warms and the sun swings over to heat up the south and then southwest slopes (most of which are grass or rock in the Siskiyous), even bigger thermal-towers form. As the sun leaves the last slope of any gulch, and then the whole canyon, the towers collapse and reverse the airflow, dumping a tumbling breeze downslope into the gulches (the *fluffy zephyr*).

16 Airflows

Wildfire on this sort of landscape is pushed around by these sun/wind/mountain/canyon dynamics. But eventually when a catastrophically big wildfire starts marching across whole ranges, throwing burning debris ahead of itself and building up its own clouds, the fire-storm overwhelms the topographic airflows. The immensity of the smoke volume and the vigor of the massive updraft spin off side weather, such as relatively cute pasted-on thunderheads with superfluous lightning and sometimes strange black-rain.

As long as Fire stays on the ground, there is some hope of stopping it with a fire-break or active wildfire-fighting, mainly by scraping more fire-breaks ahead—and to the sides—of a moving front. But if there is a way for the fire to climb into the tree-tops, that is called a *crown-fire* and can get really big, really fast. The ramp of fuels that carry a ground-fire (in fine fuels) up into the tree-crowns is called a *fire-ladder*. That could be shrubs above grasses and below tree boughs. That could be dry branches low down on a tree-trunk. Removing the fuels that could carry Fire up into the trees is the next priority job in the fire-sector, after the fire-breaks.

What else is in the Fire-sector? There might be forest, and it might be good to manage that forest to be fire-resistant, by either keeping a *closed-canopy* that prevents sunlight from reaching the forest floor— thus slowing down the growth of understory ground-fuels—or by thinning the forest to reduce the potential of Fire moving from one tree-top to the next or by creating alleys in the forest that are grazed or mowed to keep the fine-fuels down close to the soil.

The alley-cropping pattern in forests is called *agroforestry or silvo-pastural-systems* (Douglas, 1978). Perhaps there could be a series of treatments, with different land-uses between the buildings clustered at Home and the ignition-source, on the county road? Perhaps that could include irrigated-pasture or orchards? Perhaps there could be a pond or wetlands, just southeast of Home-central?

WILDFIRE IN THE NEIGHBORHOOD

*Rural home development
in the dry west
is a disaster
waiting for a solution.*

Second homes and exurban sprawl have gentrified vast stretches of used-up (*hammered*) forestry and ranching-lands. These are now areas difficult to protect and likely to burn.

Making local progress on isolated parcels is not enough. Drainage-basin cooperation will be necessary to bring these lands into tending. A daunting idea in these times of Transition. Perhaps most of these cheaply built McMansions are crap anyways and cannot be Retro-fitted to low-energy warehouses? Perhaps we will salvage the houses and take down the fences? *Re-wilding.*

The types and arrangements of buildings we keep on fire-prone-landscapes improve the chances that infrastructure could survive Fire, either passing by, as part of yard-keeping, or coming on fast. Our strategies and tactics will be site-specific. Non-flammable walls and roofs would be best.

The governments of fire-prone settlement-areas usually give out lots of advice on how to protect homes. They like to preserve their tax-base. They like to force economic-activities, by requiring that fuels are reduced to their specifications. They usually recommend heavy thinning of fuels along roads and driveways, removing as many trees as possible and mowing everything else.

Buildings are encouraged to have non-flammable roof-materials, metal eave-flashings, mowed-lawns, all shrubs and trees removed nearby, wall construction that is fire-resistant, and fire-safe utilities such as buried electrical services and isolated propane gas-tanks on concrete and gravel-pads. Trees and landscaping plants are selected from lists of fire-resistant species. Some trees or shrubs do not burn well, if they are kept well-irrigated, and some non-irrigated native-species resist fire for a while, with their thick bark or leaves.

Your local farm extension office (Oregon State University Extension Services) or Soil Conservation District office will have lists of locally appropriate native fire resistant vegetation. Until we clump our housing into defendable hamlets, local governments recommend that isolated forest houses remove **all** vegetation on the acre surrounding the house and build the house with non-flammable materials. These houses require lots of energy to build, cool, and heat. And they are ugly.

WILDFIRE IN THE HEADWATERS

The ecologically beneficial wildfires on less developed lands skip around and leave a mosaic of less burned areas, with some hot-burned pockets. Some wildfires stay on the ground and even burn downhill at night. These are the wildfires to study. There has been a push by industry to build new roads and cut partially burned trees in salvage-operations. These after-burn-interventions have been shown to be very detrimental to soils and reproductive forest bounce-back. Forest stewardship (tending) may include walking into a burned area and doing some falling of unstable snags—as log-mulch on Keyline-patterns—or doing some seed-scattering to hold soil. These interventions are contingent on the season, approaching rains, or the depth of the burn into the soil-structure.

Standing snags are valuable to birds, who bring in seeds, and raptors who control rodent eco-spasms. White Oaks and Madrones are able to sprout from surviving root-systems. These two hardwood species are also associated with many *mycorrhizal-mushrooms*, such as Morels, and these encourage the survival of germinating tree-seeds. The importance of holding legacy-hardwoods, by thinning invasive-conifers, includes this ability to sprout from burnt-stumps and to hold the mushroom-hyphae that enable revegetation. The highest species-diversity on mid-elevation slopes is found in Oak/Pine savannas and co-evolved with regular burning.

Wildfire can leave smoldering-pockets that persist for weeks or months and then re-emerge. Mop-up operations, after wildfires, usually include seeking out these pockets and isolating them with local fire-lines or digging them out and dowsing the embers. Wildfire can travel underground along dead tree-roots and come up later, and a ways away. Wildfire can go down into heavy carbon-soil such as dried out peat-bogs, or rotten stumps and logs, or deep beds of needles and leaves.

Pitch-filled Pine and Fir logs, lying on the ground, can burn like candles for days or weeks. These sappy-dense logs have been ignited by hikers, taking a smoke break, only to be found days later alight, and sometimes these fire-brands can start a moving wildfire after a long delay.

MOVING BACK TO FIRE TENDING

To repeat: an ultimate goal of *Social Forestry* in the forested mid-elevations of the Pacific Northwest is the return of regular burning. This will encourage ecological values and reduce catastrophic fuel build ups. We know from tree ring cores of old snags and stumps that First Nation burning was every three to five years on much of these mid-elevation PNW woodlands. After the removals and genocides of the First Nations, the periodicity of wildfire in the Little Applegate canyon appears to be every twenty years. Some meadows, basketry copses, and seed fields were burnt yearly in the old-times. We have lots of research, remembering, and experimenting to learn how and how often to burn the specific aspect slopes and flats that we have to work with.

The Ultimate Vision
is re-inhabitation.
A whole new way,
from a mix of old ways.

Our villages, hamlets, manor-house farms, and seasonal-camps, should all be sited and built with *relative-location* in mind. We have the clue from TEK that winter-villages, the more permanent occupation sites, were above the valley-fogs and below the heavy snow-packs in the mountainous regions. This usually means between 2,000 and 4,000 feet (600 to 1,200 meters). These village-sites are best located on southeast-facing slopes (warm up early morning, cool off early afternoon), for best gardening and solar-capture. These are usually on benches next to perennial streams or near trusted springs.

The Siskiyous have a pattern of south and southwest slopes being grasslands, as they get too hot for conifer-survival, and perhaps because these slopes were burned regularly, up until almost two centuries ago. The pattern may be a persistent remnant of burning-practices. So are the legacy-hardwoods. Many mid-elevation slopes in the Siskiyous have dead or dying giant hardwood trees with spreading branches being overtopped by Douglas Firs and Pines. These remnant giants are clues to the burning practices of the past; they show that most mid-elevation slopes were burned regularly and that these big conifers that are shading out these legacy hardwoods have grown up since regular burning stopped. See Chapter 6 for managing and preserving these relics.

From the colonists' journals and anthropological research, we know that a lot of broadscale-burning in the PNW was done in late summer. If we were to try that, we would risk giant catastrophic landscape-scale conflagrations. So we are pushed into later fall and early-winter to be safe. Since colonialism involved parceling out the landscape in arbitrary grids, we do not have rational drainage-basin-scale burning-areas to work with. We have to start with small parcels surrounded with excellent fire-breaks and do the fire-tending with good-sized back-up crews.

The problem with late fall season burning is that some reptiles and small-mammals have already slowed down and gone into *hibernation*. This leaves them vulnerable to incineration and injury from our unseasonal-burning (not what their ancestors were accustomed to). The good news is that most wild-flowers, bunch-grasses, and seeds-lying-on-soils are dormant and resistant to fire-damage.

The birds are not nesting in mid-winter. The ground is already wet, so the cool-burning almost floats, above the wet-soil and any emergent-young-seedlings. Mid-winter has proven a decent time to practice cool burning, but it is sad to find the occasional Alligator Lizard cooked in some ashes. We have several generations of work and culture-building before we rescue the landscape enough to be broadscale-burning in the traditional late summer and early fall.

A big advantage of the earlier drier and warmer traditional-burning seasons was that smoke was quickly dispersed by winds. Most of the regular burning went fast and did not leave a lot of mop-up, as described just above. By the late-fall and early-winter, we get periods of atmospheric-inversion and periods of no-prevailing-winds, nor canyon-effect-winds. The valleys fill with fog and smog, and our controlled-burn smoke can settle into that soup and make it even worse.

First Nations called these hemmed-in, mountain-surrounded valleys "valleys of death." We should be removing all that infrastructure from these big bowls and restoring wetlands, savannas, and tree crops. Move-up, out of the fog, and quit producing so much pollution that makes for smog. Again, governments generally try to limit upland-burning for just these reasons. In the winter, they declare burn-days, only when there is enough wind to keep the valleys flushed-out.

Wildfires can burn hot and pump a lot of nutrients into the sky. Regular, carefully timed cool-burning, as done by the First Nations, recycles a lot more of the understory-carbon than does erratic wildfire. Cool-burning grasslands leaves behind significant fine charcoal, which is quickly incorporated by the soils. Making charcoal from overstory and brush-conversion, on the way to being able to do broadscale-burning, captures carbon, with some control of the types of gasses released (see Chapter 8).

The First Nations burning practices kept catastrophic-fuel-buildups minimized and soil-fertility stabilized, especially in dry-season climates like the Pacific Northwest. Big forest fires volatilize a lot of woody-fuel. And soil-humus. The soil is often baked-deep-down. Sterilized. Only the Madrones and White Oaks have root systems that can re-sprout. Lots of on-site nutrition is lost to the atmosphere, where it does not help with climate-weirding.

The ultimate strategy of resilience-restoration at Wolf Gulch Ranch, is to return understory burning to the landscape. But we have inherited a build up of standing woody-carbon-fuels, accumulated since the original-people and their wild-tending were removed, in the mid-1800's. We are in dire danger of massive *stand-replacement* wildfires. We refer to some brushy-slopes as thermo-explosive-events, waiting-to-happen.

In order to get back to regular broadscale-burning, controlled and cool, every three to five years, we first convert *senescent* and crowded standing-fuels to firewood, charcoal, ash, and mulch. This is delayed-maintenance work and is intensively hands-on, way more work than seasonal downslope-burning. We have to do this initial phase of restoration on the critical areas of the ranch, so that our first broadscale-burns are not going to turn catastrophic, by escaping our control.

Reduce the threat
before we take the chances.

The opportunity this phase delivers is access to a lot of carbon-materials that we can use to build fertility in our Horticulture and use as fuels for our Home-fires. Lots of lovely artistic pieces (spotted by Bodgers) can be put aside for woods-crafts. Some herb, root, and mushroom harvests may present, as we wade into the wall-of-carbon. We are making three types of charcoal from more than six species of wood.

Manzanita and Buckbrush are the bulk of the understory thicket. Both are co-evolved with wildfire. The fire-parched Manzanita seeds are ready to germinate in the winter rains, and if they find resident Morel mycelium, their tiny, woody roots will link, and up goes the seedling. Buckbrush is a nitrogen-fixer associated with bacteria. Both are explosive-fuel and burn hot in a wildfire. The old Manzanita is often laid down, from snow load, but with parts still living.

The Buckbrush, as it tips down slope from snow-load and maturity, sends up new vertical sprouts and forms tangles of divicarate-thorns. Some of the Buckbrush might re-sprout (coppice). Some of the Manzanita may be well formed to shed snow and can be left standing for nectar-flow and berries.

What gets revealed are Oak clumps and the occasional Pine. The Oak-trunk-clusters are fire-coppice, from the above-ground tree getting fire-girdled, and the roots re-sprouting in a circle-around-the-old-stump. These White Oak clumps might be re-sprouted repeatedly following hot-wildfires. The bigger Pines have fire-resistant-bark and the first limbs are high off the ground. They like underburns but their needle-beds might need mop-up for a few days.

The seeds of wild-flowers, bunch-grasses, and geophyte bulbs are still remnant in the brush-field. After thinning and clearing these *chaparrals,* we are eventually gifted with open-grasslands, under scattered Oaks and Pines: a desert-savannah. Full of opportunities and lovely to Human eyes.

Above the canyon, serpentine-canyon-ridges with adobe-flanks are steep, but they still have some of the classic land-form-clues that the Keyline method looks for. As we prepared to treat the brush-fields, we surveyed fire-break/access-trails and we found ridge pads (also called nick-points) on the ridgelines and laid-out wing-trails, rising one-in-fifty, up-gulch from the pads, along side slope benches, for access and materials-movement. These trails may never carry water, but they do connect the landscape in useful ways. The ridge-benches are good kiln and campsites with awesome views.

The trails allow wheelbarrows-with-loads to be easily pushed in both directions. Trails are easier to build than roads, with less impact, but they still need massive-investments-of-labor and maintenance-forever. Thus trails should have *multiple-functions*. We use them for getting to remote parts of the gulch, as fire-lines, as emergency-access, and as observation-transects. These trails define the broadscale-burn areas and provide control-lines during burning. They also make moving logs and harvest-bundles to kilns and sheds relatively easy. The only petrol-motor we use is a chain saw, to handle the first clearing-phase. Good hand tools do the rest.

HOW WE LEARNED TO BURN DOWNHILL

It was a warm day in January. There was a guest/intern, Zyara, staying in Little Wolf Gulch. My herbalist-friend Heron was visiting herbalist-Harlan, over the ridge at Wolf Gulch Farm. The savanna-restoration clearing-process had gone on for several years now. We had over two acres cleared and were on our third burn-pile location. Vesta, the charcoal-kiln, had reduced a few loads of Manzanita logs. The Little Wolf Gulch branch of the Flat-Trail was freshly re-cut, and we were all burning brush below that good fire-break. We did have boots and heavy canvas pants but were mostly stripped down to tank tops in the lovely weather. The burn-pile-site was tucked into a northwest facing bowl and was somewhat protected from the canyon-airflows that chimney up-gulch toward Little Bear Butte.

At some point, with four-of-us standing by, Fire started to run up-hill, towards the trail, under a couple of vase-shaped Manzanitas. We let it go, just to watch. No problem. Cool. The fire skipped around a bit but did not run away uphill, it just creeped. "Ummm!" we said. We let it burn out, in the sparse-fine-fuels, and returned to finishing-off the burn-pile job. Then we decided to try to burn-downhill the next day.

The next day, we got our tool kit together. A couple of shovels, a McCleod, back-pack sprayer and hand-held one-gallon-sprayer (both full of water), gloves, good clothes, hats, snacks, and drinking water. There was already another small went from the third burn-pile back to Vesta, one-hundred feet below, and parallel to the LWG trail. The slope ratio was steep, about one-in-six.

We tidied up the fire breaks and stripped dry grasses from above the LWG trail, widening the fuel-break above our proposed burn-area. So we had a good-box-of-a-slope to play in. The previously massive fuel-load had been burnt in piles, and the remaining Oaks and Pines had been high-pruned. We inspected the slope and moved any remaining small branches and non-embedded logs onto contour so they would not act as *wicks*.

We took the grasses we had stripped from above the top firebreak, and we twisted them into bundles that we nicknamed "Fire dollies." We had a small can-stove with charcoal, as a Fire source.

Then we burned a two-to-three-foot strip just below the top fire-break (the trail) and lo-and-behold, some of the line-of-flames moved downhill. The Fire also moved up to the trail and went out, but the bottom of our first line moved down, majestically. We were stoked! We used our *fire-dollies* as torches, to move Fire around, and soon had a continuous flame-line, moving downhill, about one-to-two-feet tall. Ooo! Ahh!

The mop-up had a problem. Just above the bottom fire-break, there was a good-sized, two-foot-dbh Pine, and the needle-bed under its *drip-line* was hard to put out (*quench*). We had to dig out the glowing-red-needles, a foot deep. This is an adobe-desert-savannah, and there is generally very little topsoil or fine fuels litter so we were surprised. We also found a dead-cooked Alligator Lizard. Whoops. Sorry! Otherwise we managed to water-dowse (quench) a few hot-spots and had it mopped-up good enough. "Let's do it again tomorrow," we decided.

The next morning we convened with tools and had a close look at the previous day's burn. Tears-in-our-eyes! The bunch-grass-clumps, mostly Fescue, were suspending a lace-of-black-carbon! And under that delicate-canopy, we found intact green-emergent seed-leaves! We were on our hands and knees, with 10X hand lenses. We did not want to walk all over the site, as it seemed like a work-of-art to us. Ephemeral art! Surely the carbon-lace we found would soon fall-down or wash-down and be incorporated into the thin *humisphere*. Oh, yes!

The third day we burned an acre. It was awesome. We scratched in an internal firebreak, to surround a special clump of *chaparral* that we had not previously cleared, to preserve a refuge and study-plot. We moved much more confidently this time. But we still had a pine-needle-bed deep-smolder to deal with, and this one took two days of checking and short periods of digging to finally have no smoke or reignition.

These beginners-luck savanna under-burns were very encouraging. They look great now, several years later. Lots of native-grass-diversity and a good mosaic-structure with Oak snags and unburnt clumps. Vesta has been fired up several times since then. After another year of burning we eventually treated about four acres.

HOW WE DO IT NOW

Broadscale-burning—of fields full of invasive European species—and under-burning of Oaks, with more natives present, has become the star-attraction of the Social Forestry winter-camp course. This is what folks show up for. Workshops and trainings in controlled-burning, by conservation-organizations and by government-land-management-agencies, are way different than what we have been doing. So far we have been lucky, to not have any injuries or big-fire-escapes (Bauer 2014).

As we are using fire by principles to recycle nutrients, invigorate photosynthesis, break seed dormancy, and reduce excess-senescent-fuels, we are also taking risks. This is a process of *reciprocation* in that we are giving Fire back to a landscape-of-ecosystems that has co-evolved with both wildfire and regular broadscale-burning. Our biggest risk is that our small, cool, understory, on-the-ground burning can escape our fire-breaks, and we lose control of the Fire.

Because the surrounding hills and brush-fields have not been cleared or tended for well over a century (except by the 1987 wildfire on the west boundary), we are making the first small steps toward the return of regular-burning. Our hope is that we learn how, and that we increase the vigor of native-species and biodiversity generally, while we reduce the monopoly of invasive-species, such as Star Thistle and Foxtail Barley. Of course, as we remind ourselves continually, we are foolish and silly. None-the-less, we are moving forward with expectations of the return of Salmon and the return of Human cultures-of-tending and appreciation of Place.

SOCIAL FIRE TENDING

This means that there is a lot of preparation that we do before we actually put fire into the dry-grasses. Everyone gets oriented. Everyone has a place in the procession-of-considerations. We use ritual, song, parade, and ceremony, to keep us intent-on-the-process and aware-of-our-surroundings and conditions. There is a lot of training necessary to learn how to modify wildfires (see above), that are applicable to controlling cool-burning. We must be prepared to control any escapes over our firelines and we must be ready to make decisions moment by moment, as the weather changes, or the fire-behavior is not what we expected. Lots to learn and best we try within real Time-and-Place.

By watching the clouds and winds and our weather-instruments we can predict a few days ahead, if there might be dry-enough-conditions with low-enough-winds, to dare to try to burn. The morning of a proposed burn follows a day or more of preparations. Once we have identified a candidate area, we need to reduce-fuels around and in the area, to prevent any flare-ups climbing over our fire-breaks or up into brush and trees. We are only burning in winter, so far.

*The Mothers Guild
let us know
a couple of days ago:
the grasses snap.*

OK then. Here we are on a clear-calm-day in January, and after breakfast we have gathered in council-lodge to get organized. The Sawyers-guild has already scraped fire-breaks, around the field or savannah, that we have identified as ready-to-burn. The Fire-crew (Mothers) has scoped the area and bundled fire-dollies in piles (Bodgers) from fire-fuels adjacent and outside the fire-breaks. The herb/root/bark guild (Mothers) has tagged special-spots and asked for internal fuel-breaks.

The Rangers have scoped the larger area to help orient everyone to the big-picture and joined the Sawyers on-the-lines. The Elders have sat-at-dawn, on the ridge above the area, and walked the fire-breaks, while talking to the birds and denizens, to tell them what is about to happen and to listen-carefully to any clues-and-messages that may have been missed. All the guilds in the Fire-crew have reviewed their tool-kits and preparations. We have all sat-with-each-other and reported on all-our-observations.

By coming-into-alignment with the tasks, and risks, and opportunities-at-hand, we are calm, grounded, focused, and ready. The Elders of the Fire-crew have walked, singing softly to the site, and have tested the fine-fuels, once more, for *perfect timing*. They send back a Ranger to tell the-whole-crew that all looks and feels good enough to go. In guild groups, we now have a procession from the altar-of-our-intentions, the central-hearth (with hot-coals in can-stoves). We all walk, senses alert, promises in mind and take our positions.

After the Sawyers have moved down wind (with the wind) to ready for escapes (or actually with the breeze up slope at most) to the uphill firebreak, and the Rangers have deployed to walk the wider area, alert and attentive, and the Elders have found comfortable places upwind (against the wind) to observe everything, the Fire procession comes, carrying the sacred-hearth-fire in the fire-baskets and ember-nests. The Mothers are singing the welcome-songs and the Sawyers and Bodgers may be drumming, and dancing, to let the ground-dwellers know something-big is about-to-happen.

With our still-being-birthed culture of *Social Foresters*, the Elders look around, call the directions and spirits-of-Place to attention, and the Mothers light the upwind and top strips of fine-fuels. The Sawyers and Rangers, on the fire-break, guard against any escapes uphill and quickly-suppress-fire trying to move outside our lines. The Rangers and Elders watch the weather and the wildlife-messengers. The flame-line begins to move downslope and the Mothers fill any gaps in the line of fire progress with the fire-dollies and sometimes a toss of loose-fine-fuels.

Some Fire-crew folks are already mopping-up, by quenching still-burning-wood with water sprayers and sprinklers, scraping out any embers that try to embed in stumps, logs, or needle-beds (*duff*). The cool-burning moves steady and soon eats a lot of grasses, and twigs, and leaves, on-the-ground. The Mothers move around the already-burned-areas, where the ashes are still warm and scatter seeds of nitrogen-fixers, and bunch-grasses, and geophytes.

OPPORTUNITIES, RISKS, AND LIABILITIES

There are a lot more social and ecological-opportunities that may present themselves, and with practice we are learning to be ready. So much to learn, and so few Elders with us who have done this before. We do get clues from Wildlife that this work is appreciated. We do see later, in the recovery and in subsequent seasons, that the grasslands and meadows that we have burned change in their species composition and sometimes in available harvests. The challenge is to do enough of this cool-burning to secure and improve much ground.

Wolf Gulch Ranch is kind enough to facilitate our learning, but this is just a small piece of the larger-landscape. We are using large-crews. As told in the story above, about how we learned to burn, four people is the minimum and we are burning very-small-parcels. Larger gatherings of cooperators can do this safely, under the right conditions and with perfect timing, to extend the areas burned. This is labor-intensive and social/cultural in a way that industrial-demands do not even consider; this is not efficient. This is ceremonial and edifying.

Our experience so far shows that the most challenging aspect of how we are burning is the danger of unexpected fire-behavior and our under-prepared abilities to combine cultural burning with wildfire-suppression techniques. As the consequences of uncontrolled burning can result in property-damage and liability-litigation, we are constrained in our enthusiasm.

A lot of experience is going to be accumulated and shared, before we can roll out these cultural-practices in a larger context of re-localization and changing settlement-patterns. Meanwhile, we continue to need back-up from government emergencies-services, standing-by, on call.

In *The Control of Nature* by John McPhee (McPhee 2011), we learn that Old English Law principles (and the letter, and spirit) hold the last Human to make changes in Nature liable for subsequent damages to private-property. Never mind the pre-conditions wrought by industrial-extraction. If you lit the match, you caused the damage. And can be sued. If a broadscale-burn escapes fire-breaks and enters adjacent private or government property and requires wildfire-suppression support, whoever lit that burn is liable for the costs of control (bulldozer-scraped fire-breaks?), mop-up, and any damage to taxable-infrastructure. Folks have lost business-licenses and insurance-coverage after small-escapes cross fence-lines.

If a rural-property is paying into a "fire-control-district" fund, the local volunteer fire department will respond to any ignition that starts on-site or threatens your assets. This is a sort of insurance, but usually does not involve corporate skimming and government taxation. The neighbors pool funds to support fire-suppression equipment, training, and outreach. Rural Fire Districts also get some grant help from government agencies during Transition. There are networks of cooperation between many levels of fire-suppression organizations. Some wildfire-workers are paid professionals and hire out to various agencies. Some are permanent staff in forestry and natural resources agencies.

Neighbors are also forming community burn associations, where liability is negotiated in advance across fence lines, and the community cooperates at a series of controlled burns. So far most of these events are organized to conform with the standards of wildland forest fighting used by the federal agencies, subcontractor businesses, and conservation trusts. This means water trucks and fancy industrial gear. That way, if a community burn association needs backup, the protocols and accustomed inter-agency expectations are in use and control is seamlessly passed up the hierarchy.

The traditional Hicksite Quakers had advices against insurance. We called it "gambling against Spirit." Better that we are very careful and do not take unnecessary risks. If thee cannot control a horse-less carriage (Horses are dangerous enough), why does thee persist? Just because the insurance agent promises *"convenience, license, and privilege"*?

Don't try broadscale-burning at Home, without the cooperation of neighbors, community, and culture. Do a lot of training and preparation. Live someplace long enough to know what could happen and has happened in the past. The 1987 Cantrell Gulch wildfire ate our northwest corner. We burned off several acres of Star Thistle in 2000. We had a field fire escape in November, 2003 that ate forty acres (in our main Fire-sector) and cost us $2,000. The neighbors complained; they wanted a fire like that, and we got off cheap. Still, it was really scary and burned up to the foundations of the plastered straw-bale walls of the main house. Since then, we have burned several smaller plots safely with only one blow-up (see above).

At Wolf Gulch Ranch we can still see the remnants of First Nations Horticulture and learn from those patterns. The Oak/Pine savannas revealed by our brush removals still hold bunchgrass and wildflower soil-seed-loads and quickly recover enough for us to reintroduce cool burning. Fire-managed Basketry was a big part of Dakubetede culture. Cool burning goes hand-in-hand with tending-the-Wild, sharing the bounty, and giving back.

*We are proceeding
with humility,
appreciation, gratitude,
and reciprocation.*

The Charcoal principles include

38.
When wood burns, the heat of combustion cooks the unburnt wood and distills out gasses that feed the flames. If we burn up all the wood, the remains are ashes. Ashes are mineral remains of wood. Wood loses (out-gasses) carbon, water, nitrogen, phosphorus, sulfur, and other micronutrients, through combustion and pyrolysis.

In most open fires, some components of wood do not directly burn and outgas as large or unstable molecules (carbon monoxide, hydrogen, methane, turpines—all together known as wood-gas or brown-gas). Wood that combusts completely becomes ash (minerals), water, carbon dioxide, nitrogen and sulfur oxides, and several other simple gasses.

Pyrolysis is the distillation of wood. In an oxygen-control container (a single or double-retort kiln), combustion can be quenched just-in-time to capture charcoal. The flue-gasses can also be filtered and used as biofuel or condensed for varnishes and turpentine.

39.
White smoke is water vapor. Brown smoke is complex-gasses. Blue smoke is simple-carbon burning. Smoke from an open fire has all these colors. Brown smoke is the most toxic, unhealthy to breathe, and contributes to atmospheric-greenhouse-effect-heating. Different colored flames are usually from different mineral-compounds (imagine a Shaman, dusting Fire with color-flashes) or mixes of complete and incomplete combustion.

40.
Charcoal is stable, nearly pure carbon, the remains of partially-combusted wood. The vessel and tube, and sometimes bark, structure is still visible. Charcoal contains less than half of the carbon found in unburnt wood. Charcoal is seldom consumed by Life, but offers a haven and refuge to soil micro-life with its range of tube sizes and textures.

CHAPTER 8
Charcoal

THE PARKING LOT WE LIKE

*Bio-char
is a three-dimensional hotel
for soil micro-life,
easing extremes of drought,
heat, and drowning.
Bio-char is charcoal
that has been charged (rooms filled)
through composting.*

Charcoal is amazing stuff. We can see the water-vessel-tubes and the fibrous-rays, the carbon-structure of the wood still visible. The empty tubes held together by the ghosts-of-fibers, after burning away so much else, are almost pure carbon, with traces of minerals. If we did not quench the Fire at the right moment, to save the charcoal, and let it burn, we would be left with mineral-ash.

Hardwood ash is valuable. We can make alkaline-brews, soak Maize (Dent-Corns) in ash-water to make it nutritious, adjust garden pH (acidity) by dusting thin layers of ash on the compost layers, spread it up-slope to return mineral fertility, or add it to natural-building clay-plasters. If we burn all our wood-fuels to ash, all the carbon and volatile carbon-compounds outgas to the atmosphere. How we burn the wood leads to different mixes of gasses (clean-burning wood stoves deliver plenty oxygen to combustion), but we are still adding greenhouse-gasses to the already heavy-industrial-load that is melting the ice-caps.

If we burn carefully, we can capture up to half the carbon as charcoal, resins, and fuel-gasses. If we bury the charcoal, after charging it with bio-brew or layering it in compost, we build *bio-char* soils and *sequester carbon long term*. If we practice smart *carbon-farming* we will also accumulate *humus* and sequester more *net-carbon*, while improving crop-nutrition, easing irrigation requirements, reducing soil-erosion, and building *soil-bio-diversity*.

The heavy clay adobes—rich in magnesium and calcium, but alkaline (pH as high as 9)—at Wolf Gulch Ranch are especially susceptible to bio-char amendments. Bio-char introduces stable carbon-matrixes that act as small nutrient-sponges, packed with soil-micro-organisms, ready to keep the soil alive through extremes of drought, while mellowing the mineral alkalinity.

Net,
Sink,
and Loop.

Non-charcoal carbon, such as compost or mulch straw, is quickly consumed (eaten-up) by living-soil and plant-roots, and although this encourages growth, much of the carbon-attached-nutrition is outgassed as volatile-compounds of nitrogen, sulfur, carbon, hydrogen, oxygen, and traces of other minerals. Not only are we adding more greenhouse-gasses to the air, we are also losing fertility.

Tilling the topsoil with a plow or rotavator-implement, in order to stir in fertilizers and compost, and in order to add air by fluffing up the soil, increases the water absorption capacity and also encourages rampant-bacterial-composting of nutrients in the topsoil, making the fertility readily available to annual (one-season) market-crops. But this bio-stimulation also accelerates the outgassing.

Hot-compost-piles also outgas a lot of nutrition, as they quickly digest the rough feedstocks (crop-residues) and cook-out the weed-seeds. The farmer can grow vibrant-fast-vegetables and take advantage of petrol-fueled-machinery but fails with tidy-closed-nutrient-loops, atmospheric-stability, and conservative-soil-management. Farmers-who-plow have to import a lot of carbon and fertility often, from off-site (an imaginary territory).

21 Just Sink Carbon

The Ultimate-soil-management on farms is no-till, with moldering, cool, fungal-dominated, low-outgassing *compost*. These deep carbon soils are best dominated (overstory) by perennial-woody-plants, interspersed with diverse-understory-plants, and at a much smaller farm-scale than previous industrial-agriculture. Lots of small pocket-fields with hedges, orchards, and windbreaks. More like Horticulture in the Wild. After a period of transition, we can reach at least as much productivity-per-hectare, with much more pollinators and wildlife presence, as long as farms can handle all the complex harvesting, sorting, and processing.

Think Oak/Pine Savanna-like farms, with berry-patches and grasslands, *mob-grazed* and occasionally burned, full of geophytes (edible bulbs) and seed-crops. Bacteria clustered-on-roots (*nodules*) fix nitrogen-from-the-air to carbon-from-photosynthesis: *solar-carbon-capture*. Micro-nutrients are re-cycled and re-distributed by mushroom-networks in the soil. In dryer climates, regular, cool, broadscale-burning and charcoal from fuel-management and forestry—sequestered as *bio-char*—help with landscape-resilience from drought and wildfire.

Fields and gardens close to housing-clusters would be managed intensively, with top-dressing compost-applications and deep-mulch, supporting mushroom-digested carbon-recycling that is laced with bio-char (in the top-dressings) from pile-composting manures, crop residues, and shredded wood. *Carbon-farming*. Carbon in many forms is skimmed from surrounding forest-restoration work and concentrated near Home to support the workers.

29 Loop Law

22 Nutrient Cycles

Farms and forests in wetter regions have the potential of year round broadscale, mushroom-based carbon-recycling, which can break down woody materials and hold a lot of carbon-attached-fertility and micro-life. Big, downed-logs can slowly rot for a long time in shady understories that seldom experience Fire, and thus sequester significant-tonnage of carbon, often for centuries. Leaching of nutrients by rain and un-diverted runoff would be the drain: the losses to "off-site" (there is that other place-idea again, *there is no away*) to watch out for. Close open loops!

THE BURN-PILE CHARCOAL MAKING LAYOUT

At Wolf Gulch Ranch, we are living and working on a fire-accustomed landscape. Wildfire has been coming through every twenty years or so, since the First Nations' burning was ended. They were on a three-to-five-year periodicity of Human-managed and timed burning. We do not yet know what the new periodicity might be.

We only get 10 to 30 inches of rain per year. The canyon is an *ecological-mosaic* with lots of complex-edges and a range of vegetation-types, from grasslands and savannas, to fire-prone *Chaparral* brush-fields, to hardwood riparian-strips, and northeast-aspect conifer-forests. Mushrooms here are seasonally-active, following wet-seasons. Before treatment, the slow-growing forest overstory and brushy understory carry a lot of standing-dead poles and dead lower branches (*fuel-ladders*). The *carbon-fixation-rate* from *photosynthesis* is probably not keeping up with the carbon-outgassing from *senescent-vegetation*.

Wildfires burn hot and pump a lot of nutrients into the sky. Regular, carefully timed cool-burning, as done by the First Nations, recycles much more of the understory-carbon than does erratic, and hot, wildfire. Cool-burning grasslands leaves behind significant fine charcoal, which is quickly incorporated by the soils. Making charcoal from overstory thinning and brush-conversion, on the way to being able to do periodic broadscale-burning, captures carbon, with some control of the types of gasses released.

*We are making
three types of charcoal,
from more than six species of wood.*

On these serpentine (wiggly and alkaline-geology) canyon-ridges with steep adobe-flanks, we found pads (knick-points, small flats) on the ridge-lines and built wing-trails, up-gulch along benches, for access and materials movement. The ridge-pads are good kiln and camp sites with awesome views. The trails allow wheelbarrows with loads in both directions. These trails define the broadscale-burn areas. They also make moving logs and harvests to kilns and sheds relatively easy. The only motor we use is a chain saw, to handle the first clearing phase. Good hand-tools do the rest.

Kilns are located where they are useful for years, located in between the fuel-loads being converted, and the roads and farms where the charcoal is used. Burn-piles are temporary locations but still need to be reused a few times. Burn-piles cook the ground, restarting pioneer-plant-succession. We do not want too many dead-spots, left by too many burn-piles. Burn-piles also leave some ash and charcoal that wildlife comes by to sample. Recently used burn-piles are good animal tracking sites.

WHY WE DO NOT BURN GUARDIAN OAK

In a wildfire, the smoke from still-green—or dormant but still-alive—*Rhus diversiloba* (a native Sumac) will cause or trigger, in most people, constricted lungs and throat, a reaction that leads to breathing difficulties. This allergic-asthma can be very dangerous and is treated with cortisone-shots in an emergency. Extensive treatment with cortisol-drugs can lead to side-effects later in life (arthritic-pains and joint-immobility).

Skin-rashes can result from casual-contact (brushing up against leaves and stems), unless one has built up immunity. Oil, from the plant (rhusinol) can persist on boots and clothes, and be contact-transferred by petting dogs. We all benefit from careful cleaning and *Rhus*-oil avoidance, so that those who are more susceptible are not inadvertently exposed and we all avoid contact-transfer. Wash with cold water and strong detergent. After working in *Rhus*, take off, jacket, boots, and pants; hang-up and put-aside work clothes that will just be used again; and then wash hands before taking off underclothes. Take a cool shower with soap, before putting on clean clothes, or climbing into bedding.

It is important to avoid dragging juicy Guardian-Oak (we use polite nicknames) into the burn-piles. We do not want to breathe the smoke. We try to leave it standing, or lay it down carefully (if mistakenly cut or broken), where it will not be mistakenly-dragged and burnt, while green. Standing and leaning remnant-stems often fail-to-thrive, perhaps re-sprouting from the roots after they are exposed to new conditions by removing the brush around them.

If live *Rhus*-stems are still there when we winter-burn, Fire seems to move past them fast enough that it does not volatilize too much oil. We have not noticed any problems from broadscale-burning dead-stems. The cool broadscale-burning seems to fire-girdle any stems still-alive.

We have seen this ecologically-important pioneer-Chaparral-species diminish as the grasslands re-establish. Perhaps, we are accelerating natural-succession?[1] Many situations allow the brush-field to be cut and sorted without having to cut the *Rhus* vines and

---

1 The ontological development of plant communities from beginnings such as floods and fire to maturity, usually expected to result as old-growth forests in the northern hemisphere; the orderly succession of life forms that build and allow increasing carbon capture and biodiversity.

clumps. The *Rhus* we have left on ridges seems to thrive and produces good berry crops, persistent over-winter, and very important food for over-wintering Thrushes and Wrens.

WHY WE BURN IN PILES

Burn-piles and slash-piles are the tactics most land-stewards use for fuel-hazard-reduction. High-pruning (to take down vertical fuel-ladders) and thinning (to break continuous, horizontal brush-fuel) produces many small-diameter limbs and brush. Even after cutting-it-up and sorting, taking useful pieces to caches, there is a lot of twiggy and dead carbon to pile-up or lop-and-scatter.

Sometimes, fuel-management is done industrially and slash-piles (very-big burn-piles) are bulldozed up to be burned later. Sometimes, brush is clearcut and laid down for hot-forced burning, months or years later. Sometimes, many slash-piles are built, scattered across the slopes, and are all burned on some later day by a crew with torches, moving from pile to pile. The piles may have plastic or paper covering part of the pile to hold a dry section and help get the pile to burn. Perhaps, the crew turns in the butts, and often this is all done in a rush. No sorting.

Our burn-piles
are small
and made for drag-and-feed
during clearing.
Slash-piles
are built after logging or clearing,
sometimes covered
with a bit of plastic or paper,
and ignited much later.

We prefer fewer burn-piles and more brush-dragging. The soil-seed-bank, along with near-surface tree-roots and bulbs, are not impacted much by careful dragging, but are reduced on hot-burn-sites. We hope to use our piles more than once. We also like to start with a smaller, dry-brush starter-fire and *then* add fuel, so as to control how clean the fuel burns and to build a "hot center" first that allows workers to take breaks from cutting and dragging and then easily re-kindle the burn-pile. Drag-and-feed allows careful burn-pile-management and we see that very little brown-smoke (the greenhouse-gas turpines) escapes. We see white-smoke, which is mostly water-vapor, and clear or blue-smoke, which is carbon-dioxide and water—fewer heat-trapping vapors.

We strategically locate our burn-pile sites so that we *stack-functions* and *integrate-flows*. Burn-piles are accessible by wheelbarrow, near trails. We want to reuse the site, so the location is available for several days' work and thus several burns. Our spread-out-sites—unlike the multiple, or giant, slash-piles from industrial-processes—are few and scattered. They will be reused, but not too-many times. Most will go through a new plant-succession, starting with lichens, algae and mosses, and eventually allowing grasses and fungi.

Ultimate nutrient net-sink-loop, on the way to broadscale-burning, is done by capturing the tip growth leaves and twigs, and getting them into mushroom moldering piles or passing them through Goats (or other browsers). The Goatherds Guild might be looking for good winter browse and could work with small herds to keep up with the Sawyers, Bodgers, and Charcoaliers. The branch-tips (*ramial tissues*) could be cut and sorted for fodder, bundled and carried to predator-safe pens (near Home) to feed dairy producers.

The small branches left from cutting back the tips are mostly carbon and contain fewer micro-nutrients, except in their bark. We can burn these branches to reduce fuel and make charcoal. Or we can use them for brush-mulch on tree seedlings (Nut-Pines?). The abandoned and cooled-off burn-pile pit could be used for mushroom composting. In the absence of machines to shred the material, perhaps we can pack cut-up tips and leaves into the pit and turn the subsoil, ash, and remnant charcoal shreds on top, followed by spreading the original topsoil/seed-bank (and some tree seeds?) over the top. Might need mushroom inoculation?

> *Cut,*
> *untangle,*
> *buck-up*
> *for tips and logs,*
> *sort, pile,*
> *cache, glean,*
> *feed, burn, sleep.*
> *Repeat.*

Drag chutes for pulling cut-brush down-hill (on diagonals) and alleys (near contour, often along fire-break trails) should converge on the burn-pile. There will be un-burnt materials (charcoal-logs, Bodger pieces, firewood, branch-tips, and drum-quenched charcoal) to carry-out in wheelbarrows on the trail/fire-break system.

MOUNTAINSIDE BURN-PILES

Slopes greater than one-in-four are way too steep. One-in-six might work. We do not always have flat-pads to burn on. We do not want to have burning debris cascading downslope. We want to be able to build a hot-center coals-pile, so that we can harvest some charcoal at the end of the day's burn.

There is a trench to build that catches coals and prevents runaway rolling fire brands. After a rare-species-check, the on-contour-trench is built by first scraping the seed-load and top-soil down-hill into a small berm. The crescent-shaped small-swale is then dug out of the subsoil, and the subsoil is turned down-hill to bury the seed-load-berm, which hopefully will protect any live-germ-tissues from too-much heat. This sidelong pit is now ready to build the burn-pile, on the upper slope of the scrape, so that the coals settle to the trench, fuel can be kept uphill as it burns, and very little burning debris rolls downslope past the berm.

The glowing-coals—the hot-center—piled in the kick-trench at the end of the burn, is shoveled (see below) into oxygen-quenching drums-with-tight-lids. These have to be brought-in on the trails and then wheel-barrowed out, full of charcoal, after cooling off. The view should be good, as we sit around the remains of the heat, at the end of the day, and drum-and-sing to the sky and Fire.

SIDEHILL BURNPILE PITS

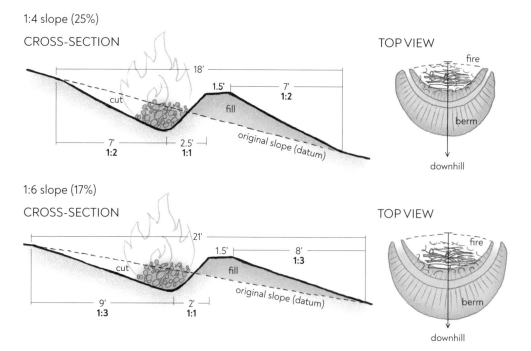

*Burn-piles
can be ignited
on cold and even damp days.
Wax-wick fire-starters
and tucked-in dry-kindling.*

This burn-pile size is site-specific. Steeper slopes secure smaller burn-piles. The kick-trench catches almost all the coals. We are feeding the pile-burn from the sides and above, pasted-on-a-slope. We have pre-positioned brush in windrows, up-wind and up-slope of the burn-pile. Then we can steadily feed fuel, keeping the burn-size manageable, as more fuel is cut and dragged to the burn.

*Butts out,
tips in.*

The trench holds the hot-center, where we turn the butts in. If we do not want to harvest all the charcoal, we can leave a small pile banked-up (smothered) with ashes, to stay-live until the next morning's relight and feed. Still-burning-butts can be put aside—off the top of the coal pile, while shovels feed drums—and then replaced to provide a small campfire at the end of the long day. Charcoal quenched in the tubs cools off by the next morning; the charcoal can be bagged up and the drums freed-up for another quench.

A crew of four, working a half day cutting Manzanita and Buckbrush, can cut, drag, feed, and quench 110 gallons (two drums). Packing the hot coals in the drums with a McLeod allows maximum harvest.

The size of the burn is relative to the workers available, the fuel already cut, the ease of working on the slope, and the heat (from a too-big burn) that makes loading fresh fuel daunting. The available number of quenching-drums may limit the capture of carbon as charcoal.

*Harvesting the charcoal
is the way to skim
some carbon-sequestration,
off of a pile-burn
that otherwise
would leave only ash.*

After the savannah is restored and under-burned, these small terraces (the old burn-pile-sites) can be flattened into tent, yoga, or observation perches. Hopefully we will not need to use these sites again for burn-piles, as we will be broadscale-burning again as soon as the grasslands re-establish enough fine-fuels, such as grasses and twiggy, pruned-or-shed branches.

ONE DAY AT THE BURN-PILE

The first day of a pile-burn, we may already have a small-dry-pile ready to light, or we gather dry branches to build the starter-fire. We might have built a moderate pile of branches the day before, leaving a small cave with a bark-covered dry-twig-nest. Perhaps we have some tricks in our bags, including prepared fire-starters: Beeswax kindling, fine-split "fat-wood" from standing-dead Firs and Pines, pitch-soaked grass-bundles, or flashy Juniper-twig-sprinkles.

Once the first small-pile of coals is built-up, we have our "hot center." Then we can even feed green branches, butts-out and tips-in, on the hot-center. After feeding the bonfire-stage of the pile-burn for a while, it is time to take a break, and we can flip the unburned-butts onto the hot-center, using long-poles (two or three meters, two or three centimeters diameter) or long-handled shovels.

Tips-in, butts-out,
lay those sprays of branches
like fans,
on the hot center.
Whoosh!

And proper clothing! We need way more protection from heat at burn-piles than we do while cool-burning. Wool-felt hats actually work fine, if kept damp, and we watch each other for sparks. Heavy cotton pants and shirts also only smolder if a spark parks, so we can again watch out for each other. Hats and clothes, and even packs, will have burn holes from floating embers, a mark of distinction among Charcoaliers. Put unused and untended packs and gear up-wind and up-slope to avoid holes from hot-floaters.

Leather boots are best as they resist burning. Leather gloves are preferred, for sure. Leather chaps, sometimes used by Sawyers and Blacksmiths, can be hand crafted, will resist some fire, and are good enough for controlled burning. All the modern petroleum based "sports fabrics" are liable to ignite too quickly, give off evil gasses, and sometimes even melt onto the skin beneath. Fire-retardant fabrics are generally non-natural fibers, soaked in chemicals. And they are expensive. We like the old style "tin hats" for fire-resistant hard-hats. The plastic hard-hats are flammable. Imagine the ultimate helmet: perhaps leather with metal-grommets? Wire-sewn bark?

As the day of feeding the hungry burn progresses, we look to salvage some limb wood and butt logs from the perimeter of the burn-pile to cool-off, dry-out, and be used in the charcoal kilns some other day. The butts that we turn-in are cap-fuel for the hot-center and (after a break) we can resume dragging and loading again. This cycle of loading, burning-down, sorting, turning-in, and re-loading, can go on for hours, as long as we have enough brush to burn. Eventually, the pile of glowing coals, draped in ash, is as big as we have quenching tubs to fill. We have done a good day's work, with multiple products, and values gained.

There are lots of details here. We are handling Human-scale brush and branches and poles; the forest-type dictates the arrangements. We are usually using sharp handsaws and broad-knives; we need to work-safe, and watch out for each other. We may be using a small chain saw (usually only the days before the burn), but that is not very social and requires thicker and more encumbering layers of protection: ear cuffs, face mask, heavy chaps, steel toed boots. Chain saws also scare away most wildlife, so we miss any messages from visitors.

Smaller chain saws with carbide cutting-chains can be handled by our shorter people, in thick brush-cutting, with some practice and grace. The brush is often small enough in diameter to cut from below and shoulder the cut-branches backwards, as we take another step, get a good stance, and cut again.

If we can hear each other, and move easily, unencumbered by heavy protection, we can maintain a culture-of-safety and mutual-care. If we do use chain saws, we send in that crew separately, and earlier, to do some large-diameter cutting, some firewood-bucking, and first-cut laying-down of log-mulch. See the guild-sequence story, below. Chain saws are inelegant short-cuts and do not actually save energy, if we look at all the manufacturing, marketing, and mining that goes into their "convenience." Efficiency can be the cloak that hides the loss of multiple-values and careful-attention.

The kit of hand-tools and the crew of guild-cooperators are scaled to the ecosystem we are working in and to the concerns of safety and ergonomic-stamina (can we do this day after day?). As in mountaineering, the minimum safe-crew is four—one injury requires one nurturer to remain and protect and two emissaries to head out for help. A crew of four works well for social reasons too: two pairs of workers, in close coordination. Enough voices to sing good harmonies at the end of the day's work.

Fire management, both for burn piles and broadscale burning, can use a fire-tender, a backup-shovel, someone cutting, along with someone sorting and dragging. The shovel-person could be taking a break and standing by, and all can rotate positions during the process. The steepness of the slope, the entanglement of the brush, the aspect of the slope, and the fuel-conditions, all contribute to the social organization and group-mindfulness. This whole work-discussion informs the social-coordination of multiple-goals.

BRUSH-FIELD CONVERSION FOR MULTIPLE-VALUES

Brush-Fields are often temporary pioneer-plant succession. They want to be overtopped, shaded-out by conifers and hardwoods that they have nursed. Brush-fields were often open grasslands, until they were over-grazed or a burning regime ceased. Brush-fields grow thick in logging clear-cuts and after hot wildfires. There is a complex implication of wildlife and brush-fields. They contain important bedding-grounds, berry-patches, nesting-tangles, cow-no-go pocket-meadows (wildflower and native-bunch-grass *refugia*), and Human-stay-out walls.

The fire-breaks were bulldozer-scraped in 1987 and came back in a wall of brush in twenty years. Where we have cleared, on ridge lines and under remnant savannah, grasslands have quickly moved in. We did not have to apply seed; there are patches of nearby native seed reservoirs. Where we are cool-burning, some brush species have not returned.

> *Disassembling brush-fields by hand is a one-time opportunity. Converting these fuels to charcoal delivers fertility to the farm, fosters atmospheric-care, and disaster-dodge.*

Taking apart a wall of Manzanita and Buckbrush to uncover an overgrown fire-coppiced White Oak-clumps savannah, with remnant native bunchgrasses, proceeds with much attention. First, we consider the list of "60-*plus-considerations* before we cut a tree" (see Chapter 6). Then the logistics of handling and sorting several classes of salvage, doing it all with adequate safety, and the protection of special features (such as bird nests and embedded logs with salamanders).

All disruptive-interventions at Wolf Gulch require ecological-inspection. We do not want to reduce the biodiversity here, which includes many rare species. The alkaline-desert theme of the landscape means that conditions are *brittle* and we should work-smart, so that we build resilience and fecundity, not set eco-succession back.

Manzanita is often tipped over by snow loads; the six-to-ten-inch diameter trunks are almost lying on the ground or are hung up in other trees. Often they are still partially living, sometimes with turned up root balls. The *pit-and-mound topography* from the tipped-up-ones is valuable; we cut off the trunk and leave the stump-pit. We are selecting the still standing stable ones for keepers and preserving the upright-ones with rounded-crowns for their flowers, berries, nesting sites, and wonder.

The tangle of laid-over-brush is taken apart like a puzzle. The dry branches are sorted out for wood stove fuel and kindling. The heavy trunks are cut into charcoal logs and either propped up on stumps or stacked in Oaks to dry. They will weigh much less in a few months. The crown-branches are cut first, leaving the trunks standing bare. The branches, which are not so heavy, can be pulled away, making it easier for the saw to cut the dense, hard, big Manzanita logs. Some standing stumps are left tall for bird perches and late-winter firewood scavenging. Manzanita does not coppice. The stump dies and dries out.

Branch-tips and bark
are concentrated nutrient materials;
saving these is daunting.
It takes a Culture of Place.

The stripped-branches are dragged to the burn-pile by Charcoaliers and Bodgers but watched for special curves or twists that could be used in natural-building and crafts. Logs and branches are stacked up against Pines, Oaks, and retained Manzanita to be sorted-out (tagged). If Sawyers can, they lay choice-cuts out, for easy sorting or dragging. After craft-pieces are separated out, bundled, tagged, and cached, all wood that could be charcoal-logs is cut to the kiln-length preferred. Vesta likes fat, freshly-dried Manzanita, one-to-six inches in (3-15 cm) diameter and two-to-four feet (~1 m) long. The Fire Pig in Hestia requires two-and-one-half foot pieces (<1 m) to fit in the drum-retort. (See below under **Kilns**.)

CHARCOAL FROM THE BURN-PILE

After a day of gleaning, cutting, sorting, propping, dragging, and feeding, we are left with a big pile of glowing coals. Perhaps the other Guild workers have headed back to the central camp-kitchen, but the Charcoaliers are still there at the fire.

Perfect timing
is essential
to successful charcoal.

The last butts, enflamed or smoldering, can be shoveled to one side, to keep a small campfire for light and warmth. The coals are shoveled into the steel-drums. This is a crew-rotation move, as transferring more than two shovelfuls in a row is enough before a personal cool-off. Long-handled shovels are nice. The coals are so hot that even with a long shovel, a broad-brimmed wet-felt-hat, and a wet-wool face-mask, it is necessarily a social-dance, with folks lined up for their turn and others standing back. When the drum is full of coals, it can be shaken down to add more coals. Then, the lid is tapped on with the back of a shovel, over the drum-lip.

Drum-quenched small-wood charcoal from pile-burns has the advantage of not needing to be dried-out or ground-up to be ready for composting. The mixed size classes of this charcoal are appropriate for diverse soil-organisms.

These are dedicated charcoal-quenching-tubs and have no trace of gasket or paint left on them. A locking lid is not necessary, as the oxygen-quench is fast. The coals eat up the little oxygen trapped under the tamped-tight lid and combustion ceases. The drum is very hot and should be set on mineral earth or up on rocks (before filling).

Theoretically, the drum could be carried (on a metal litter, or with a chain-net?) to a lean-to or into a structure, and can heat up the space if it is not outgassing any toxic residues of paint. The charcoal can re-ignite if the hot-lid is pried off; oxygen rushes in, and the coals are still hot enough to combust anew. Usually we wait overnight before opening or moving the drum.

INDUSTRIAL FUEL REDUCTION AND ITS WOES

Most private-land stewards and government-land-managers do different methods of fuel reduction than we do, described above. As the use of petrol-fuels-fed-machinery peaks, many of the industrial-practices will change. When contracted-professional-crews in heavy moon-suits, wielding long-bar chain saws, lay down brush, they may exempt Oaks and big Pines, but the sea of cut-brush left down is often neglected for a year or more, left to dry. The plan is to return and burn the heavy-continuous downed-fuels in force-burned strips, starting from an uphill fuel-break. This hot-burning eliminates many plants and small animal species, especially if the remedial-burn is done without perfect timing. Paperwork gets in the way of bureaucracies.

Alternatively, the fresh-cut brush is dragged and piled, with perhaps a small tarp-patch to protect a pocket of dry fuel for later pile-ignition in winter. A different crew comes back much later and moves from pile to pile, with a drip-torch using bio-diesel-fuel (!) to light a whole slope of bonfires.

With the good-enough timing of dry-enough piles on a wet-enough landscape with little risk of wildfires, a lot of fuel is burned in one day. But not much charcoal is left behind. And several deep-burnt-soil-circles stretch across the slopes. The good news is, if these piles are well-built and tended well-enough, they burn fast and clean. The longer-chain terpenes and carbon-compounds are oxidized down to water and carbon dioxide. Not much brown smoke is seen; these pile-burns do not tend to smolder for long.

That is much better than the classic big, bulldozed, logging slash-piles that can smolder for months and even re-ignite in later winds, spreading into wildfires in the wrong season. Big, slow-burning piles, with lots of dirt mixed in, emit multiple partially burned brown-gasses and pollutants and often even have obnoxious-trash buried in them (by lazy-loggers) that produce dioxins and other dangerous toxins. There are better ways to do industrial-logging without cutting corners just to save a dollar. The business-like directive to "externalize your costs" has decimated our commons of the atmosphere and soil and reduced the quality-of-life on this planet through inappropriate resource-extraction.

TOP-DOWN SMALLER-SCALE SLASH-PILE AND RICK BURNS

If during Transition, we have access to medium-sized trucks and tankers, we have the option to do water-quenched, top-lit pile-burns. This method leaves a wide-range of charcoal sizes and combustion-completion. The large pieces are best broken up (sometimes by driving over them with truck tires) after they have dried enough. The burn-piles need to be accessible to water from long hoses connected to thousand gallon (5,000 l) or larger tanker-trucks. Burn permits are required during wildfire season.

A rick is a cross laid cubic pile of same-sized wood, leaving lots of air between sticks. When the top is lit and the whole rick is in flames, the water is turned on to quench.

One of the oldest traditional ways to make single-species, specialty charcoals is the *rick*. A rick can be built to many sizes as long as there is enough water to quench it at full flame. The charcoal is dried and sorted. Most pieces come out consistently combusted. These ricks are made in a Charcoalier yard, fire-safe, and perhaps using off cuts from Bodger crafts?

Some small-woodland stewards (rural land-holders) have learned about making charcoal while burning slash. Truck hauled water-tanks are required for industrial-logging operations, especially in wildfire season. If the water-tanks can be placed close enough to the slash pile-burn, they can quench the flames with water pumped through big hoses. These burn-piles work best when the slash is stacked loose, without dirt (as in *ricks*), and with plenty of air spaces in the pile.

Knocking down the bonfire with hoses leaves wet, steaming heaps of charcoal that can be crushed later and eventually charged with bio-life to make bio-char. These somewhat randomly stacked slash piles, with their ample air spaces, are similar to the ancient water-quenched rick method of making charcoal. A rick is a cross laid stack of layers of consistent diameter fuels with common dryness that is lit from a top kindling nest. The whole rick, with its open air flow allowing plenty of oxygen penetration, quickly makes a huge, clean burn. Just as all the wood in the rick reaches combustion, the rick is doused with hose water. The charcoal is knocked down by the water and the fire is extinguished. The wet pile of hot charcoal is left to cool and perhaps dry out. Crushing the charcoal comes later.

At the Wolf Gulch pocket desert, we do not have easy access to water. We do not have an extensive road system to get trucks to burn piles. Our slopes are steep and (so far) the brush is thick. We do not want to build lots of erosive roads and have to maintain them. We use trail systems that allow wheelbarrows and walking crews (see elsewhere). Thus we choose oxygen-quenching instead of water. This limits the scale and size of our pile burns. We also drag and feed our fires instead of building large slash piles to be burnt later.

There is another method of making charcoal that does not require any quenching. One can use a double-retort charcoal-kiln (see below). If we can capture and burn the brown-gas phase of wood-pyrolization by cooking the logs in a contained and vented metal drum (the Fire Pig), the brown gasses in the inner chamber are forced through a burner under the drum, and when the burner goes out we have high-grade charcoal left in the drum. There is no need to quench, and the charcoal finishes itself. We wait until the drum has cooled off enough, to avoid re-ignition by opening the drum lid prematurely and letting in oxygen. The still-hot-charcoal will spontaneously re-ignite if we open the drum lid too early.

MEANWHILE BACK AT THE SAVANNAH

The Charcoaliers, sitting at the coal-heap in the fading-light (crepuscular-glooming), are looking around as they sip their brews. The brush field has been thinned, the Oak clusters and scattered Pines revealing the legacy-savannah. The Charcoaliers are intent on multiple opportunities. Charcoal work follows the previous Guilds-work and supports the emerging Social Forestry. The evening crew, warmed by the coals, are reviewing the string-and-web.

What is the sequence of adjustments and interventions that will move the community-of-tenders and Wild-denizens toward broadscale-burning, *water-detention*, and net-carbon-gain? The brush-field likely had ongoing net-carbon-loss, with lots of dead and senescent-vegetation and very little active carbon-fixation per area. Still, full of Life and habitat, just ready-to-explode.

1. The Charcoaliers are collecting limbs and logs for the burn-piles and kilns. This task follows branching and converging nets-of-opportunity.
2. The selected green, heavy, logs and limbs are propped up to dry against Pines (or Oaks) on the sunny-sides. The Bodgers have tagged and bundled goodies in Oak-trunk forks and bark-covered caches.
3. The old, partly decayed, ground-contact logs that did not get moved are *refugia* for soil-dwellers during drought-season and may get stake or flag-marked by the Mothers, or Rangers.
4. Big Pine and Oak snags that have fallen recently or came down during the clearing-work are bucked-up by Sawyers so that they can be oriented just-off-level, Keyline-pattern, to collect carbon-litter and slumped-soil. They become sponges, like the old embedded-logs just above.
5. These log mulch swales resist upslope or downslope ground-fires. And they catch some rolling embers and wood, holding them back from live tree-trunks.
6. The non-embedded logs and fallen-snags, lying oriented up/down slope, are *wicks* for wildfire and need to be cut-up (bucked) and re-oriented. Sometimes this requires digging a shallow kick-trench (to hold them back from rolling) or staking the logs to the slope so they can stabilize (while they settle-in).
7. After the savannah has been rearranged and disturbed by several guilds, the Mothers and Rangers have the moment to scatter seeds and rearrange dragged carbon-litter and soil. This is also done on pattern, net-and-pan, or perched bio-swales made of slash and litter, staked on the slope.

So there we have one sequence that focuses on moving directly to broad-scale-burning. Another sequence moves us toward the kilns.

1. The Charcoaliers are the carbon-workers, they assess and plan to meet multiple goals.
2. After Sawyers, Bodgers, and Mothers have salvaged some opportunities, before and during the cutting, sorting, and dragging of brush-limbs to the burn-pile, the Charcoaliers do their log and limb-gleaning for the kilns.
3. Big heavy logs are propped up to dry. This saves work; move them later after they dry out enough to be lighter to carry. Use south aspects inside Pine drip-lines.
4. Small diameter, short, but still-green limb and branch wood is sorted by species so that specialized charcoal can be facilitated. These bundles are tagged and cached,

or they are light enough to move to the kilns where they can be dried out faster, through more turning.
5. Pruned White Oak branches, not taken by Bodgers, are sorted by size-classes and sent by wheelbarrow to Hestia and the double-retort Fire Pig. They are stack-dried for later pyrolization to make tea-ceremony charcoal. See below.
6. The dried (after one year?) Manzanita logs (5–15 cm, 2–6 in) are moved to Vesta and stacked to dry for that single-retort kiln process. Bigger Manzanita logs are bucked up for firewood by the Sawyers.

During these coordinated-flows, the newly opened savannah is revealed for under-story-burning, and the explosive-load of tangled-fuel has been creatively cut and sorted by multiple-cooperators while singing, drumming, and laughing. The Charcoaliers are thinking about the kilns.

THE KILNS

The old, simple word *kiln* refers to a contained combustion chamber where materials can be heated to high temperatures. There are brick-kilns, pottery-kilns, board-drying kilns, and charcoal-kilns. The idea is to be able to control temperature, drive-off moisture or brown-gasses, and cook clay-and-glazes to pottery. An oven for cooking or drying food generally does not heat up more than 500°F (>200°C). Kilns can go much higher for longer periods of time. Pottery can take all night.

> *Kiln*
> *can be sounded with the n,*
> *or not.*
> *In Dutch,*
> *a kill is also a river.*

A retort means a vessel that holds a chemical-reaction. A single-retort-kiln contains Fire inside the main chamber. The fire is lit and is kept burning through a controllable oxygen-intake-port. The smoke from the gas-outtake flue (the chimney) is watched carefully for color changes, and the *pyrolization* is shut-down by capping the intake-port and the chimney-port.

A double retort is a chamber within a chamber, with the inner chamber heated by combustion in the outer chamber. The smoke from the inner chamber is watched carefully and combustion is quenched in the outer fire when the inner chamber vent smoke turns clear or blue.

SINGLE RETORT KILN "VESTA"

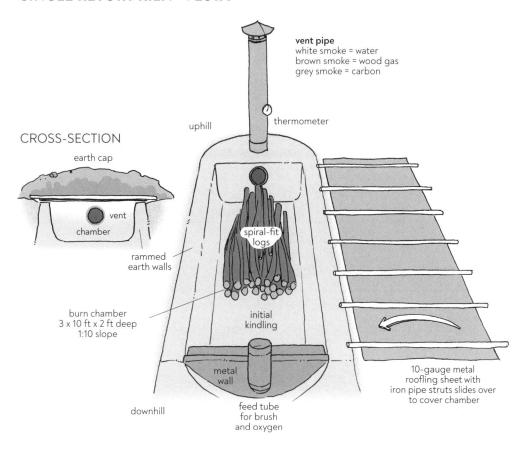

DOUBLE RETORT KILN "HESTIA"

FIRE PIG IN A PIT

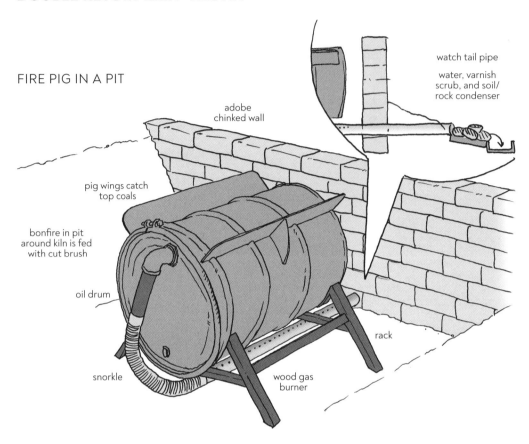

Loaded with 1.5- to 2-inch round sticks.

Bonfire burning around the Fire Pig.

LOCATION, LOCATION

Both Hestia and Vesta are kilns built to use portable-by-wheelbarrow parts. The Fire Pig drum in Hestia comes apart, and the bricks and pipes can be moved to a new location someday. Vesta is a simpler but larger kiln with a big sheet of metal-roof that can be rolled-up and moved, various portable stove-parts, and a down-slope metal wall. These are both small-kilns that are nonetheless moveable as fuel-reduction proceeds. What we will have to do is dig new pits and trenches on the new site, somewhere out on the newly extended trail-system.

Hestia is a one-half cord (64 cubic feet or more than 2 cubic meters) pit, built in an old bulldozed cross-bar drain, left by the 1987 firefighters on the ridge between Wolf Gulch and Little Wolf Gulch. This north-rising ridge was bulldozed in 1987 and has grown back in thick Buckbrush and Manzanita. Twenty years ago, when we first started tending the ranch, we could walk up and down this steep backbone ridge, following wheel-ruts on easy, open, ground that standard fire-prevention wisdom would keep open, as a big fuel-break.

There was brushy Oak savannah here before the big bulldozers scraped the 50 to 100 foot (15 to 30 meters) wide ridge top down to subsoil. The Oaks got dozer-coppiced, and yet, their deep roots re-sprouted. The exposed subsoil gestated a new brush-field that is way more dense than the original.

This presents the challenge of a-lot-of-work, to clear the brush and re-establish the savannah. We have been slowly converting this new-brush to charcoal, while watching the grassland and wildflower recovery. Our 30 centimeter (one foot) tall Buckbrush coppice-stools have mostly failed to re-sprout. So much for the cut-forage/Goat-browse option. Instead, where we have cleared the brush, we now have wildflower-filled native-grasslands and widely paced Oak clumps. And yes, we have begun to winter-burn the grasses. So far so good.

At Hestia, the kiln-site on the ridge, the pit is protected by a wind-wall built of salvaged-bricks and on-site adobe-mortar. In the pit sits the Fire Pig. This is a 55-gallon (280-l) steel-drum, laid on its side in a steel-cradle, to lift it off the floor of Hestia's pit. The drum-lid has a reverse snorkel, coming off the high-middle threaded-hole in the drum-lid and bending down and under the cradle.

The two-inch (five-cm) pipe, under the drum, has forty one-quarter inch drilled-holes on a staggered one-inch grid-pattern, with the holes facing up toward the drum side. This is the vent-pipe for the inner-chamber of the double-retort. This is also the brown-gas-burner that ignites only after enough of the water vapor is driven out of the wood inside the Fire Pig (the inner chamber). The outer chamber is the pit, Hestia-herself, ready to hold Fire.

To pyrolize the Oak logs or limbs inside the drum, we burn Buckbrush and Manzanita brush (from the ridge-clearing-work) laid alongside and on top of the Fire Pig. The Fire Pig drum has two wings, welded as a wide V on the top, to hold some hot-coals from the brush-burn-pile in Hestia. As soon as the brown-gas-burner ignites, we drag some of the hot-coals away from both sides of the Fire Pig and shovel them into the oxygen-quenching-tubs, just as we do with the burn-pile method described above.

This brush-charcoal does not have to be ground up and when cooled off can be bagged and wheelbarrow-delivered to the farm-operation compost-piles. Once charged with bio-life and nutrients this brush charcoal becomes *bio-char* and is added, with the compost, to the food production-fields.

Hardwood-charcoal
makes the best cooking fuel.
Softwood-charcoal
is good for composting.

The charcoal from the inner chamber Fire Pig is exported as artisan-charcoal for special uses, such as barbecue, drawing/marking, and Japanese tea ceremony. All the brush-charcoal is kept on farm to close our carbon-loops. The principle here is: *Export no more than the surplus carbon fixation which can be estimated from photosynthetic carbon fixation per area.* And we do mean estimated. This figure is hard to quantify. We are expecting that keeping all the brush-charcoal on farm will increase our food-production, and thus the carrying-capacity of workers. Carrying capacity is also hard to estimate, but we are trying to do our best, on principle.

29 Loop Law

21 Just Sink Carbon

26 Water Cycle

THE SINGLE RETORT KILN

Hestia is the Greek name of the goddess of the hearth. Vesta is the Roman variant for the same goddess. We use Hestia to name the pit that holds the Fire Pig, and we use Vesta for the slot kiln.

Vesta is a one quarter cord (32 cubic feet, more than one cubic meter) trench, cut into a slope so that it can be covered and vented. The trench is about three feet across, one-and-a-half feet deep (less than a meter by less than a half meter) and less than ten feet long (3 m). The slope is at a ratio of one-in-ten rise. The walls are tamped-adobe that was pounded into wooden forms, and the top is a ten-gauge sheet of galvanized, flat metal with three pipes as ribs on top.

When the trench is filled with wood, the lid is lowered and covered with up to 6 inches (15 centimeters) of dry soil. The cap-soil is a non-flammable insulation and comes from two storage pits on either side of the kiln-trench. The 2-foot (60-cm) down-draft feed-tube is at the down-slope end of the trench and enters the trench through a 6-inch (15-cm) pipe-elbow at the bottom of the downslope end-wall. The chimney is a five-foot (150-cm) length of galvanized stove pipe, set on another six-inch elbow, set into adobe at the up-slope lip of the kiln-trench.

The chimney and the downdraft feed-tube can be pulled off the elbows (with heavy gloves and pads) when it is time to shut down the kiln and cut-off oxygen to the interior-combustion. There are two stove-pipe caps, of the sort that is used to close an unused input-hole in a brick-chimney. When the *pyrolization* is complete and the stove pipes are pulled off, the elbows are capped and set tight with a mallet to prevent smoke-escape. Then extra cap-soil is added where there are smoke-escapes and tamped down lightly to seal up Vesta all-the-way.

This combustion-system is a hybrid of a downdraft rocket-stove and a single-retort kiln-box. The scale is appropriate for two workers, or one very-competent operator. The lid can be lifted off (after the dirt cap is removed), once the kiln cools off after about a day. We must be sure the fire has gone out before we dare to unseal. We could lose a lot of charcoal to a small oxygen input.

Vesta burn chamber loaded with spiral-fit logs and initial kindling.

CHARCOAL CAN BE BAD

Most portable charcoal kilns are scaled larger and take up to two days to finish driving off the water and brown-smoke. There are many single-retort kiln styles. The simplest, and perhaps oldest, is a *rick* covered with a soil-cap, with a hole in the top of the soil-cap and several holes (4-8?) in the soil cap along the base. The kiln can be lit from the lower ports, all at once, and the smoke from the top-port is watched for the colors. When done, someone has to get soil up on the pile to close the top-vent (a death-defying move on a big-pile).

The industrial scale extraction-system, which uses a bulldozer, pushes whole Mesquite shrubs (root ball and all—no coppicing then) into a pit and then covers them with a soil cap. I guess they pre-locate input and vent pipes. The pit is dug out to harvest the charcoal.

The Mediterranean basin was clear cut for charcoal for bronze metallurgy thousands of years ago, leading to civilization-collapse, from the ecological fall-out. New England and New York colonies were cut-over several times in the 17th and 18th centuries for charcoal-production. When fossil-fuel extraction of coal and oil industrialized, charcoal production lessened. We could imagine a resurgence of greedy fuel-taking during some phases of Transition. Charcoal needs to be made carefully, following principles, to be worthy of respect. Only wildfire-hazard materials should be used as we return Fire and Salmon to the landscape.

*Charcoal
is a valuable fuel,
as it is light to transport,
holds heat value,
and burns clean,
with less indoor pollution.*

Vesta is scaled to the fuel-flow that is a product of the pace of our savannah-restoration. We are not trying to produce commodity-charcoal for export. Charcoal has many uses and values, but it can be just as ecologically destructive as any of the petrol-fuel industrial-consumptions that have filled the atmosphere with greenhouse gases and reduced the water-holding capacity and net-carbon-sequestration in our forests and soils. We can justify some charcoal production in the mix of multiple-outputs that follow the work of tending-the-Wild.

We cannot make charcoal our single-product goal. Multiple-forest-products are best used on-site to support the work of ecological-restoration. Charcoal as just-another-fuel is unethical to market as a commodity (during Transition), with low wages for workers. All our handling-and-sorting is proper to Social Forestry and inefficient for mass-production of market-commodities. We all need to replace global-consumerism with modest-local-sufficiency, while we Transition to whole-landscape-resilience and the return of Salmon and Fire in Ecotopia.

THE CONTEXT AND TECHNIQUE OF OUR KILNS

Vesta, the single-retort kiln, is not as tidy and efficient as Hestia, a double-retort. See above. This single-retort does allow more capacity and simple construction. The best net-carbon-capture, in the form of finished-charcoal, is seldom more than half the carbon of the feedstock-wood. Hestia offers a cleaner burn, both by after-burning the brown-gasses from the inner-chamber and by drag-and-feed burning the outer pile-burn, so that very little brown-smoke is released by either combustion or pyrolysis.

The big oxygen-flows and loose-structure of the burn-pile method described above assure complete-combustion with good-carbon-capture (see elsewhere). Vesta captures as much net-carbon as charcoal but releases a lot more brown-gas to the atmosphere. Vesta allows the processing of long-log-fuel, which needs less cross-cutting and is heavier to move long distances. Once an area has been thinned and cleared, enough to return to traditional cool-burns, the kiln-site will be moved closer to the next brush-field challenge.

Hestia's pit will stay on the ridge for a long time, as we have so much fuel that can be delivered down the ridge to the kiln-site. Vesta will need to be rebuilt elsewhere sooner. Vesta is set up near the cabin in Little Wolf Gulch and is accessible by wheelbarrow via the west-wing of the Flat-trail from (and to) Hestia, and also from a set of multi-purpose Keyline-trails, stacked up the slope, through the savannah-restoration area on west aspect slopes.

The bulldozer scraped fuel-break, built during the 1987 Cantrell Gulch fire, cuts directly uphill from the bottom of Little Wolf Gulch and down from the BLM access-road (that runs up the west side of Little Wolf Gulch). North of this now-grassed fuel-break, you will find Oak/Pine savannah that under-burned downhill near the end of the 1987 fire. This pattern allows observation of a wildfire-fuel-reduction event. We have been thinning the savannah south of this old fuel-break with crews and tools to mimic the wildfire effect.

Let us not forget that the First Nations broadscale burning practices, developed over thousands of years of experience on this landscape, created a forest structure that was way more open than what we are dealing with today. We have an emergency-situation and a lot of work-to-do before we can re-introduce extensive broadscale-burns. Our work, at first, is a lot of thinning, clearing, and carbon capture through charcoal and log-mulch. We are using chain saws, while we have them, and we are skimming the first material flows that are associated with this initial fuel-reduction process.

Eventually we will have a different set of tools, harvests, and opportunities. We are just beginning to try to recover the resilience and catastrophic-wildfire-mitigation that the First Nations facilitated and the colonists squandered, as they did not know what they were looking at—and didn't ask. The best and most effective way to intervene in complex-systems is with a change of mind-set, or paradigm (Donella Meadows, 2008). We are trying to wrap-our-minds, around-the-corner, towards an Ultimate-restoration-strategy.

A BURN DAY AT VESTA

We did the permit process last year, when we were late getting ready for the big day. It was late spring; they actually showed up to see and GPS the operation. The permit process with the state department of forestry was good to go through. The nice young men that came out were supportive and glad that we were applying for the permit. They said most folks do not bother and go ahead, inappropriately. The inspectors approved that we had a 500 gallon water tank on the hill above the kiln and hoses laid. They were glad to see our backpack water-pump and our tool set of shovels, Pulaskis, and Mcleods, but advised us to move the tool package closer to Vesta. We had the tools parked up on the fuel-break/trail, above the kiln.

This time it was a mid-June morning. Fire season was about to start. The do-not-burn signs had not been put up on the county roads yet. We could tell by the changing weather that this was going to be our last chance to burn safely, and without jumping through all the permit hoops.

The advised safe-window was early in the morning. It was a Tuesday morning and even though only one worker was on-site when we needed two (at least), the fire was lit by 6 a.m..

The safe-window closes at 10 a.m. Although not quite light enough at 5 a.m., tea was made in the cabin and thermos, radio, and treats-bucket were assembled. 6 a.m.—the match was tossed into the downdraft tube. It took a few matches to get the Manzanita twigs to ignite.

Fanning with folded paper got the smoke to make-way through the twisted pile of logs inside to wisp out the chimney. Then, with updraft pull, the fire and smoke were sucked down the feed tube to start the burn. It was a half-hour of feeding small wood into the downdraft tube before the chimney-smoke got strong and we got a tea break.

The early morning light, the cool shade of the Oaks on this west-aspect slope, and the birdsongs in the savannah were lovely. At first, there was not a lot of smoke and all of it was white: mostly water-vapor driven out of the wood by the heat of the downdraft-tube feed.

The Manzanita logs, fitted like a puzzle, a twisted-bundle, were cut in mid-February. We had rick-stacked the small-logs under a big Jeffery Pine, big pieces leaned up against the southwest-aspect tree trunk.

Dried five months, yellow twisted cracks
of the shrinking wood
peeked through red bark.

This charcoal-batch is for the local on-farm-store. Barbecue season and the neighbors are eagerly awaiting the new batch of Manzanita charcoal. Our charcoal has an excellent reputation. It is known as the best charcoal available, and folks like to know that it was made righteously and with love. The price we ask for artisan-charcoal does not cover the total work. Much of the value is in fuel-reduction, intern-opportunities, ecological-resilience, joy of lovely mornings, and the fruition of labor. That we can use the kiln-product in trade, and receive appreciation, is the icing on the cake.

We had to trap the Ground Squirrels, who tunneled the walls. The Squirrel-quarried subsoil, filling one end of the trench, is useful for the dirt-cap that insulates the burn chamber. We mixed up mud-and-sand to patch the holes. We had to mix mud anyways, to patch any smoke-leaks, when we shut down the combustion.

The wood-in-the-trench-pattern allows plenty-air-space; the smoke and heat move through the pile. We lowered the metal lid onto the lip of the trench walls and *chinked* the heat-warped-lid edge with small rocks. This was Vesta's twelfth firing, the ten-gauge metal, black underneath, discolored and rusted on top, warped and sagged between the three top-ribs. The lid fit tightly enough by chinking, side piling the soil cap on the lid edges, and tamping-down to prevent smoke-escape. After the edges were secured, we spread dry-soil from the side-pit-hampers onto the top of the metal-sheet-lid. Our goal was to have four-to-six inches of insulation. We learned to hold our palms over the insulating-cap-soil to find hot-spots and add more soil to even out the insulation.

WAIT FOR PERFECT TIMING

The week before, we laid a plastic-sheet on the soil-cap to keep things dry and allow sun to dry-out the soil-cap. We placed the chimney-pipe and downdraft-tube on their embedded-elbows and gathered up Manzanita branch and twig-fuel to use in the downdraft-tube. With the backpack-pump full, the wildfire-tools arrayed, and the fine-fuels (grasses, twigs and dried-up wildflowers) scraped off—all around Vesta (ten feet, three meters, as a fuel-break)—we were prepared.

This cool morning, only the Charcoalier-elder, tea-in-hand, standing at the feed-tube to get Vesta hot, hot, hot. On the chimney, just above the dirt-cap, the fire-tender can see a magnetic flue-thermometer, as used on wood stove flue-pipes. It shows up to 900 degrees Fahrenheit (450 C), with a burn-clean zone in the middle, and creosote-warnings on the cool and hot side for a clean, less-polluting, and less chimney-clogging burn.

One advantage that Vesta has is *ESI* (essential-scientific-instrumentation). Hestia's steel drum is in a bonfire, and it would be hard to see the dial. Besides, the magnet loses its magnetism when the flu gets above 900 degrees F. On Vesta's flu, we have a loop-of-wire around the pipe, to hold the thermometer. The color of the exhaust-smoke is the key ESI, but the thermometer adds information and wonder. *Pyrolization* goes through a series of chemical-reactions and temperature migrations (fluctuations), pushed by the progressing heat.

The first-smoke, from the chimney (the tail pipe on Hestia), is white. Water-vapor is driven out of the wood. Even well-seasoned stove-wood has a moisture-content. Milled-boards are cured in a shed, often called a drying-kiln, with small-sticks between layers of stacked up boards, so that sap and water can evaporate over-time and boards stabilize, flat and straight. A charcoal-kiln drives all water out of the wood. No matter how dry our feedstock, white smoke for a while.

Vesta's flu-thermometer sticks-a-while at 200°F (100°C) to evaporate all the moisture. As long as water can still evaporate, any heat in the kiln is absorbed in driving it out. On this fine June morning, with freshly-dried Manzanita, even with a variety of diameter-sizes, this takes a while. The mass-of-logs heats up slow, with the downdraft-tube stuffed with long Manzanita branches. The kiln was lit 6 a.m., 6:50 the flue reaches 150°F. By 7:50, 200°F.

Water boils at 212°F at sea level (100°C), and we are at 2,400 feet elevation above sea level. Water boils at a lower temperature at higher elevations.

At 8:05 the temperature starts to jump up, first to 350°F, and then at 8:10 to 400°F. The smoke is still white, but *pyrolysis* is distilling-off the first rush of carbon-molecules. When the internal gasses start to ignite, we stop feeding the downdraft-tube. Then it slowly climbs to 500°F by 8:35, 550°F by 8:50, and 600°F by 9:10; the brown smoke becomes obvious at 9:35, and 650°F.

When a wildfire crowns in a forested-canyon and the fuel-moisture is low, in an extreme-fire-season, the trees do not burn, they *pyrolize*; they out-gas their hydrocarbons and terpenes and the air ignites. This is called a *blow-up* and is very dangerous for wildfire-suppression-crews; oxygen below the burning air is quickly used up, and asphyxiation is probable for any animals in the *understory*.

Similarly, when the heat "boils" wood-gasses out of the wood, inside the single-retort kiln, the space between the logs no longer has any water-vapor, and the gasses ignite inside the kiln. Then the brown-smoke becomes visible. In the slanting-early-sunlight, the brown-smoke is visible in the shadow cast by the billowing flue-gasses. The shadow of the smoke on the ground turns red-brown.

Inside the kiln, hot-and-vigorous, there is not enough oxygen through the downdraft-tube to burn all the long-chain-carbon compounds, so brown-smoke is a mix of partial and complete combustion. Burn-pile smoke is mostly clear, or blue, or white with water-vapor, when we lay on fresh-cut green-brush. The drag-and-feed burn-pile has lots of airflow and makes for a lot of water-vapor and carbon-dioxide, not so much brown-smoke.

BROWN-GAS AS FUEL

An advanced single-retort kiln-complex would capture the brown-smoke and use it to heat water with a direct-smoke burner (as is under the Fire Pig), or would filter the brown-smoke and feed an internal-combustion-engine or steam boiler. During war-time, the Britons made charcoal in the back of their cars and ran a pipe up to their carburetor to fuel the engine, as they did not have access to petroleum for gasoline.

Brown-gasses need to be filtered for particles and long-chain-carbon compounds, such as turpines. This can be done through a water-bath: the smoke is bubbled through (leaving a floating oil sheen), then up through a compost or horse-manure screen-tray (to scrub more carbon), and then what hydrogen and carbon-monoxide is available. This is the actual fuel that runs the engine.

Carbon-monoxide is toxic and so are long-chain terpenes. One does not want to breathe brown-smoke. Unfiltered brown-smoke is a serious atmospheric-pollution. Running Vesta is not the best for the air. Thus we prefer the Fire Pig or the well-oxygenated open burn-pile. What Vesta gets us is a lot of charcoal for a short period of pollution, with less complication than scrubbing and after-burning the brown-smoke.

THE HIDDEN BURN GOES ON

By 9:45 a.m that fine June morning, Vesta has reached a flue temperature of 700 °F. Now things get interesting, as the temperature starts fluctuating. The temperature goes up and down between 700 and 750°F until 9:55. One might suspect that we are seeing evidence of a complex-combustion-chaos, with a pattern that we can sense through our ESI. By 10:00 the temperature starts to climb above 800°F, and the brown-smoke is thinning out.

Vesta has driven-off all the brown-gasses; the almost-pure residual carbon starts burning. At that point, the charcoal—the skeleton of the wood we put in the chamber—starts burning, and the smoke clears or has a faint blue tinge. Time to shut down the combustion as we are now losing our charcoal. The smoke from the flue stack is now a lot less polluting but there is no need to cook the wood any further. Our charcoal is at its most smoke-free and highest fuel value.

The early down-slope airflows are replaced by up-gulch canyon-effect breezes and smoke-drifting-around-clouds turns to a directed plume.

By 10:30 a.m., quick-moves capped the pipe elbows and sealed-off any smoke-leaks. First, the downdraft-tube is pulled off the feed-end of the kiln and laid-aside on bare-soil. The cap is fit onto the elbow. Then, with very heavy leather-gloves, holding another pair of gloves as hot pads, the chimney is snatched off its elbow and laid safely on bare-ground. Hot gases are pouring out of the vent-elbow. Now, we have to cap the top-elbow, without scorching our face or clothes. This took a few tries. At last it slipped into place, after some clearing of dirt around the crimped-top of the elbow. Whew! That was a hard move.

After the intake and exhaust elbows have been capped, smoke-leaks become evident around the edges of the metal lid, through the dirt cap. A shovel of dirt here and there, tamped gently-tight with the back of the shovel, suppresses those leaks. The inside of the chamber is quickly using up any oxygen still-in-there and combustion is slowing, but any smoke-leaks suggest that air may also be sneaking-in.

The downhill front wall of Vesta is a heavy frame of metal, fit into the rammed-earth walls with rocks and adobe-plaster. This is where any further leaks can now appear. With a trowel, a water-sprayer bottle, and a heavy pair of rubber-gloves, the stand-by adobe from the bucket is slathered onto the front wall of the kiln and poked into the leak zones. At last, after quite a bit of hustling, no more smoke is escaping. The kiln is sealed.

Time to sit down, drink any tea still in the thermos, eat some snacks, and hang out a bit, to see if any new leaks of smoke show up. After a quarter-hour, down to the cabin for a break, more tea, and treats. All the heavy-boots, and pants, and shirts, come-off. The day is warming up fast and the work is almost done. A few more trips out to check Vesta for leaks, and by mid-afternoon, time comes for a nap.

SECRETS REVEALED

Two days later, a couple of neighbors come by to open up Vesta and see what we got. The dirt-cap is shoveled off the metal-sheet lid and into the side-pits, the small stones are pulled back from the edge of the lid, and the lid is carefully lifted to prevent any soil from falling into the charcoal. There it is! The logs have been reduced to a pile of charcoal on the bottom of the trench. The feed-tube end of the load is mostly ash. We were careful to not load good charcoal-logs too close to the feed-tube. Small-brush had been put there and it is all gone to ash.

The rest of the trench is holding a mass of rainbow-colored Manzanita charcoal in a wide range of chunk sizes. One can still see the shape of the biggest logs, but they have slumped and cracked so that when we gingerly pick them up they fall apart. Now we bag up the charcoal with our lightweight gloves, oohing and ahhing. This time we got over 40 kilos of finished high-grade barbecue charcoal and a small bag of crushed charcoal mixed with some adobe shreds that we can put into the bio-char compost.

One-half feed-bag of charcoal stays at the forestry-camp for outdoor-cooking in our rocket stoves. The other six bags are wheel-barrowed a half-mile (1 km), on the Flat Trail to the main ranch drive, and down to the parking-lot and Boat Barn for processing. This is where the bigger chunks are broken up or saved for a blacksmith's-forge, and the rest is weighed, bagged, and labeled. If we charge $15 for two-kilo bags (about 5 pounds) and we end up with 18 bags, the retail take-home will be $270.00. We are not going to get rich selling our charcoal without a lot more forestry-work and infrastructure-improvements, but we have some happy-neighbors and we export a high-grade, value-added product, that may be within our net-carbon-budget.

Most of the charcoal made at Wolf Gulch Ranch is from pile-burning brush and all of that goes to bio-char for sequestration in our fields. That carbon is staying on-site for a long, long time. It is very stable. The barbecue-charcoal burns very clean, without smoke, and gives off water and carbon dioxide for its clean exhaust. Still, greenhouse gasses—but not the worst. We are increasing the carrying capacity of our farmlands and reducing the risk of catastrophic Fire. Charcoal rocks!

The principles of Gifts and Yields include

41.
The gift goes around and comes back. Opportunities are offered us that benefit from discretion, so that we are gifted-again. Reciprocation is expected and appreciated. We get what we need, without fear or worry, when we do not take too much. Host rituals of gratitude at every effort.

42.
Skimming is accepting the gifts of Nature that we notice as part of our tending. When there are surpluses, and All Sentient Beings are content, we receive opportunities. By converting only by-products of eco-restoration, to our cultural-resilience and participant simple-comfort, we bless the gifts with use and treasure the goods with care and maintenance. Skills are our contribution to transformative-work (offerings)—we carry the lump of clay to the wheel, and make a beautiful pot.

43.
The Ultimate gifts of Nature, to her tenders, are Beauty, Serenity, and Wisdom. We receive these through forest-bathing, cooperative-work, and cultural-continuity. Our cultural-reciprocation is art through ornamentation, thrivance through Potlatch, and wondrous-appreciation through drumming, dance, and song.

CHAPTER 9
Treasures from Thickets

JUST ENOUGH ABUNDANCE

*The yields
of a complex ecosystem
are endless.
We cannot imagine
many of the gifts that flow.*

There are values produced in the forests that we will never notice, as they are beyond our ken. And there are values that we do notice, and they are the foundation of our material and spiritual comfort. A purposeful simplicity that shines with beauty. The gifts from the forest allow us to discover how we live, where we are. We build our pride and longing in Place through our vernacular art and skills. First learn the where and what of a drainage basin, then sit in council with All Sentient Beings and envision a way-of-being.

*Potlatch
is the Pacific Northwest coastal
abundance-tradition
of giveaways;
wisdom-status comes
from moving-gifts-on.*

The basic skills and materials, used in handcrafts, are common to the mid-latitude northern hemisphere of our only Earth. Please see the Bibliography for the many sources of hand-skills with natural-materials. Then review the poster "Prescription," to see an abbreviated inventory of materials, skimmed in restoration-forestry work. This is a local, site-specific poster. Other drainages will have different materials to work with.

Prescription

Let's spend some time exploring baskets, fibers, rod-mats, lashed-poles, herbal-medicine, high-density-nutrition, compost, brush, and wood. We discussed charcoal, in Chapter 8. Some fuel-load materials will be surplus during Transition, as we bring back tending. Then, as we learn to roll with the cycles, we will have seasons of surpluses. Storage and maintenance of our goods will bridge gaps in abundance. Feast and celebration will celebrate times-of-plenty.

FIBER BASICS

Materials themselves have homes. We can speak of our fiber-basin, or morel-shed, and think of ways to live and be. The fiber-basin of the Little Applegate is mostly on the farms: Hemp from fields, wool from Sheep pastures, and we could grow Flax. There are some other fibers from crop residues we could process.

The other introduced-fibers nearby the farms are Blackberry vines and milkweed, and perhaps *Vinca* (Dogbane family) and English Ivy. The handling of these materials is well known, and cloth, string, and rope can be made by hand, with carders, spindles, spinners, knitting needles, crochet hooks, and looms. A very handy settler-technology is the foot-powered-treadle sewing-machine. There are ways to put wool-spinning-heads on foot-power.

Up-basin and in the Canyon eco-mosaic there are many sources of fibers and vines. These can require local-knowledge to make good string, twine, and rope. To make the best native-plant fiber-crafts, one would use Nettles, but we do not (so far) have surpluses. Hemp is done the same way: the long, selected-stems (after processing for other yields) are soaked in a long-trench or trough. The stem tissues rot away from the fibers in the water. This is called *retting*. The by-product bio-brew is odiferous (very stinky) with outgassing nitrogen and sulfur compounds (fast composting). Capturing this nutrition requires large amounts of empty carbon-materials (such as charcoal).

Goats for hair are possible. We have an eco-spasm of *Torrilis arvensis*, which we call mat-burr for the seeds' awesome clinging-architecture. Fills up socks and wooly-animals. This too may pass, but we now have a hard time cleaning the wool of woodland-browsers. Perhaps a hairy rather than wooly-fiber is appropriate? And there is Human hair?

FIBER PLANTS

Top to bottom: A. narrow-leaf milkweed; B. nettle; C. dogbane minor; D. Himalayan blackberry; E. vinca.

Rope is very useful for moving logs, tying up animals, lashing poles to make structures, tying off loads, and raising walls. Knots are key to using ropes and learning knots and rope work is a great social activity. Winter by the Fire. We use the clove-hitch, the figure-eight, the tautline-hitch, the square-knot, and many others. Sailing a small-boat is knot-thick. Putting up a pole-tripod tipi-cone, hanging the liner, pegging out the cover, and controlling the smoke-flaps will keep you good with rope. A lot of fancy rope-work (twirling, lassoing, tying-off) comes from hauling and herding: Horses, Oxen, Cows—wagons, harness, skidding, and corralling.

Keeping ropes handy and intact is also a hearth-craft. Splicing is used to tidy up a frayed rope-end, or weave a rope end back into the rope (to form a loop), or fix two ropes together with a smooth conjunction (to extend a pulley-rope). Whipping is done with a thread or string, to hold the strands of a fraying rope tight so that there is no more fraying and the rope fits easily through loop, knots, and tackle.

BASKET BASICS

Baskets are essential to our life and work. They can do many jobs: carry, store, winnow, sieve, cradle, dry, and even cook (clay and pitch sealed—stone boiling). Global trade degraded local basketry arts by substituting cheap deco-baskets with no obvious specific use. The baskets we have found most useful in Social Forestry are burden baskets (baskets in a harness, worn as a backpack), gathering baskets, storage hampers, and small gift baskets.

Advanced Cultures of Place produce beautiful, tightly-woven, pattern-decorated gift-and-show heirlooms. What we have learned to make are *rough-baskets*. For those, we have plenty salvage materials from fuel-reduction work. To get the fancy-grade materials, we have much to learn.

As some materials for baskets come from upland harvest, the materials are stripped of bark (medicinal yield), and dried for bundling and packing down to winter-camp sheds. Willow and some other withies are split three ways with teeth and skilled hands. Hazel and Maple are vulnerable to Borer Beetle larvae who eat out the cambium under the bark and make way for fungi and subsequent Beetle mining. Each species has a different Borer.

If we do not de-bark the larger withies (> ¼ inch, 7 mm), they will leak sawdust wherever we place the finished basket, weeks later, and weaken the structure. Hazel bark in midwinter,

just at bloom, can be stripped by hand and pocket-knife (or scraping-rock), and after the leaves appear a month later, the withies can be heated over coals to ease stripping bark.

There are such details for each different withie species. The end result of lots of handling knowledge is better baskets, more lovely and lasting. Basketry withies can be facilitated by burning (clump-base burns or patches), or by *coppicing*.

COPPICE FORESTRY AND CRAFTS

> *A copse is a thicket of shrubs or stump-sprouts.*

Coupe in French means a blow or a cut. To *coppice* (or *pollard*) a woody-shrub or tree, we cut the main stem(s) back to the ground, or back to the central-stems, in order to force (stimulate) new growth. This is usually done when the plant is dormant, during winter. Spring sap-flow brings nutrients up from storage in the roots, and the cut-back branches, or stump, grows new buds under the bark that then sprout through the bark to vigorous lengths.

These long skinny-twigs are good materials for a wide range of Human artifacts. They are called *withies*, and bundles of withies are called *faggots* (French), or *fascines* (Italian), or *brindles* (German).

The stump is called a *stool*. As the new branch-buds are *epi-cormic*—forming under the bark after the upper branches have been cut-off—the stumps and remaining stems should be tall enough for new sprouts to form. On Oaks, one foot (30 cm) will do. For Hazel, working multiple-stemmed stools, better to leave twice that. Pollarded trunks sprout mostly just below the elevated cuts.

10 Social Forestry Terms

12 Hedgerows

Many other simple-words are associated with the multiple-product type of forestry and land-management, often called Standards and Coppice Forestry. The *standards* are the over-story or timber-trees, left-standing for their good-form and genetics, or forest-structure reasons. Traditional British Standards and Coppice Forestry maintains a patchwork of spread-out tall trees, letting sun into the understory, where fresh cut coppice allows grasses to grow while the new sprouts get started. Grazing animals could be moved through until the grass is shaded out. These Royal Forest Lands allowed prescribed forestry activities with lots of specific job-descriptions (Maitland 2012).

A log
is too heavy to move easily.
A post
is a log-upright,
holding a beam-log
and sitting on a sill-log.
A pole
is a long, straight, small-diameter log,
for bracing, rafters, and tents.
A staff
is a strong and Human-scaled small-pole.
A rod
is a stiff straight branch for mats and walls.
A withie
is a flexible long-twig for baskets.

When the withies are thinned on the stool for baskets, the remaining sprouts are thinned further, by the Bodger Guild, to stimulate new fine-withies and to select fast growing *hurdle* rods to weave into the vertical staves (*shores*).

*A hurdle
is a section of woven small-wood.
Can be fencing or wall-panel.*

The finest traditional baskets are made with select plant-materials, mostly harvested from carefully burned brush-patches. One can *coppice* with cutting, burning, or browsing (as with Goats). Local-practice and careful observation are essential to achieving the best crafts-coppice. Before there were plastics and metals, we had pottery and baskets. Many parts of a small farm were made from coppiced small wood.

The animal fences, called *hurdles*, are woven onto *shores* (or *sails*) which are stout, straight *staves*, with *withies* or *long-rods*, twined or woven into panels (*hurdles*) that can be carried by one person and tied-together in rings to corral animals. These hurdles were also used to panel buildings in-between posts and beams (from the *standards* and *logs*) and then plastered with adobe or clay/lime to make a *wattle-and-daub* wall.

Bodger Guilds are the crafts-folks, who turn small *poles* and *staves* into stools, benches, chairs, and furniture. The poetic-terminology of coppice-craft goes on and on. Luckily we have some good resources from the English Isles and from early American wood craft (John Seymour, 2001, and Eric Sloane, 1973, among others). We at Wolf Gulch Ranch are envisioning a Social Forestry that reconnects Human-culture to the forest and brush-lands, so that ecologically necessary tasks and a vision of lovely-local cottage-culture can co-evolve to enhance a vibrant, persistent, and appreciated desert-gulch eco-mosaic.

ROD MATS

Rods are too small to lash, too big to bend for baskets, and just right for weaving. They can be used in *hurdles, wattle-walls, back-rests* and *rod-mats*. We sort materials in fuel-reduction and when we have enough rods we set up a *field-loom*. We often have Hazel, Ash, Maple, Mock-Orange, and Service-Berry rods that can be stripped of bark (to slow down those Beetles), sorted to be all-one-length for rod-mats, or into tapering-lengths for back-rests. A *back-rest* hangs on a *tripod* (see "Tripods," below), while a rod-mat makes a great sleeping-pad or can be used for awnings and privacy. Both can be rolled up and carried, or put up in rafters after use: nomadic camp-furniture.

The *field-loom* consists of four stout-stakes (from *staves*) driven firmly into the ground in a rectangle, one-half meter by three meters (two by ten feet). A *heddle* is the horizontal loose-stick—about a meter long—that has the two longest twines (wire, vine, or small-rope) tied one-half meter apart on the heddle, over to the front two stakes, and about a half-meter up in the air. Another two twines are tied loosely from the front two stakes to the back two (½ m up). By lifting the heddle, one worker opens a four-twine jaw, where another

FIELD LOOM FOR ROD MATS AND BACK RESTS

LOOM BOX SETUP

— four stout stave frame posts driven slanting out and away from center
— two frame warp strings
— adjustable clove hitches

BUILD TO SIZE OF PROJECT

WEAVING

— string heddle warp long, adjust with clove tails on heddle-stave
— start right side — lift for second rod
— shift heddle left and down for third — up left fourth — right down fifth — and so on
— nudge rods up warp and pull out slack
— tie off with clove hitches, exit loom
— adjust as rods dry and bundle to keep straight

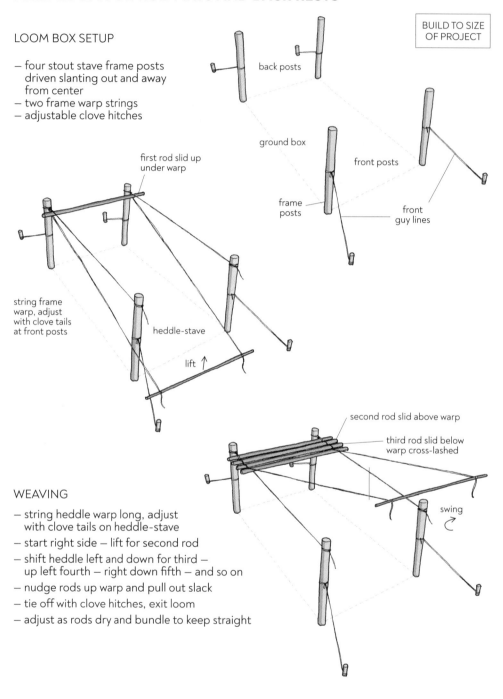

TREASURES FROM THICKETS 257

worker puts in the first rod. The heddle then drops to the right-side of the back-stakes and traps the first rod.

Next, a second rod (the weft) is placed under the warp-twines, and the heddle comes up high. A third rod is now placed over the warp-twines, and the heddle comes down to the left of the two back-stakes. This way, the loom captures the rods and ties them in, by going down to the right and then up and down to the left, and up-and-down, side-to-side. The rods are tapped forward with every new weft-insert to cinch up the mat. The warp-strings and heddle-strings get tighter as the rod-mat uses up the slack. Maybe leave long-tails past the clove-hitches on the heddle and back stakes? So that more warp can be fed to the fat weft-rods? The end result is a carefully tied-off-of-the-loom rod-mat, with back-woven twine-tails and lots of clove-hitches.

A rod-mat foundation, with only a bit of bedding, is a surprisingly comfy bed.

MEANWHILE, BACK ON THE RANCH...

Our Little Applegate Canyon was the home of the Dakubetede tribe, Athabaskan speakers, and famed for their California Hazel (*Corylus californica*) baskets. These relatively recent Indigenous arrivals, still having been here thousands of years, are part of a migration through the Yukon territory, and include Apache, Navajo, and other language relatives with names that include Dine or Dede or Tede. In spite of one hundred and seventy years of imperial-extraction and genocide, we can still find the shade-growing stools of Hazel in all our gulch bottoms. Several other local *ethno-botanical* species still persist and suggest extensive Horticultural use of Fire and disturbance-regimes that maintained a vibrant and productive landscape. The local Salmon runs were very large, as reported by the first colonists. Thousands of Beaver (for their pelts) were trapped out of the valley and then gold was discovered—now, there are few Dakubetede survivors. The landscape was cut up into farms, ranches for cattle, and sheep scoured grasslands. The timber was cut from the ridges and valleys, where roads could be built.

The work in Little Wolf Gulch, thinning riparian-corridor Doug Fir mid-crown overstock to improve wildfire-readiness and rescue remnant old-growth overstory, delivered *poles* and *logs* to the cabin. The work of reducing fuel-hazard, by *coppicing* the wildfire-coppice Buckbrush, delivered the best *brindles* to fill the cross-diagonal-cribbing (see Chapter 10). Whippy Douglas Fir pole cribbing allowed us to plaster the walls, inside and out, with on-site adobe. Thus we used regenerative forestry-products and on-site materials.

TREASURES FROM THICKETS

REVIEW OF RANCH COPPICE EXPERIENCE

The woody-species that we are working with are listed below, with cultural notes:

Symphoricarpus mollis (Snowberry) is a Honeysuckle family shrub that is very-flexible and grows in straight-rod clumps and thickets. We are experimenting with one-quarter inch rods for basket hoops. Burned clumps or coppiced thickets rebound from the roots. We have built brush-walls and hurdles with Snowberry brindles.

COPPICE SHRUBS

Clockwise from upper left: A. snowberry; B. silk tassel; C. buckbrush; D. deerbrush; E. hazel; F. spoon willow; G. red willow; H. grey dogwood; I. big-leaf maple; J. Oregon ash; K. mountain mahogany.

Garrya fremontii (Silk Tassel) is a shrub or small tree (in its own family), with evergreen waxy-leaves. The red, stout stems from wildfire-coppice make strong hoops. Silk Tassel is not common here, but patches up near the old-mining-ditch were burned by the BLM in 2007 and then cut back by trail crews in 2010. We were able to salvage some withies from the trail-work and they show much promise for baskets and rod mats. The bark did not need to be removed. We did not find the borer holes and dust we see with Hazelnut.

Ceanothus cuneatus (Buckbrush) is in the Buckthorn family, a nitrogen-fixing-shrub with small evergreen leaves. This is the common shrub of our Chaparral. The wildfire and cut-coppice regrowth have very-flexible branches, but with spiky spur-twigs. Although it is listed ethno-botanically as a basket-material, we have yet to learn how to strip it for withies. We have used it for brindles in our natural building walls and for erosion-control, brush-mulch (to protect new plantings), and charcoal. Old growth Buckbrush is sometimes a small tree, and is very flammable. This is our concern-species for catastrophic-fire. The fuel-breaks that were bulldozed in 1987 are almost continuous Buckbrush. The root is bright red and is used as Redroot in herbal-medicine.

Ceanothus integerrumus (Deerbrush) is much easier to handle. Although also a nitrogen-fixer, it does not have the thorny nature of its cousin, Buckbrush. The twigs taste like Wintergreen, and we browse them all seasons. We have made "root-beer" by fermenting a decoction of late winter twigs which delivers an immune system tonic. We have used long-stem coppice for *hurdle* weaving. We have used the twiggy long-flexible coppice-sprouts for *wattle-and-daub* walls as well.

Corylus californica (Hazelnut) is our premier basketry-withie. Remnant *stools* from Dakubetede coppicing are found in all our gulches. The best withies are shade-grown and new (one to three years old). We have been able to practice with these to make solid, lasting baskets of varying sizes, utilitarian and rough, using the older branches and twigs from salvage, as we thin the stools to grow more fine-material.

One chance observation is that long stems that are pinned down to the ground by a snag-fall send up a lot of withies of good quality. We have cool burned some clumps, but the new-growth response was poor. We have a lot to learn.

Salix scouleriana (Spoon Willow) is a common roadside and clearcut, disturbed-ground pioneer Willow. Although it is "snappy" (brittle, not flexible) when green, we have discovered that the green stems turn red, in drying, and when soaked months later, work great for baskets. This Willow is heavily browsed by Deer and Elk, so we *pollard* to keep the new-sprouts out-of-reach.

Salix sitchensis (a red Willow) is found only in a big patch in Little Wolf Gulch. We are not sure of its identification, but it seems to be a local variety. This is an excellent basket-Willow and has a perfumed flowering-period in the spring. We have less than one hundred wildfire-coppiced stools to thin and are piling brush-mulch to reduce browsing-damage. This red Willow grows on seemingly-dry sites and its shade-grown withies are weak. We hope to propagate from root cuttings and bring it into cultivation. We tried to grow cuttings of various European basket-Willows, and they all died in the drought of 2010. This is our best native candidate.

Cornus stolonifera (Red-Stemmed Dogwood) grows almost vine-like, in the shade of our *riparian-corridors*. It is supremely flexible, and we have made several baskets. It is especially good for wrapping handles and edges. Some wildfire-coppiced stools put up straight rods that we have used for hurdles and rod-mats. Very-excellent material with a medicinal bark used in pipe mixtures. The Deer love to browse off the tips, and this reduces the basketry values of the first flush of withies after coppicing. Later, second branching seems to be less browsed. Brush-mulch again?

Acer macrophyllum (Big Leaf Maple) grows only in our gulch-bottoms. It produces a good Maple syrup in wet-years, from larger trees that are right-alongside creek-flows. We are using shade-grown coppice and pollard for hurdle-withies and for wattle-and-daub. They are snappy when green but soften to "crunchy" after a week or so. We are looking forward to making baskets, but are making hurdles. The bark needs to come off of fat withies, staffs and rods, and especially logs, as the Borers love Maple bark (don't we?).

We are trying to remember how to make whistles, such as are made from Striped Maple (Moosewood) in New England, loosening up the bark by tapping-all-over on a short piece of young rod, slipping off the bark-tube, carving the wood for the air channel, and re-fitting the cylinder of bark over the shaped wood-core.

Fraxinus latifolia (Oregon Ash) grows in our gulch bottoms and is responding well to coppicing in *shady-riparian*. The long-withies are pithy (with a corky center) and make good tubes for pipe stems and blow-tubes after reaming with hot wires (or patiently with hot coals and blowing). We hope to be able to split the fresh-cut withies for *basketry-splints*. They will probably need to be dried and later water-soaked to become flexible enough for baskets.

Cercocarpus alnifolia (Mountain Mahogany) is an occasional savanna-understory-shrub on southeast aspects, a nitrogen-fixer in the Rose family with medicinal-qualities, another of our immune system tonics. The bark is another candidate for our "root-beer" trials. We have also coppiced it in our fuel-reduction work. The stave-sized stems that we thin from multiple-stemmed-clumps are very good quality "rosewood" that we hope to use for musical-instruments, digging-sticks, and decorative Bodging.

SUMMER COPPICE REPORT

This is a story of out-of-season coppicing of multiple-species near the forestry-camp cabin. Let's review the conditions, impulse, and tactics chosen, to pretend to do something-useful to protect infrastructure. We should have done this fuel-reduction in winter.

It is two weeks past Summer Solstice and two late-spring soaker-storms (five and three inches, 7, 12 cm) soaked enough water into the adobe clay subsoil to support awesome grass-growth, tall wildflowers, and exuberant brush-sprouts (lots of fine fuels!). Now we have tinder-dry standing-dead annual vegetation, and the wildfire danger is high. There are thunderstorms on the way.

Pile-burning summer prunings is out-of-question: too-dry, too-explosive a landscape, and illegal to burn in this season. If I cut-and-pile to burn later, the crackly leaves and twigs in the piles are a bomb waiting for a live-fuse. A ground-fire, meeting a pile-bomb, can flare-up and throw fiery-brands down-wind or climb into the Oak/Pine canopy and start a forest-crown wildfire-storm.

Lop-and-scatter spreads cut-up fuels more widely and keep the wildfire on the ground. It is tedious with all the cutting but at least there is no slash hauling to burn piles. The twigs and annual plants end up close enough (six inches?) to the ground to potentially keep a fire on the ground and out of the crowns.

This time, in this hot-scary season, I chose to bundle (lassoed-brindles) two *Ceanothus* species (*C. cuneatus, C. integerrimus*) and left the Lacquer Bush standing. This avoids the brush-pile-bomb, compacts the fuel to reduce flammability, and allows the brindles to be relatively safely stashed.

The Oak branches and the Manzanita trunks were real work, but I cut small-enough-pieces, spread them flat (ground-contact) and on-contour as brush-mulch, on the drip-line of a young vigorous Ponderosa Pines, southwest of the cabin, in good-position for future late-afternoon-shade. Some of the Manzanita was put aside to dry for kindling and some of the more gnarly-beautiful Oak branches were sorted for crafts-material. The remaining three-dimensional larger pieces are put on a more isolated burn-pile for late-fall pile-burning.

The greenish (at least some moisture is still there) *Ceanothus* withies proved still-flexible. Especially the Deerbrush. The brindles tied tight-and-long. Still soft to handle, and at the best-stage to stuff into wall-cribs. That will have to wait, since this is summer *quick-fix* with elegant-timing, we hope. The Buckbrush has *divaricate-branching*, almost thorn-like, that makes it hard to roll up and tie and pokey to stuff-in-walls, but holds the plaster well.

This summer-coppicing of *Ceanothus* is unlikely to re-sprout. It seems, so far, that if we cut too close to the base (less than 20 cm from the ground) and do not leave a live-branch standing, the stool fails. The bristly dead-stools will still hold slope-soil for a few years. The roots and the nitrogen-fixing nodules will break down, adding to carbon-sequestration and soil water-holding capacity. New seedlings are to be expected, where there is still-exposed subsoil. Someday, this straight-up-the-mountain cat-scrape, south of the cabin, will be grassland again, but so far the brush keeps coming back. We also hope to cool-burn past the cabin someday, if we can get the fuel-loads reduced by Human-labor.

The *brindles* are 10–15 cm in diameter where the twine constricts. I sat on the porch in the shade and tied about 50 lassos, with a figure-eight-loop, at one end. As I worked my way across the slopes, I left the brush I cut in rows to be sorted into brindles. The long straight-rods (good for rod-mats, not brindles) of White Oak and Ceanothus were bundled-to-dry in the shade. Break-time, tea, and sit!

To make brindles, branches are sorted by species and laid tips up, butts down, along the slope, with three-to-ten stems per pile, all about the same length. With leather gloves, the butt ends are lassoed together and pulled tight, with a half-hitch tie-off at the figure-eight loop. The twine is pulled up the twiggy-brindle to the next opening, to loop a second half hitch around the brindle. Roll-and-crush the brindle down the thigh, against heavy pants-fabric (14 oz canvas), to compact it enough to hold the brindle together and run the next loop up, near the tips. Next, the twine goes to the tips and pulls together all the thin twigs, and another loop gets doubled to form a clove-hitch. There may be enough extended-tips to tie one more clove-hitch, and then the loose-end is left dangling. Got that? Check out the Bibliography!

The next day it rained a half-inch, and the temperature moderated. The relief was palpable, and the desert-forest looked and smelled great. The insect-frass (insect-processed wood and plants), pollen, and dust had been washed off the Oak leaves, now shiny. The hot-spells in the Siskiyous often portend a weather-change. After this scary hot-spell, we had no local wildfire start-ups and a significant rain-event came mid-summer. Three weeks later, hotter slow-weather (temperature inversion high-pressure) has oozed into the canyon, the red-Sun veiled in smoke.

The brindle-piles of two *Ceanothus* species dried fast. The *volatile turpines* wafted into my upslope shade-shelter. Delicious, slightly fermented, and very-much-like *Artemisia* Sagebrush. The desert has a special set of odors. The leaves had shrunk and cupped, and curled or twisted, and easily shook off the brindles.

Then, predictably, the hot-spell broke, with new weather moving-through. No rain or wildfires, but the mornings were cool-again, and I took a couple-of-mornings more, to reduce-fuel on the west and northwest sides of the cabin. This time, I made Oak-brindles from the most-flexible-and-straight White Oak coppice (from previous years cuts) and low branches. There are two building-projects that need wall-filling-brindles, and I want to experiment with plastering Oak brindles with leaves dried-on-the-twigs. Re-localized-innovation?

Both Black Oak and White Oak have robust branches and twigs, which are harder to crush and roll. At least they are less stickery than the Buckbrush. Another half-cubic-meter got lassoed, tied-up, and piled. One wall of the new-woodshed received three species of brindles, stuffed-in-cribbing. The odor of the crushed Oak-leaves is rich and good: yet more perfume for our efforts.

Later, after they dried, we plastered our mixed-species brindle-wall with the first-filling-coat. The plaster/straw hung better on the Buckbrush and Deerbrush, but the Oak leaves added to the filling-effect (we call "papering"). The woodshed wall, with the lower-half papered, definitely blocks the down-slope airflows. Someday: the cool-fire-resistant adobe coat.

To review our materials-harvest: one cubic-meter *Ceanothus* brindles, one-half cubic-meter of Oak. The small twiggy-branches that were lopped-and-scattered on-the-slope amounted to less than half a cubic-meter. A bundle of twenty rods, cleared of branches and leaned up to dry. They are one to five centimeters in diameter at the butt and one half to one meter long. This may be enough to weave a privacy-screen or rod-mat. Some Oak branches were diverted to the ear-cuff-bench, if they had dark figures in the core of the cross section. Some Oak-branches were put aside for tool handles. None of this went to charcoal. There is a small amount (probably one bucket, broken-up) of dead/dry small-branches and twigs (especially Manzanita), for stove-kindling.

The perfumes were an ephemeral-extra and I do not remember the same smells in winter-work. The cabin is safer now and closer to being able to be burned-past. I relieved some anxiety in a scary wildfire-season by working in small-windows of weather-opportunity.

POLES, LOGS, AND ADOBE

We are running short on long poles on the ranch. There were only so many pole-thickets, grown up after logging and fires. There will be more poles, growing tall in the shade sector or legacy-hardwoods. They won't be the fifty-year-old, two-inch (5-cm), twenty-foot-long (9-m) Doug Fir wonders found in the riparian understory. Dense, flexible, pitch-filled, long lasting (kept out of weather, after debarking) and perfect for cribbing ribs on post and beam frames, to be filled with brindles, and slathered with straw-adobe. More on that in the next chapter.

Poles are so valuable—we should treat them with respect. Leave a low stump to avoid trip stakes, cut the limbs off tight to the bark, take off the bark as soon as possible (to slow down the beetles), cut the tree in the late-winter (when the bark comes off best), keep the bark for inoculated bio-swales, tip the poles up in the north shade of a well branched Pine, and visit monthly to turn and straighten. A pole left out in weather or used in tents and tipis, even of this high grade, will only last ten years, although it can then be recycled to other tasks: cut up for braces or stakes and eventually stove-wood.

Logs are big reservoirs of carbon and water. Lying on the ground, they invite fungi. Good way to grow Turkey Tails. Log mulch, laid just-off-contour, is the quick-fix for taking down just-dead, dying, or suppressed trees. Review "60-Odd Things to Consider Before We Fell a Tree" in Chapter 6. When we want to use a log, we could process it on-site, where it lays, to take only the wood we want and leave the bark, limbs, and some side-slab wood.

Pit saws can slice a propped-up log in a tree-fork or tripod (see below). If a log is to be seasoned first, it is put up off the ground on "sleepers" and debarked. A dry log, and some green logs, can be split from-end-to-end by *riving*. A riven-board is rough and not so even or flat, but can be dressed further at least on one side with an adze. A log, green or dry, can also be cross cut repeatedly to an even depth and then the short slabs can be chopped off with an adze or ax to make a rough, flat slab log. If one side is left round, legs can be inserted into holes bored by a bit-and-brace. Then we have a bench.

Peeled logs can also be used as sill-logs or log-buildings. Log cabins are made with notched and stacked logs; the walls are massive, with not-so-good insulation, low specific-heat (for mass-storage), and need chinking where the round logs touch. Lots of tactics in the Bibliography books for tightening up a log wall. A better log-wall is called a "palisade": the logs are stacked vertically on flattened sill logs and under flattened header-beams. The palisade-logs are cut to fit up-and-down and then a rough saw (or small chain saw) is used to chase the vertical crack between the logs, over and over, until enough diameter is reduced on adjacent logs, that they meet tight.

The lower log of dead-standing-snags or wind-falls can be clear of knots, as the young tree lost those lower limbs first. These "butt-logs" can be split for *shingles* and *shakes*. The *shake-bolts* that are salvaged from top-logs are usually dried to make the splitting process easier. The froe is a side-handled splitter that allows a bit of leverage to "pop-off" the short riven-boards. Cedars and Sugar Pine (a Blister-rust threatened species) are the favored shake-bolts. Other Pines, Juniper, and some hardwoods will split well-enough. Doug Fir is more splintery and needs careful handling with gloves and thorough-flattening (planes and sanding) to prevent slivers from sitting.

The landscape dictates the architecture. Here at the ranch, we are blessed with building resources. The challenge is wildfire. It is almost impossible to build a blow-up proof cabin. Even if the walls are adobe, the roof is metal or living sod, and the eaves are metal-flashed, a catastrophic-conflagration will eat the cabin. Heavy adobe-houses in New Mexico still failed when the beams, covered with adobe, burned away inside the wall. We do the best we can and treat the fuel-load, hoping for a ground-wildfire to burn past.

Most drainage-basins contain some water-sorted clays, perhaps in pockets on the stream-sides or revealed on cut-banks. Once a Culture of Place has found the best clays for plaster and pottery, those clay-pits become shrines of gratitude. With enough searching, we should be able to find sand, clay, colored minerals (such as red and yellow-ochre on beaches), and the best fibers for integrity. Straw is traditional and there are other fibers, as mentioned above.

Adobe clay at the ranch is "ready-mix." The clay contains its own "grapple." The clay wants a rough sand, and some clays already have broken-down grit. We haul in straw in wheelbarrows and have water tanks near the building sites. Once we have put up the timber-frame and filled in the walls, the adobe-plaster covers all the logs and brush, seals out the wind, and holds in the heat.

All clays, quarried on-site, need some maturation. If there is significant carbon-material (such as roots, grasses, humus), the clay will not stick together well. We need to "sour" the clay by soaking in lined pits or barrels, with some nitrogen added, to push composting and the outgassing of the carbon. Clay becomes more sticky two ways: souring out the carbon and "wetting" the clay by soaking it long-enough. We have clay in buckets that has been "developing" for years. And we brought slack-lime ("builder's-lime") in years-ago for white-wash-paint to set-up over-time.

TRIPOD OF POLES

To make a tripod that can carry lots of weight and hold canvas, materials, solar-showers, or roadkill for processing, we choose three poles (tall enough for the job) and good, strong lashing rope or wire. Rope is easy to reuse until it rots and wire will be around as salvage material for a long time. Different materials and ways of lashing. We will talk rope here. A tripod can carry a lot of weight and is stable when built well. With a hand-winch or block-and-tackle (ropes through pulleys), we can move boulders or very-large logs.

Here is the recipe for a moderate tripod for stacking staves, small poles, and rod-bundles against: three smooth, strong, Doug-fir poles, three meters tall (ten feet), are laid down on the ground with two poles alongside-each-other crossing the strongest pole, a half-meter down from the tips. The angle is about sixty degrees (one sixth of a circle). Using a 1.2 mm (½ inch) rope, that is about tie a clove-hitch around the junction, leaving a half-meter "tail" (*bight*). Take the rope around the three poles, making a *lashing* with the rope-wraps snug-against each other. We now have a tube of rope holding the poles together. Tie the long end of the rope to the bight (or tail) with a double-half-hitch.

The lashing should be snug but not-too-tight. With one foot against the butts of the double-pole legs, pull the tripod up to nearly vertical. Now with the tripod balanced-upright, walk the outside pole of the double-leg away from the inside pole, and set-up the tripod: standing on its own. The lashing should have tightened up as the outside leg was walked out. If the lashing is too loose, take the leg back to double, let the tripod down to the ground, and re-tie the lashing tighter.

The Lakota-Sioux tipi is a conical-tent (Laubin 1970), framed by several poles set into a tripod foundation-frame. The set-up involves much ceremony and orientation, with the door-pole east and the rafter poles set into the tripod in a careful order so that the walk-around rope-wrap lashes the tightest cluster, to allow the canvas-cover to fit tight and keep out weather. The half-circle canvas cover, with door-hole cut-outs and sewn-on smoke-flaps, is tied to a last raising-pole, unfurled around the slant-cone (a half-circle folded into a cone is slant), and pegged together at the front and to the ground all-around.

The clincher-move, for awesome-comfort, is to hang a three-piece liner on a perimeter-rope, tied to the rafters and foundation tripod. The liner is folded under floor mats, all along the inside of the pole-cone, to seal off a boat-like interior with in-slanting reflective-walls. An air-trench is bark-roofed (from under the cover in the northwest), to bring a draft into the center Fire-pit; oxygen and airflow push smoke up and out the down-wind stake-tied smoke-flaps.

The space between the cover (which is not pegged tight-to-the-ground) allows airflow, and the liner between the poles also allows a constant airflow that sweeps out insects and smoke, leaving the tipi-interior warm and comfy. Managing the air flows and comfort

TRIPOD RAISING

Simple lashing with clove hitch

First, lay out one pole across two poles on ground and tie clove hitch with half-hitch lock and stop knots.

1" to 4" diameter poles
½" to 1" rope

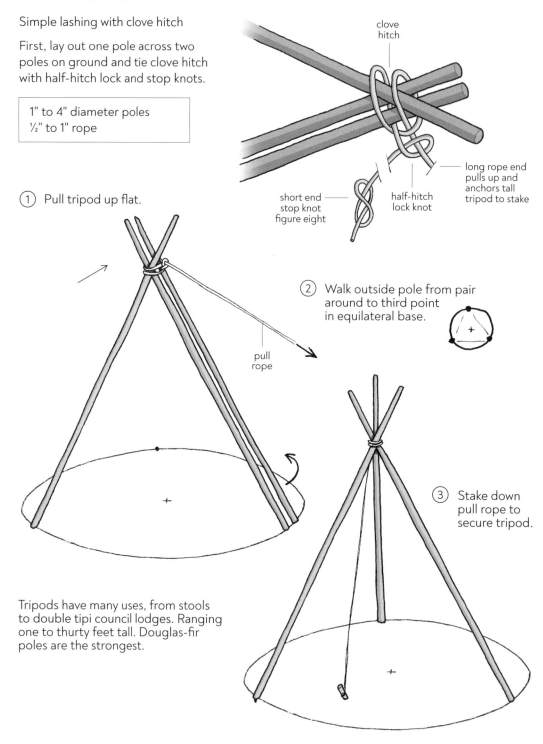

① Pull tripod up flat.

② Walk outside pole from pair around to third point in equilateral base.

③ Stake down pull rope to secure tripod.

Tripods have many uses, from stools to double tipi council lodges. Ranging one to thirty feet tall. Douglas-fir poles are the strongest.

of a tipi is like sailing a boat; tattle-tails on the pole tips show wind directions, and the smoke flaps can be adjusted to prevent back-smoke and hold in warmth, while keeping out most rain.

Lots of details!
Vernacular local living.

A tripod stool or chair can sit firmly on most rough floors or ground. A four-legged stool, bench, or chair needs a flat surface to sit solidly. Three-legged-seats tip easily, and are called "suicide-stools." They seem good, but don't be too tipsy, or over-you-go. Most milking-stools are three-legs. This is appropriate-furniture for up-drainage living, we just need to learn to use them.

CRAFTS

To replace consumerism we will need to re-skill, big-time. The essential crafts will cluster around shelter, clothing, tool-making, nutrition, and art. We already discussed baskets, above. There are many other containers that will be in demand. Tight wooden-boxes that keep moths out (made of Cedar, or True Fir); pottery of all sizes; tightly woven bags; leather and many other animal-products; and caches—big baskets perched on boulders; hollow Oak trees, sealed up with clay and wax; and rope-lashed pole-towers with small cabins on top and metal-cones on the uprights.

The essential craft-skills that children are taught are Fire-keeping and lighting (dangerous stuff), fiber-crafts (discussed above), and sharp-blades (inevitable blood). Burnt and smoked-out, tangled and tied-up, cut and bleeding, we all learn truths: Hot! Get-down! Be careful!

Whittling is slow but exact, shaving by shaving, but bowls, figurines, tools and ornamentation emerge. Can be done at the fireside, makes useful shavings (for hearth-kindling), and teaches how the material wants to be something that we have to discover. Arrows, pipe-stems, chest-plates, ornamental panels, handles, spoons, and much else, with a small sharp metal-blade. A pocket-knife is every-day at-hand. Valuable and often lost, colorful ribbons or handles help keep them around.

Small axes and hatchets are common accident-facilitators. The handle radius is tight-enough to leverage body-blows. We have thumb base and index finger hatchet-scars from learning to split stove-kindling. The weight of the head has momentum and balance is tricky. A long-handled ax is safer, once learned. The broad-knife, used in coppice-craft, is a good hand-blade to learn: split kindling, shave bark (as a draw-knife), de-limb small

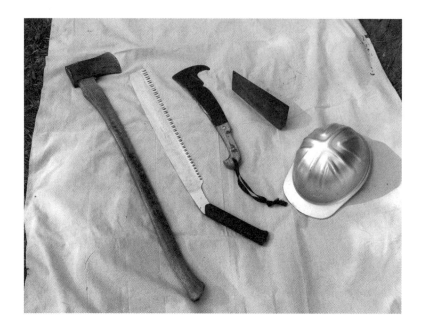

poles, reach up and lop off branches, hook-pull withies off stools. The dull tip can catch some soil while de-limbing without dulling the blade.

Knives for cutting rope and twine, for carving points and scoops, for cutting leather and fur, are the metal-tools we will be able to salvage and re-make, over and over. Blacksmithing we can do with charcoal. We can make handles and sheaths. Sharpening metal edges is constantly needed—we can learn to use magnification; file-guides (to keep consistent angles); and mirrors and saw-vices to keep the best edges.

*A dull edge
is a dangerous edge.
Don't push it,
let sharp do the work.*

Sharpening stones can be locally quarried in many drainage-basins and are likely to be trade items forever. Working obsidian, jasper, agate, and salvaged, thick consumer-glass is called *knapping*—a chipping tool is used to dress the glass-sharp edge, revealed by the chunk-striker. There are old midden-heaps of stone-flakes all over the Great Basin. We imagine lots of sitting and chipping, while drums-talk, overlooking vast-views.

Sharpening saws is advanced metal-care. Usually there are only a few with this skill. Many old saws and tools are made of excellent metal and are worth the attention to

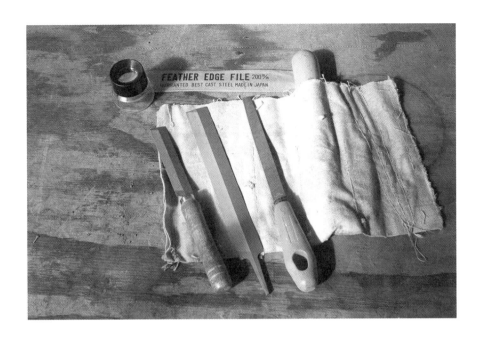

refurbish. Saw files are fine-grooved and thin. The saw-vice is long enough to hold a lot of saw, and has excellent light—magnifying lenses for sure. Pull-saws (traditional Japanese tech) have three surfaces on each tooth to sharpen. European saws cut both directions, which makes a less smooth cut. The gentle for-stroke of the pull-saw pushes out sawdust, and the pull-stroke cuts clean, if we use our whole-body, swiveling at the hips and shoulder—takes some guidance and practice.

The two-person bucking-saw (frame-saw) is thousands of years old. A take-apart frame of wood is tightened with a turn-buckle (or twist-stick sinew-cable) to stretch the saw blade. The two workers take turns pulling and relaxing on the return-stroke. The saw-blade cuts both-ways but the effort is not against the wood, only with the pull. Makes a smooth cut for less work and is social: There are sawing songs, as there are canoe-paddling songs.

JEWELRY FROM FUEL-REDUCTION OFF-CUTS

Our ecotopian marker jewelry is made from White Oak (not surprised?). We tried other woods but she is the best. We use very-fine toothed pull-saws that are very hard to sharpen; we hope they hold up a while! The ear-cuffs slip on over the tip of the ear and slide down into the cup. They do not require piercing and can be taken off easily, but should not fly off while dancing.

A seasoned Oak branch of the desired diameter is clamped to the work-bench, debarked, and drilled out with a half-inch (7 mm) borer (bit-and-brace). The discs are then sliced cross-grain off the branch. If they have not split and broken, we carefully cut a 3 mm slot in the rim. With pin-probes, we clear the sawdust out of the Beetle tunnels. Then we sand the end-grain patterns to a gloss. Pendants are made from larger diameters (4-7 cm).

REMOTE DANCING IN THE SUMMER OF THE PLAGUE

June 2020 at Little Wolf Gulch

Yesterday was cool with showers. I dressed warm and draped some regalia over my heart. Our yearly round dance, centered on the Pine Tree, was in the greater mind of all-of-us, the community of dancers spread out across the Pacific Northwest. We were unable to be together during this plague, but our hearts reached out to each other.

We had agreed to dance in place. All in an afternoon. We had this time to think of the circle, center our breath, and align with spirit, to find what spirit has, as a way for us. Our prayers were thus released and offered through the communal practice of dancing around the memory of the Pine.

I decided to keep my presence focused, without the help of the usual crowd gathering, by doing art. I thought of how the heart of art is finding what the materials want to be. The artist comes into alignment with the world at hand. My chosen material is White Oak. I live in an Oak/Pine savannah. I make pendants and ear cuffs from lower branches that were cut off Oaks to prepare to bring fire back into the grove. No harm meant and great thanks given for the gifts that are revealed by the pruning saw.

At the workbench, the Oak branch is sliced cross-grain with a fine-toothed saw to find figures and patterns. The compelling pieces are then poked, sanded, filed, and polished, revealing the complexity of the end-grain growth rings, the colors of fungal stains, the perforations of Beetle larva mining, and black tyloses deposits. The Oak moves minerals to vessels near heart rot to contain the hyphal spread while the limb is alive. Thus we find black-hard heartwood that polishes wonderfully. The beetles drill in after the limb dies or as it sits on my stockpile. The blue stain of the *Fomes* shelf fungus shows the progress of decomposition, as wood returns to earth.

I clear the Beetle holes of sawdust, left behind by the mining of the larvae, with pins stuck in twig ends. This reveals the tunnels to let light through. The bark is carefully removed and the reticulation, laid down by the inner bark in a network of raised complexity, emerges. Thus, the pendant, step by step, is revealed with great beauty and meaning in my hand. Lots of careful attention with small files and probes. Much smoothing through four grades of sandpaper. Polishing Oak end-grain is a lot of work, as one smooths over the edges of open tubes, the water vessels in the growth rings.

The Oak guides me. I am patient for clarity. The beauty emerges from the dance. We, the remote dancers everywhere, are standing with our prayerful intentions, shuffling from foot to foot. The weather surrounds us with its music and the wildlife comes by to see what is going on. Prayer is a matter of clearing, to be open to wonder. The dance distracts our busy thoughts and allows presence. This is how we walk with spirit.

(continued)

My bench is on the west porch of my cabin overlooking the gulch bottom. I have breezes and bird song, feet on ground, eyes on the Oak in hand, and moments of looking up into the Pine to carry my attention into wholeness. After four hours of this dancing artwork, I am filled with love for my community of dancers and for spirit. I feel fulfilled and connected. Burdens are lifted.

Eventually, when the Beeswax and Orange oil have soaked into the end-grain tubes of the shiny Oak pendant, the colors come out strong. The pattern of the wood is revealed with all its elaboration. Beauty blossoms. We see meaning in the image. We have come into appreciation, through alignment with our work, by dancing.

I am filled with gratitude and wonder for the strength of our remote dancing. I feel the heart strings that connect us to each other and to the wonder of creation. Nature blesses us if we pay attention and let ourselves be transported into the wonder of life and spirit.

All things have spirit. We are all part of the great wonder. Our intention is to join in the compassion of that immense love. Blessings flow with all of this. I am humbled by the deep and wide inclusiveness of dancing. "The spirit of the people is equal to the power of the land."

WHITE OAK

EAR CUFFS and PENDANTS

MADE FROM OREGON WHITE OAK *QUERCUS GARRYANA*

===}}}}ACORN WOMAN'S TREE

===}}}}GAELIC WHITE GODDESS / TRIPLE GODDESS TREE

===}}}}ECOLOGICAL KEYSTONE TREE

(*HIGHLY IMPLICATE!*)

===}}}}ECOTOPIAN-PRIDE ORNAMENTATION

===}}}}SLICED, CARVED, AND POLISHED FROM DEAD LIMBS AND PRUNING

===}}}}CUFFS STIMULATE THE PANCREAS AND LIVER POINTS IN THE EAR

===}}}}BEESWAX AND ORANGE OIL FINISH

===}}}}THESE CUFFS ARE HAND CRAFTED FROM

SOCIAL FORESTRY OAK – PINE SAVANNA RESTORATION

WHITE OAK EAR CUFFS DETAILS

ALL EARS ARE DIFFERENT. TRY FIT—SLIP DOWN FROM TIP OF EAR INTO EAR CUP. CAN BE ADJUSTED WITH FILING.

DIAMOND NAIL FILES WILL SMOOTH EDGES. 1/8 INCH ROUND FILE (SMALL CHAIN SAW FILE) IS THE STANDARD GAP

OAK WOOD SWELLS AND SHRINKS WITH WEATHER AND AGE.

CROSSCUT END GRAINS ARE DETAILED AND LOVELY, BUT CAN GROW ROUGH WITH AGE.

CUFFS FROM SALVAGED DEAD LIMBS HAVE FUNGAL DARKENING AND BEETLE DRILLED HOLES.

(continued)

BLUE STAIN IS FROM *FOMES SPS.* FUNGUS AND BLACK IS FROM MINERAL DEPOSITS CALLED "TYLOSES".

CUFFS HAVE BEEN SOAKED IN AN ORANGE OIL/BEESWAX FINISH AND CAN BE BUFFED WITH SOFT CLOTH.

CUFFS CAN BE RUBBED WITH BEESWAX AND BUFFED.

SANDED 600 OR 1500 GRIT: CAN SAND AGAIN AND WAX.

BROKEN CUFFS MEND WITH GLUE AND FILING. USUALLY BREAK ON FLAW IN WOOD AND REPAIRS LASTS.

Little Wolf Gulch Social Forestry also makes barbecue charcoal and Hazel baskets, teaches advanced Permaculture courses, and demonstrates deep local living and ecological restoration.

CHARCOAL STOVES

A rocket-stove is an insulated short-chimney with a small fuel chamber; the pot is perched on top.

A small stove using selected fuels can boil water fast. These are quick-heat stoves. We prepare all the cooking-to-be-done and then light the stove, moving the jobs through in sequence: first a kettle, then the hot-pan stir-fry, followed by the vegetable steaming and last, another kettle for dishes and tea. We use salvaged restaurant cans, as learned at campouts in the 1950s Adirondacks. We upgraded the craft. With salvaged wire, we pierce the can wall to hold the charcoal; we use a "church-key" punch to open air-holes along the can-base; and we also punch after-burn holes along the lip of the can, making a 'trivet' of wire to hold the pot about an inch (24 mm) above the can-lip.

A skilled woman, with three stones to hold the pot and selected fine-fuels, can boil water faster than any made-stove.

The charcoal is loaded around an internal wire-mesh air channel, stuffed with kindling-twigs. Matches or coals, dropped into the kindling bundle, get a fast column of flame shooting up, and the charcoal lights from the bottom. Once the kindling burns out, the central airflow directs heat right at the pan bottom. If we wait until all the wood is burned out, our pans will not blacken. The charcoal burns clean and hot. Get ready for the next pan-change!

A rocket-stove can be made with rocks and adobe. The burn-chamber is at the bottom and is fed from one side. The pan perches on three rocks embedded in the top of the thick chimney-sides. Not as portable as the can-stove. The can-stove, once going, burns clean-enough to carry the stove, still-burning, into a shelter for warmth and cooking—the air quality is good-enough (Japanese paper-panel houses with charcoal under the quilt-draped table).

NUTRIENT-DENSE FOODS

We already talked about Maple sugar and Deerbrush root-beer. Let's not forget that weeds in the Home-garden are more nutritious than the vegetables—we just have to know how to use them. Wild-foods have density, both energetically and nutritionally. We do not need to eat as much food if we are eating dried, strong-tasting leaf-teas, roots, barks, nuts, berries, mushrooms, and seeds. There is some taste orientation for eating the bitter, sour, and astringent. But those tastes are on the wheel of molecular diversity, delivering enzyme-precursors and complex-vitamins. We can learn to love the special-foods of the season and the carry-foods on the trail and quest.

Much of the diversity of nutritional and medicinal (what is the difference?) native-foods on this landscape was scoured out of existence by starving, sick, desperate-miners and their hydraulic-giant-guns. All the blowing-away of stream-sides and hill-sides washed away the wild-foods (actually tended-foods) with the rich topsoil. All of that floated and flooded down-river and out-to-sea. A huge flush of accumulated-carbon and rich-ground, gone for a long, long-time. We do not understand what the carrying-capacity of this valley was, before the water, soil, timber, animals, herbs, and bulbs were mined-out (extracted) with the gold.

By taking the Beaver first, the river flooded and exposed gravels that the prospectors stumbled on and kicked-off the gold-rush. The benches along the river that did not have gold were cleared for farm-fields and pasture. These complex, metamorphic soils do not hold carbon very well, and the farmers and ranchers exported all the fertility they could to serve the mining, logging, and ranching (and pay the mortgages).

These farmed-out soils need lots of bio-char and love. They can be re-built. The water-tables will be harder—we have to start at the head-waters. The river-side bulb-grounds will not replenish until we temper the flash-floods off of denuded and compacted industrial-extraction disasters.

There was a summer early in the new century when a crew of Hoop-culture folks came to the Applegate and did several walk-abouts, inventorying Wild-food. They could not find much at all. All over the west, the back-country old mining-districts, some of them now considered wilderness, are barren of useful plants.

This was not the case before the mining. We will need to collect seeds and bulbs from remnant patches and build nurseries to propagate the best remains. And we will need to learn what is here that offers skimming for high-quality nutrition, while we rebuild run-down farm-fields to grow staple crops to keep us going.

The most available Wild-foods are tree-needles and leaves. Douglas Fir and Ponderosa Pine young-needle clusters are full of vitamin C and essential-minerals. No calories, fat, or starch, but great fasting-support. We may need them some harsh-winter famine. Young

Maple leaves, also maple blossoms in the spring, and perhaps Alder and Poplar leaves are non-toxic and are said to make fermented leaf-curds—very-dense nutrition.

We hope to experiment. Various inner-barks are nutritious: Pine (Sugar and Ponderosa), Douglas Fir, Maple, and Birch. These can be used in fermentation, soup stocks, and herb-baths (nutrition-through-skin). Other edible-barks here are medicinal and need some knowledge to use them safely. Mountain Mahogany bark contains cyanide-relatives that are not deadly, but can clear a harsh-throat. Also, in the Rose family, is Choke Cherry bark, with cyanide-relatives.

The most dense-nutrition traditional foods on the Pacific drainages are dried-Salmon or dried-meat with pounded-in dried berries (*pemican*). Dried berries; dried bulbs and mushrooms; and processed seeds. *Pinole* is a diverse-nutrition mix of seeds, rinsed and then soaked in a small bag to become gelatinous (sort-of-like Frogs-eggs). This fermenting and germinating (sprouting) mess is super-food. One small handful will carry a Runner (special Ranger?) all day (Runners are very fit and trained to deliver messages over long distances by foot). Miners paid gold to have some Pinole. The seeds usually included Mustards, grasses, Chia Sage (Mint family), Buckwheat, and Thistles.

Wild-beans and peas are tricky, with many different toxicities. The study of nutrition includes deep knowledge of poisons, irritants, toxicities, and chronic malnutrition. This is call *Phyto-pharmacology*. There are textbooks and sources in our Bibliography.

We need to honor the wisdom of the Mothers and the Elders, with Healers, clinics, and good backup. All of us, as children, can learn the basics of bites and stings, scratches and scrapes, burns and bruises. Lots of help (and accidents) out there in the thickets.

Acorns are the most basic Wild-staple on these mid-elevation drainages. Black Oak was preferred for lasting longest in storage, and White Oak was left for *others*, except in times of abundance, or where there was no other choice. Live Oaks and Tan Oaks were valued for their high oil content and special tastes.

Acorns are astringent, with lots of tannins, and need to be rinsed with water to leach out the raspy-taste. This can be done with fresh or dried acorns. There are many songs and rituals that go with making acorn-mush (stone-boiling) and acorn-bread (in pit-ovens). Lots of local-knowledge to learn. Not every year is a *mast-year*. Keeping three to five years of acorns in storage was common.

Walnuts, Hickories, Hazelnuts, Pine-nuts (Pinyon, Grey, Jeffery), and even California Buckeye are dense-nutrition. They need careful storage-tactics to prevent the oils from going rancid. Buckeyes are Horse-chestnuts and are poisonous, until carefully water-leached to remove all the nerve-toxins. Do not try this on-your-own! The Buckeye-nuts are also used to stun Fish in creeks! Nut-trees take decades to come into maturity and start producing yields. Many traditional-cultures treasure their nut-trees in lovely tended-savannas.

Berry picking asks for logistics and perfect timing. Special baskets and berry-rakes (modeled on a Bear-claw), drying baskets, and berry-storage baskets are essential-tools. Berries make Bears fat before they hibernate and there are many stories about Human berry-pickers becoming Bears. Mutual-respect was kept in the berry-bushes. Dried berries last well and make berry-soup, or mush-flavoring, or pemmican: all loved and valued.

Berries come from many different plant-families: Manzanita ("little-apple"), Madrone, Huckleberries, Wintergreen, and Cranberries are edible Heathers. There are some poisonous Heathers: Sheep Laurel, Rhododendron. Yew tree red-berries are edible, but the seed is poisonous: spit-it-out! Blackberries, Black-caps, Salmon-berries, Service-berries (Saskatoon, *Amelanchier*), Thimble-berries, Strawberries, Rose-hips, and Choke Cherries are all in the Rose family.

California Coffee-berry and *Cascara* are in the Buckthorn family (with *Ceanothus*, which does not have berries). Some Dogwoods have berries that can be cooked as preserves. Gooseberries and Currants are in the genus *Ribes*, which is a co-host for Blister-rusts. Some Honeysuckle berries are almost tasty, but always medicinal. The Southern Cascades host almost thirty varieties of berries. These are critical Horticultural yields that beg for nursery-work (selection and reproduction), cool-burning, and coppicing.

We send you again to the Bibliography. The sequence of becoming People-of-Place starts with learning identification: knowing something of All Sentient Beings and coming into the circles of relations. After we learn the scope of our local plant-diversity, we can start to learn the TEK that keeps us healthy and safe through various challenges. Mushrooms are even more daunting to learn, but have some special treats for us. Grasses are the advanced ethno-botany to grow into.

The Principles of Forest Shelters include

44.
Staying warm in the winter, with a small heater and solar glazing, is easier than staying cool in the summer, which needs heavy curtains, mass to cool overnight, and perfect timing. Mass is as essential as insulation. Small cabins and cottages, to retreat to during cold and hot extremes, can be built from on-site materials.

45.
Most of the year we can live outdoors in open shelters. We need shade and rain-cover with wind-block on two or three sides. Most cooking can be done in open shelters. Many of our common needs can be met with shared, semi-open spaces: bath-house, tea-house, council-lodge, dance-hall.

46.
Fresh air, of high-quality, is essential to health. Airflow follows a set of patterns that move and capture air. Our shelters are located in landscape-scale air-shed flows that shape our lifeways. Midlatitude Humans have favored aspects: southeast, above fog, below snow.

47.
We in the Pacific Northwest have year-round seasons of mild outdoor weather. The work of Social Forestry often involves moving camp and returning to the winter-village. Much of our up-drainage shelter is temporary. A hoop-of-shelters is a necklace on our drainage-basin. Well-made clothes, appropriate to the season, are our most important shelter. We Humans are naked without borrowed-coverings. I told you we were silly—who else has to do this?

CHAPTER 10

Forest Shelters

A DOWNRIVER ARCHITECTURE TOUR

Imagine a mountain-ringed drainage basin. There are tea-house pilgrimage-huts in the passes between the high peaks. They are made sturdy, with steep slate-roofs or snow-dumping tiles. The walls are stone-and-log with renewable-chinking and on-site clay-plasters. There are house-boards or moveable-shutters that seal up the open-solar hearth and summer-view patio and convert the tea-house/pilgrim-hostel to a winter-shelter, with firewood piled up inside and a chimney-tower/access-ladder, or a storm-wind-swept, wall-baffled entry-porch. The crest-trail winds along the high-country, taking long-distance pilgrims towards the Great Basin and east around Mount Shasta.

These passes are trail-junctions, where down-drainage trails drop to log lean-to shelters on benches by rushing creeks. These are open-one-side, facing-the-creek, for storm-shelter, staging, and overnight-travelers. They have a hearth in front and can be closed off with peg-tied-tarps. The roofs are mossy split-shakes, or succulents (roof-gardens), to slow down ember ignitions, yet a big canyon-wildfire will incinerate everything.

As we go down the headwaters-trails, we find side-trails up/over to spur-ridge notches and pads. When we head-over from the canyon bottom trail to ridges, we pass work-camps. There we see bark-shingled roofs, brush-walls, shade-arbors, tarp-tripods, and caches. These shelters are temporary and can burn or snow-crush. All the craft-bundles, artifacts, tarps, and tools are nomadic, and move-on.

The ridge-trail, down spine from the pad, connects a series of cleared ridge-pads where we find bark-walled Bodger-sheds and shake-roofed lean-tos around the uphill-edge of the pad. There may be a rock or adobe wildfire-resistant Ranger-cabin, Charcoal-kilns, sorting-yards, and processing-floors spread about. The side-trail that took us over from the creek is gentle-enough to pack—or wheel—the craft-goods down into the upper-valley.

8 Prescription

The steep canyon trails flatten out and widen, as we make our way along the river. Here we find short side-trails up to bench-houses, above the flood, at the edge between the riparian-forest and the sunnier canyon-slopes. These are up-river houses, above the persistent valley-fog: solid, wildfire-resistant pit-houses or adobe-core pavilions. They have clay-tile or garden-green roofs to resist incoming, wind-carried wildfire-bombs. If there are big river-side meadows nearby, these upriver-hamlets and housing-clusters are hosts for seasonal-festivals and markets. A few hillside-terraces and solar-shelters host fruit trees and small gardens, just above the valley frost-flow.

As we wheel further down-river on well-built wide-trails, we pass side-valleys that open up, as hill-ringed-basins perched above the canyon-bottom or river-plain. The most well-endowed of these support small-farms and ranches, with wildfire-resistant hamlet-cores, and side-drainage water-detention. These mountain-pocket farms are stable-enough to help support the whole-drainage-basin carrying-capacity for Humans, to do the eco-systems repair and maintenance-work up-slope in the headwaters.

Down-river, perhaps where a significant side-drainage joins in, there might be a larger, permanent, winter-village. This would be best set up on a southeast-aspect, just above the worst winter fog-banks. The slopes near the village would be well-tended and stabilized with Horticulture on earth-works. Wildfire might dodge the right-situation of the village.

The buildings—the compact-jumble of rooms and storehouses—would be tile-roofed, adobe-walled, mass-heavy climate-batteries that modify outdoor temperatures with a seasonal-flywheel lag-effect. Mass, like a heavy flywheel in its vacuum chamber, is slow to change when there is a draw on the stored energy. A thick-walled mass building that warms up from the outside in the summer will cool off from the outside in the winter, leaving the indoor cycle very pleasant and seasonally appropriate. The adobe-walls are thick and warm-up over summer to keep the inside moderate during winter. The walls cool off over-winter to keep the inside cooler in summer. And so forth. Passive cycling. Lovely outdoor-rooms and patios, shaded north-walls, flat summer-sleeping roofs, and ice-chimneys for summer-treats.

COZY AT THE WINTER FORESTRY CAMP

Wall-tents, unlike tipis, are contained-spaces and have ventilation issues. They work best for short stays of a month or less. (Wall-tents are not insulated and the dew point moves in and out, leaving condensation on the canvas or tent fabric. Even when the tent is on a platform, keeping the floors and lower walls dry and clean is a challenge. Tipis do better by having active air flows.) They are best pitched on raised wood or tamped-earth (dirt-deck) platforms, so that the water running down the walls soaks away from the floor. Tents with floors sewn-in still need a foot-print tarp, or moisture barrier under them.

If a wall-tent does not have a sewn-in floor, one can build-up a temporary floor, in layers. Start with dry-bedding (straw, Pine-needles, dry leaves) on a smooth, tamped, raised earth-pad; add a rot-proof moisture barrier (that laps up the walls 4–6 inches (10–15 cm); cover that with a one-piece tarp floor (a cloth-boat); and last—tucked and fit rugs. A waxed-canvas tarp, over a conifer-twig-strewn dirt-deck, will only last until the green-branches, and perhaps the tarp, start to mildew. Very short term and it's easy to sit and work on the tarp-floor.

A strewn floor, deep mulch built up with layers of fresh boughs, is lovely and perfumed, and may even resist bugs. If anyone drops a small object on such a wild floor, it disappears. As the boughs dry, the scales or needles drop off the twigs, which are harvested for kindling. The needles accumulate to form a dry, soft, warm floor. Messy, temporary, mysterious, and local.

Consumer petrol-materials sure are handy, but toxic. Try to imagine weaving our own tarps, blankets, clothes, and tipis.

Sleeping, cooking, keeping a kettle on a stove, and sponge-baths will raise the humidity in a tent or small-cabin. Overheating the space to drive all humidity out is dangerous and unhealthy (too hot a stove or too little oxygen); venting and understanding airflows hold the solutions.

*The dew-point
is where, in a wall,
the humidity level meets
cool temperatures,
to the point of
condensation.
Wet-and-hot humidity floats,
looking for a cool edge
to precipitate on.*

    Take in oxygen, for any wood stove, with a tube from outside to the draft control. The tent should provide peak vents to outgas carbon dioxide (from breathing), excess heat, and humidity. If the tent allows, a cross-vent pair of openings, from one end to the other of the ridge-pole of a wall tent, allows "sweeping, or skimming" of top-air. When top-venting, better that fresh-air come in at the floor corner of a door, near the stove. If the tent is too-tight, cold air will be sucked-in, all along the floor/wall edge and any other leak, causing uncomfortable-draftiness. Provide a draft-feed tube under the floor or wall-vent that feeds directly to the stove draft control.

*Carbon-monoxide gas
is produced by incomplete combustion:
a smoky lamp,
a wood stove back-smoking,
the damper closed too much.
It is heavy, puddling on floors,
odorless and deadly:
keep some fresh-air flow-through,
and do not burn fuels in small tight-spaces!*

    A dry raised earthen-floor (say under a shed-roof or pavilion), with good exterior trenches for water drainage, can be strewn directly with Douglas Fir, Incense Cedar, or True Fir boughs, to make a soft and perfumed natural-floor. Strewn-floors are easy to lose small things into! They can be found later (perhaps with magnets) upon breaking camp. We can spread the strewn floor back into the forest—whence-it-came (keep those *ramial-tissues* in fast *nutrient-rotation*)—as mulch.
    Rodent and insect-access is easier to control where a tipi, a wall-tent, or a tarped-shed, has its cloth-walls tucked under the floor-mats. Otherwise, keep clean-culture, avoid

temptation, and use a lot of tight-containers. The Hoop-culture-household, nomadic most of the year, includes many types of baskets, boxes, bladders, and bags.

A *strewn-floor*, dry under a roof and left to settle, can build-up, thick-and-soft, over years. In summer-visits or the next fall, the twigs, shorn of needles, have floated to the surface and are harvested for kindling-bundles. Some lost items may still be found, but if the pavilion has been left open for months, the local rodents and birds have already picked it all over and stirred up the twigs for us. After the floor is mined, year-after-year, a bed of needles and scales builds up soft and dry. Once a decade, perhaps we shovel out the dusty-duff. And assess the floor-tunneling?

When winter-camp arrives, we go out and sing to the Cedars and Firs, thanking them for the boughs, and relieving them of lower-branch ladder-fuels. After the branches have been stripped of green-twigs and the stripped branches are piled to dry for kindling, the new green-floor is strewn with flourish. Sure smells nice.

Wall-tents and tipis can experience moisture-buildups, where the walls meet the floor. Especially on their north-aspects. The moisture settles, the hearth is far-away, and the Sun never shines there. Any storage needs to be up on pallets, such as rod-mats or basket-disks. These allow some air circulation, under the pallet, but the dew point (see above) is moving in and out of the structure.

A tent is not sufficiently isolated from the ground (55° F, and damp) and seldom has enough roof extensions (broad-eaves) that can push the outside drainage further from the floor-edges. Moisture income and capture is inevitable. A tent under a roof is good; the tent is a sleeping-container in an otherwise open-space. Perhaps the pavilion has a dry outdoor-hearth-kitchen, and wood-storage.

The warmth of a wall-tent is from reflection (infra-red radiation) more than retention (mass and insulation). A big pot of water (mass) on a stove can radiate heat, after the stove is dead. Avoid boiling too much humidity into the confined space. A tent with just body-heat will be about ten degrees warmer than outside, especially with enough ground-insulation.

Insulating-pads, mats, boughs, and straw, are more-important than insulated-walls. A wall-tent with a woodstove, does not need much time to warm-up. The most useful skill is keeping a low-smoke, small-flame, coal-building stove well-tended. A chimney thermometer will help.

A nomadic, thin-walled metal tent-stove is easy to carry but delicate to operate. Pack-stoves are best started just inside the feed door, with the chimney damper full open. The chimney-cap should be adjusted to let smoke out down-wind. Start with a fast, twig-kindling flash-fire to warm up the chimney and then make a small carefully-built fuel-crib that has bigger sticks crossed as a base and smaller sticks fan-laid so that a match or fire-starter, put to the center-front kindling, will quickly ignite the *rick*.

When the chimney thermometer reads safe-zone for burning (low creosote-outgassing), turn down the chimney damper until the blaze slows, then back-off to hear it step-up again. By watching the thermometer, we can now adjust the draft control on the feed-door to slow-down or speed-up the burn-rate. We should not have to adjust the damper again, unless the chimney is sooty and blocked, the flames are died-down too-far (thermometer is showing too-cool a burn), the wind outside is down-drafting (as just before a rain), or the damper was too tight from the start. If the stove back-smokes, opening the front-door when we feed it, look for these clues.

There is a whole art to using woodstoves so that they cause the least pollution. Lots of oxygen and good fuel-spacing; continuous attention to feeding; and raking coals forward before adding more fuel. Stove-wood needs to be dried-and-split, in a spectrum of diameters, to allow the tending the best finesse.

A tipi, in the worst winters. can be insulated with straw, stuffed down in-between the liner and the cover. Airflow challenges will occur (see below), but a small hearth-fire or can-stove may still have enough ventilation (if the air feed channel is open). Stoves-in-tents are great, in that they dry the tent out. Usually, folks try to use a stove that is too big for the tent, and that makes it harder to keep a clean-burning small-fire going.

WOOD STOVE FLUE

Chimney parts and flows

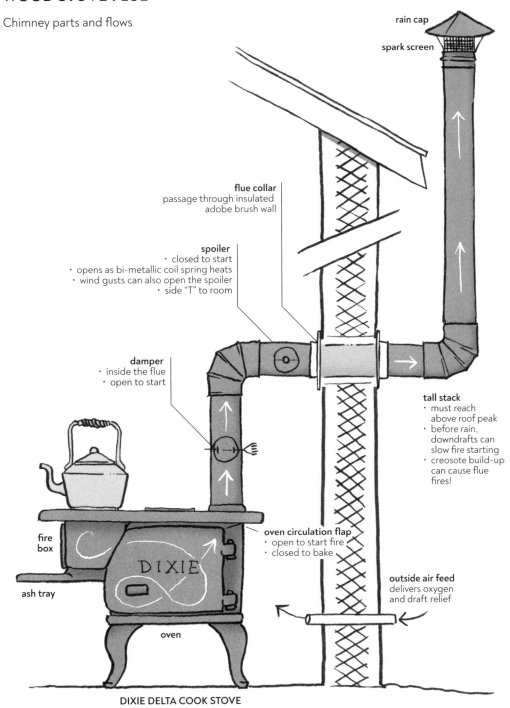

DIXIE DELTA COOK STOVE

A clean-burning wood stove shows only heat-waves lapping the flue-cap.

A tent-with-a-stove has limited ventilation-options, as folks are constantly adjusting doors and flaps to keep, or dump, the heat. A tipi can be retrofitted with a stove-pipe, safely-insulated and wrapped with the smoke-flaps, but that has lots of air-circulation compromises (the smoke-hole top-vent is closed). Stoves in tents are temporary setups. They will need use almost-every-day, just to keep the canvas from rotting, and to try to keep the inside storage dry. Better everything inside a tent is moved around a lot, rolled up, put away, stacked up, aired out...Lived in.

Tipis and tents need deep-and-wide drainage-ditches, especially if they are going to be used in winter or for a long-season. Sandy-soils drain better than clay-soils, and sometimes it is necessary to build a raised gravel-and-sand platform, to perch a tent on to protect it from damp-ground, such as peat, or near-surface, perched water-tables. Lots of other details to remember, such as staking, orientation, wind-flags (tattle-tails), boot removals (mud-entries), etiquette, and manners.

Tight-space,
warm-folks,
wide-wilderness around,
and stormed-in-times.

THE TIPI

A tipi stands as a slant half-cone of cloth, inside a cone of poles, inside a slant full-cone of cloth. The inside half-cone is called the *liner* and is hung on the inside of the tipi pole-cone, pegged tight to-the-ground and folded under the floor mats. The outside-cone is called the *cover* and does not extend all the way to the ground, but is staked tight and lays taut on the outside of the tipi pole-cone. The cover has two big ears alongside the *smoke-hole*, which opens right above the *door-flap* hole on the east aspect.

These *smoke-flaps* can be set, in various directions, to guide smoke down-wind. They can be folded over each other to cover the smoke hole in a storm. The smoke flaps can be directed with smoke-flap-poles, set in canvas-cups at the top-tip of the smoke-flaps, to prop-out the ears. The smoke can be aimed down-wind, to hurry the airflow: in under the cover, up between the poles (inside the liner), and out the smoke-hole, between the flaps sweeping along the fire smoke. Whoosh!

All the bugs that fly under the cover, near the ground, rise up behind the liner and are swept-out the smoke-hole, if all goes smooth. The inside-cone of the liner creates a bowl that holds breathing-air, with the hearth in the middle that has a draft-feed pipe or trench to keep oxygen-feed and help hurry the smoke out the smoke-hole, under the floor.

The inside-lean of the liner, a truncated cone, reflects radiation back down onto the liner-bowl mats. The bodies and accoutrement (gear) inside-the-tipi act as mass, absorbing heat. The critical-insulation is under the floor.

The light from the hearth is reflected back down, by both the liner and the upper-cover. As long as there is a bed-of-coals or warm ring of hearth-rocks, tipis are comfy-enough. In the cold-morning, as soon as the flames rise, the tipi warms-up fast. Then comes the Sun, to warm All-Creation. Only a small volume of Manzanita brush-wood is needed to spot-heat and flash-cook in a tipi.

The small open-fire needs constant-tending to keep it less smoky, which uses less wood than stoves seem to eat, as the flame is fed carefully. Dead-and-dry Manzanita burns clean and hot. When we use can-stoves and charcoal, we have fewer smoke-issues and the cooking-pot is not blackened. Once Manzanita or charcoal burns down to coals, there is no smoke, but lots of heat-radiation. When we need light, we throw twigs into a charcoal-fed can-stove, or have a candle, or oil-lamp. Keep-it-simple, and sing the long-song story.

A tipi is a *dynamic-airflow* structure, spacious enough, yet portable (with help). Our favorite size is 16 feet (about 5 meters). The poles can be carried or dragged; the cover, liner, and parts weigh less than 40 kilos. A tipi is a *dissipative-structure* that channels airflows and radiation as part of the landscape. The spoke-flaps allow open-fires in almost any weather.

Rain can be an issue, with the open smoke flaps and the pole-cone cluster dripping near the hearth. If anyone is tending-camp, the rivulets-of-water, running down the inside of the pole-cone, can be channeled by wetting-the-way with a finger, to go behind the waterproof-liner. Tipis cannot be left-up, un-lived-in. The canvas rots from lack of drying, and opportunistic life-forms move-in. Sunlight degrades canvas, and it becomes brittle. Nonetheless, tipis are great semi-nomadic shelters.

16 Airflows

A hearth-fire, inside a tipi, when well-tended, does not back-smoke. A smoldering fire forces residents to get close-to-the-ground or roll into the liner bottom-corner, looking for air. When all the flows are well managed and good-dry Manzanita wood is placed, stick by stick, on the small bed-of-coals, everyone is happy.

TIPI FLOWS AND LAYOUT

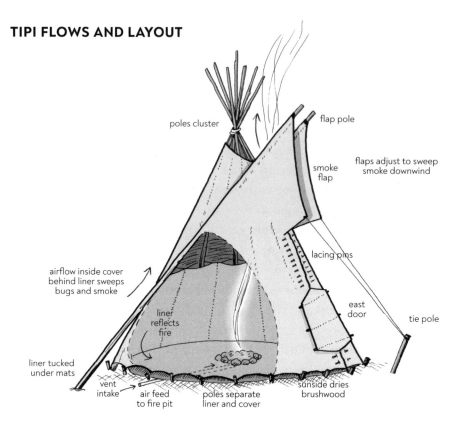

TIPI FOOTPRINT

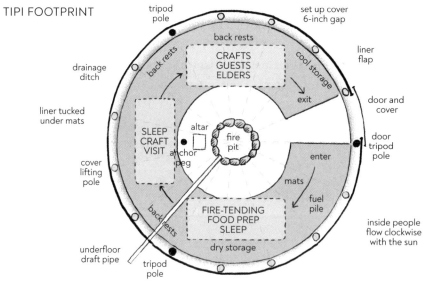

A campfire indoors! The light and leap of the flames projects a lot of radiation, shadows, and light. Flashes of light-and-smoke illuminate the smoke-hole and the top-inside of the cover. Useful during storytelling. Our charcoal can-stoves work great on the hearth. Less smoke, no extra light while cooking on charcoal. Good light when carefully fed good small wood; very efficient small wood use with good heat projection.

The traditional Plains First Nations sitting and working positions around the three mat-covered sides of the tipi are relative-locations in flows. As folks come in the east door (least-likely direction for storm-gusts), they drop-off firewood to the southeast door-side (best early solar) and food to the northeast (coolest handy floor). The area just inside the door-hole is where we take off footwear, turn around, and crouch to close the door-flap and liner; the slant of the tipi-cone is longest here. The Willow cover-pins lace up the cover, below and above the door, and up to the smoke-hole.

Folks make their way with-the-Sun (deosil, Gaelic for clockwise), around the hearth, to their seat. The first mat—near the door, next to the woodpile, on the sunny southeast side—is the hearth-tender's seat. Might be grandmother, and she can get-out-the-door quick, or greet-arrivals, or chase-out-dogs.

The fire tender holds the morning warm-side seat. Next, on the south side is the *lodge-Matron* (women-owned lodges) mats, with room for helpers and lots-of-light coming through the cloth behind them. Midday-sun-warmed backs and warm fronts are turned to the fire, doing the cooking that is done inside and managing the scene.

In the southwest is usually a back-rest or two for guests or crafts. Everything and everyone is sitting on woven grass or rush mats, laid on willow rod mats: warm and comfy. Lots of blankets and baskets.

The west side of the oval tipi foot-print is the main sleeping-platform and crafts area. The liner is steepest here and the hearth is toward the door.

The altar is at the northwest corner of the hearth-pit, with the tipi pole-cone anchor-rope tied to a stout-peg. Backrests face-in, defining the space, with storage behind them.

The north side of the tipi is a guest-space, extra sleeping, storyteller corner (northeast), and lots of cool-storage. This is the cool-side, where the floor near the liner seldom sees sun-radiation. Sun warmth does stream-in through the canvas from the south. Most of these positions are warm-enough. Next to the door gets some drafts.

Make the most of the small winter-lodges, winter-work daily warms us outside-in-the-woods.

THE BRUSHWOOD CABIN

There are two little cabins,
in Little Wolf Gulch,
just below Little Bear Butte,
and just above the Little Applegate River,
and we are having a little bit of fun,
but we do not brag about it.
And that is the modest little story of our Place.

We established the forestry-camp in Little Wolf Gulch starting in the year of the empire 2000. This canyon side-drainage comes down from Little Bear Butte in four steep ravines and combines to one narrow gulch, just above the ranch property-line. The whole upper drainage-basin burned in 1987 in the Cantrall Gulch Fire. The lower-gulch hasn't burned since the gold-rush in the mid-1800's.

The spine of the ridge between Little Wolf Gulch and Wolf Gulch Farm had been scraped-wide (20-30+ m) by bulldozers in 1987. The bulldozers used an old logging-road on the west side (east aspect) of the gulch, and then a logging-spur down into the gulch-bottom, to get up the west-aspect side, ahead of the wildfire, and up to the ridge. The wildfire burned mostly downhill and on the ground, north of this fuel-break, in the end.

After the planning process and infrastructure construction for Wolf Gulch Farm, Hazel asked for a life-lease to develop a forestry-camp to start reducing fuel and managing access to this challenging part of the ranch. The families said yes. There was now a forester/cottager for the ranch.

Hazel looked at the topographic-map and found a side-swale with slopes that faced southeast. A pencil made an X on that site. Then they went to have a look. That meant walking up the old logging road and bushwhacking to the east and south toward the imagined site. After beating through Buckbrush and Manzanita, under White Oaks and Pines, they popped out into a nice Pine-grove just north of the bulldozed fuel-break and looked downhill.

There, about where the X was on the piece of paper, was a small-bench where the riparian forest of Maples and Douglas Firs met the upslope mosaic of brush and Oak/Pine savanna The very steep bulldozer-scrape was right-along-side the pad, on its southeast edge. A Black Oak snag had fallen on the small-bench, leaving a torn, four-foot (1 m) stump on the hillside. Three White Oaks stood just upslope, and east of the small-bench. The under-burned Pine grove felt welcoming. This idea was looking promising.

A bit hard to access, but assessment now began. The gulch bottom holds an intermittent stream. Just by good-timing, Hazel found a seep in the stream-bed, just below the

FIRE SECTORS

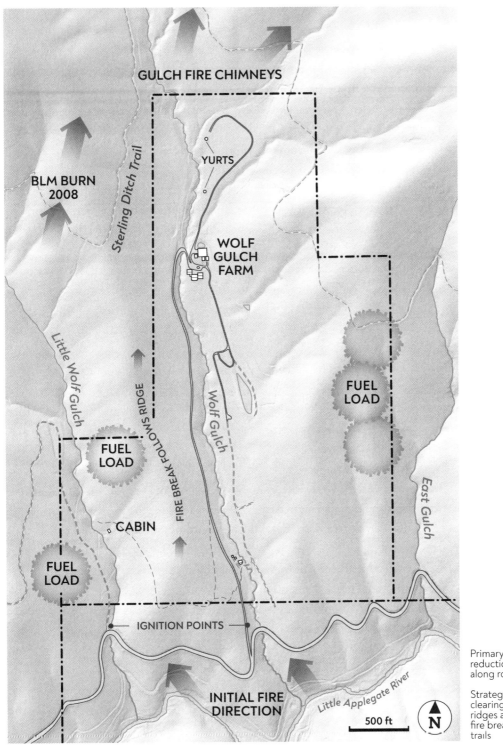

Primary fuel reduction along roads

Strategic clearing on ridges and fire break trails

discovered small-bench. There, the bulldozer had tried to climb a serpentine-outcrop and had spun out a small pad of debris, before it quit trying. The big cat then backed down the gulch, to go straight-east up the steep but less rocky slope, passing alongside the small cabin-bench.

The next few visits set up the first footprints for the forestry-camp. A tent-site was leveled, in the bulldozer spin-debris, alongside the seep-pool. As the savanna was hotter than the bottom of the gulch, the seep-side camp was best for an initial tent base.

Another pad was leveled under some small White Oaks, just north of the small cabin-bench. This Oak grove-pad, which came to be known as the *palapa*, was where the best observations could be made of the cabin-bench potential.

The Oak grove-pad then got a 60 mil 6 X 6 foot (< 4 square meters) sheet of HDPE drinking-water-grade plastic, as a rain-catchment and shade-roof. This rain-catchment tipped—with a wide V-shaped trough—to the southwest, into a 70 gallon HDPE recycled pickle-barrel with a perforated lid.

Now there was rain-water collection, shade, and a place to put a couple of backrests, to sit and think. Observations accumulated, to frame this forestry-camp idea as more intriguing and feasible. Notes were taken on a small rain-proof paper-pad and lists of on-site materials started forming alongside sketches of pole framing. But first, the question was, "What does the land want?"

The southwest forty acres (~ 18 hectares) of Wolf Gulch Ranch was a mix of microclimates/ecosystems, after thirteen years of recovery since the wildfire. In the early new century, this was a downcut dry-gulch, with riparian shady depths, brush-fields too thick to walk through, and ridges either wildfire or bulldozer-cleared, on both east and west sides of the main Wolf Gulch drainage.

The ranch wants to burn. It misses Fire. The 1987 wildfire revealed the open savanna potential. It was lovely. There were lots of native species, and as the botanical search continued, the more rare and threatened species were found. So far, the two spots we impacted had already been disturbed by the wildfire-containment and showed no obvious botanical rarities.

The bedrock, exposed in the gulch-bottom, and the outcropping, just-above the first tent camp, were both competent-greenstone, a metamorphosed sea-bottom volcanic-rock with quartz veins and glitter. It split on curved and dished cracks. There were big pieces, lying on contour, about 20 meters north of the cabin-bench.

Also found: white-rocks in the stream-bed hissed when vinegar was dribbled on them (coated with lime!). Lime-sand sat in pockets in the gulch-bed! The red-clay upslope-subsoil tested out to be awesome building-plaster material. It has angular grit, also called *grapple*, embedded in weathered-in-place adobe. This turned out to be what some, in the natural-building craft, call "ready mix."

AND SO, THE CABIN IDEA EMERGES

The cabin-bench looked good. The slope ratio was 1:6 (one unit of drop in six units of run). The side-to-side orientation of the bench-on-contour was 15 degrees east of solar-south. Perfect: early-morning warm-up and late-afternoon shade.

The cabin would be a long-rectangle, the solar-wall on a narrow southeast end. Using a magnetic-compass, orange flagging was hung at two spots: southeast across the fuel-break andat the Oak grove pad rain-shelter, the Palapa.

The rotten Black Oak log was rolled downhill and stabilized with stakes-on-contour to hold the topsoil crib. Using a sight-level, on a Jacob's-staff, green flagging was tied on several small White Oak trees, north and south and west, below the cabin-bench, all at the proposed cabin floor-pad level. This line has stayed up for years. It oriented the leveling and excavating of the cabin-footprint.

The cabin-bench had several advantages for desert-gulch comfort. Above the riparian trees and below the savannah, with cool-airflows in the evening pouring down the gulch. Earlier cool-air drifts down the east-swale.

Late afternoon shade comes from the riparian forest and the high-ridge to the west. Sunrise, near the hot summer-solstice, would be delayed, coming over the east-ridge with first-beams, lovely, through the Oaks.

The low-angle winter-Sun, rising further south down the east-ridge, at a low angle, penetrates the length of the cabin. At noon, the winter-solstice Sun skirts above slow-growing Madrone trees, south across the fuel-break. Using the materials-on-site, this cabin might be comfy with mass and insulation, using very little stove-wood in the winter and cooling off passively, overnight, in summer.

Across-gulch from the cabin-bench, is a great, old White Oak, with a tall, one-foot-diameter (30 cm) standing-dead Douglas Fir snags in its limbs and north-shadow.

Under the riparian-old-growth, *overstory-trees* were many, tall and skinny Doug Fir poles, two-to-six inches DBH (5-15 cm), as *overstocked-understory*.

The Buckbrush, burned 13 years earlier and wildfire-coppiced, stood in tall spiny clumps.

"How about a rubble-foundation, with a pitch-log sill-box, supporting a pole-barn frame with brush-bundle filled walls, slathered with adobe?" the land said.

MEANWHILE BACK AT THE FORESTRY

The forestry-work began at the forestry-camp. The riparian-forest yielded more than fifty thin Doug Fir poles, up to twenty feet long. One half-acre of overstocked forest also offered several bigger, longer, and even-tapered Doug Fir poles, for posts, beams, stringers, braces, and rafters. The bark peeled off with a drawknife at a tripod on a small flat in the gulch-bottom, where the bulldozer had shoveled in soil to get across. The peeled-poles were propped-up in trees and sorted and counted as they emerged, glistening with pitch.

More-and-more low, small-stumps appeared, and the removal of the riparian-understory revealed the old-growth legacy-overstory. The thinning work was compensating for the long-missed under-burning, while the forest moved toward wildfire resilience, and the potential of future controlled-burns.

The peeled Doug Fir-bark is useful for graywater reactors, and the branches were lop-and-scattered, or pile-burned. So-many poles from so-small a patch of the gulch-bottom.

Guests and interns helped out, and took good breaks, and discussed many philosophies and sciences. In the cool morning, we could excavate the cabin-pad. Big White Oak logs that had been pushed to the other side of the fuel-break and partially buried were retrieved and moved, with hand-winches and block-and-tackle, to be the sleeper-logs under the Doug Fir sill-log—for the west-porch posts, just upslope of the topsoil-crib.

A crib-wall of smaller, salvaged White Oak logs rose downslope of the porch-foundation logs catching the topsoil and rotten-wood, skimmed-off the cabin-bench. We planted potatoes. The porch-pad-bench was hacked-level, below the cabin-floor level-line. That excavated adobe was turned downslope, against the old White Oak trunk-sleepers, to make a flat porch-bench.

A short, off-contour but wheelbarrow-ready trail was cut, over to the greenstone-outcroppings. With a miner's wheelbarrow, big, one hundred pound plus boulders (40+ kilos) were placed as southwest and northwest corner-piers. Other big-rocks were placed between them, to form the west foundation-wall of the cabin.

That massive-wall got backfilled with rock-rubble, and the first pitch-log-sill was perched and leveled in a kick-trench against moving downhill: anchored and embedded—the first sill-log was laid.

The east cabin-pad bank, cut into the adobe-subsoil, appeared as chopped-down sub-soil back-filled against the west-sill foundation log. To chop down to the floor-level, as flagged around the perimeter, the rest of the foundation-trench got pick axed out for drainage, to hold rubble and support the sill-log box.

A big, long-pile of adobe was built-up in the middle of the cabin-pad. When we dug more than three feet (one meter) down, at the east pad bank, the adobe transitioned to more-competent bedrock, especially in the northeast corner. A drainage trench falling south, toward the fuel-break, was filled with rubble, on top of a two-inch (5 cm), perforated HDPE pipe.

Any off-site building-materials had to be carried in. There was no trail system over to Wolf Gulch yet. The bulk of the cabin-materials were from on-site resources. It would have been best, for privacy and security, if the lower-gate on the Bureau of Land Management (BLM) road did not show too-much-use or traffic, but that was the first, obvious-access.

The distance from the gate to the cabin is 1,000 feet (300 m), and the rise in elevation is about 100 feet (30 meters).

The first trail we started at the West Pond near the Wolf Gulch housing-cluster rose 1:50 to the LWG east-ridge, went down-ridge on the fuel-break, and cut steeply back, following the LWG fuel-break to the cabin-bench. Backpack only. No wheelbarrows yet.

We learned how to unload at the lower LWG gate and stash boards and water tanks; drive over to Wolf Gulch and park; pack in the groceries; then go back to carry up the hidden stashes to the forestry-camp above.

At first, we shouldered two 16-foot planks (5 m) at a time and walked them up. Then we improved the BLM access-road and used a hand logging-arch, called a Blue Ox; with the arch we could tie together three or more boards and wheel them up the steep dirt road. The 550-gallon HDPE tanks could be rolled up on their side, and some other gear was wheelbarrowed. A few years later, the "Flat Trail" was built; the BLM road reverted to a wildlife corridor, and signs of our trail-use faded.

This forestry-camp cabin will burn-down in a catastrophic blow-up. It will be a while before the ranch accommodates wildfire well-enough to reduce that risk. Any cabin built mostly of natural materials is called a "melt-down" house. The life-expectancy is one hundred years at best, not the three hundred years of a town-house. Very little evidence should remain a few hundred years after abandonment or collapse: rock walls, bits of metal, ceramic-tiles.

UP GOES THE POST AND BEAM FRAME

The Douglas Fir snags, extracted from Grandmother Oak, proved to be dense and pitch-soaked. They had survived several wildfires, pumped pitch into their wounds, and grown new-wood over pitch-layers, only to succumb to the Cantrell Gulch wildfire. When the butt-logs were cross-cut, we could see wildfire scars had been grown over. The wildfire periodicity appears to be every 20 years.

That meant heavy log-weight, but also promised long-life as rot-resistant sill-logs. The most likely threats to this sill-log box, after catastrophic-wildfires, are Termites and Carpenter Ants (only doing their jobs). Six Douglas Fir logs, each more than twenty feet long, got rolled and slid-down to the logging-spur, and then to the gulch-bottom. From the peeling-tripod to the cabin-bench, up a slope of 1:3.

That required the Blue Ox (a *hand log-arch*); the log is up on a set of wheels at the balance center, and with block-and-tackle, we pulled the log-arch directly upslope, with three-quarter inch (2 cm) Sisal rope. Fun! Two people, tying the Blue Ox up to a tree-stump with another big-rope, while we reset the block-and-tackle for another pull.

The log-sill box of pitchy Douglas Fir logs was laid on the ten-by-twenty-foot (3 X 6 m) perimeter rubble-trench. The sill-box got pinned to the slope with steel rebar pins, driven through drilled-holes, in order to tie the sill-box more securely to the slope. There is some threat of a serious earthquake here from the Siskiyou-fault, which runs east from the coast to the Cascade range. And besides, this pad is chopped out of a serious slope, on greenstone-bedrock.

Luckily, we had to chop-and-pick hard, down into the greenstone at the northeast corner of the cabin, which tied the cabin to bedrock. And with those big-stones, placed in back-cut kick-trenches, holding the west sill-log, so far so good. With an inwaled-path against the very-steep-bank (1:1) on the eastern uphill side of the cabin-bench, there was good drainage and the ground would be shadowed by the two-foot-wide overhangs of the roof deck. The trimming back of this bank continues-on, as the handy-source of plaster-grade-adobe (see below under "Plaster It").

The cabin floor and cool-storage in the northeast-corner is the most temperature-stable, thanks to the geothermal-mass in the most shadowed and cool-microclimates of the cabin-bench.

A Ranger-cabin is a sturdy-small retreat. It is built to weather periods of no occupancy and be ready when needed for shelter.

The sill-logs got notched-and-pinned at the corners with strong rebar-pins (salvaged construction-iron) set-through the corner-overlaps, so that the corner-posts could be set vertical on the pins and board-braced, upright, to begin to raise-the-frame.

The whole post-and-beam-frame was pinned-or-bolted-together, to lend-coherence to the small-pole construction. The posts and beams, as well as the sill-logs, have very-dense growth-rings (slow-growing). These are desert grown, understory-poles, with very-little taper.

All these material-characteristics orchestrate the integrity of this natural-cabin. The long-skinny-poles, of less than two inches (5 cm) diameter, often have fifty annual-rings. The one-foot (30 cm) diameter sill-logs are pushing one hundred and fifty years old. These Doug Fir poles have very dense compression-strength; they hold vertical-weight fine. But they are whippy; even after drying thoroughly, they wobble. Hard-to-break, but flexible. Douglas Fir is similar to Spruce-wood—excellent tensile strength.

The old farmhouse in Wolf Gulch was gutted and torn down to make way for the new straw-bale common-house. Doors and windows were salvaged, and some of those were found to be previously-salvaged from Army Camp White, near Medford. Two doors and three windows got carried into the forestry camp.

Single-sawed four-quarter boards (full inch thick, unplaned) from a friend's land on the other side of Ashland were hauled on the top of a 1955 International van, named Beulah. These boards came from a familiar-forest and were harvested and milled by the landowner. They are rough-boards with very little embedded energy; the energy budget they contain is mostly from the fifty miles of hauling, to get them to LWG.

The boards were carried up the BLM access road to put a roof-deck on the pole-frame and to box-in window-openings and doors. There were enough boards to also build 2 four-by-ten-foot (1 m X 3 m) lofts that tied the pole-frame-walls together, across the narrow-dimension of the floor-rectangle. These lofts act as torsion-boxes; they add stability to the wobbly-frame; and they went up after the cabin-roof got its membrane (see just below).

Even after the cross-diagonal-cribbing was fastened to the pole-barn frame (see below), and the three layers of roof-deck-boards and rafters were perched on top, the cabin-skeleton still had a lot of wobble. Eventually, a couple of pieces of plywood (construction-salvage) were fastened next to the solar-wall, and between a door frame and the window above it in the southeast corner, tying posts-and-beams together to act as stress-panels. It was only after a lot-of-plaster was put up, that the cabin-frame settled-down (again, see below).

Before we put up the cabin roof-deck of boards, we built a west-porch. We fastened a long header beam to the west cabin-frame, raised a post-and-beam porch-wall, and tied the two with rafters. With a 1:1 metal-roof pitch and a gutter, we added another rain-barrel. This covered, but not enclosed, porch offered shade and rain-cover. We put up two sawhorses and some planks, and had a work-bench. Dry tools, building-supplies, and a feeling of building-being.

Big, heavy, two-and-one-half by seven foot (75 cm X 2+ m), quarter-inch (6 mm) thick panels of plate-glass were carried up the gulch-road to build a big, glass solar-wall on the south-end of the cabin. Plate-glass breaks-up into small-cubes when shattered and contains no infrared-blocking-laminates. It is good for solar-income, but needs a lot of quilt and curtain-insulation backup in cold and hot weather.

These sheets of plate-glass had lain stacked-flat for many years and had been etched by acid-rain corrosion. The lace-like patterns left a few areas of clear-glass. This etching would not scrub-out, but no birds have ever slammed into the solar-wall (as they do with clean windows). These patterns also work as camouflage, and sitting just-inside the wall is a great wildlife-observation-blind. The bright-light outside glints and reflects on the white-corrosion and conceals the observer. There has not been a significant loss of solar-income. The cabin solar-heats well, fall through spring, while the two-foot-overhang of the roof and the slightly southeast cabin-orientation keep out a lot of summer afternoon Sun.

The salvaged doors and windows are not insulated, so they need curtains and added padding to slow down the outside weather extremes. This cabin, with its passive-heating and cooling, is appropriate to this desert-gulch, where it is harder to keep cool in the summer than it is to keep warm in the winter. The forestry-work delivers plenty-firewood. The cabin only uses a cord (two cubic meters) a year, because of the architecture.

A very-small cast-iron cookstove, a Dixie Delta—manufactured a hundred years ago in Cleveland, Tennessee, and rebuilt after it was salvaged from a wildfire lookout-tower—is the star-attraction of this small-cabin. She even has a full-circulation-oven, but only takes up about five-square-feet (½ square meter) of floor-space in the southwest corner, next to the solar-wall. The firebox is about a half-cubic-foot (< 0.02 cubic meter), reinforced with plate steel, and is capable of burning Manzanita, which would melt thinner fireboxes. One can burn one 16-inch (40 cm) stick at a time and keep a hot-oven baking.

The finished-cabin also benefits from having some thermal-stability, with adobe-mass walls and earth-floor. It stays unfrozen or unbaked during periods of non-occupation. Some visitors remark that there are few windows besides the solar-glazing, but if we want a view, we go outside. The cabin presents a place to retreat to, a small comfortable-space during extreme-weather. We are otherwise resting and working on outdoor-porches, shaded-patios, or out on the landscape, most days. Outdoor-rooms are found on all sides of the cabin; its cave-like box is a welcome-respite when needed.

AND THEN, THE ROOF—A HAT ON THE NAKED FRAME

Through local contacts, a business that lines landfills with very thick, textured, HDPE was found, near Medford, and they had "scrap" sheets of 90 mil. One piece was bought and cut-to-size, 16 X 26 feet (5X8 meters), which would cover the roof-deck and wrap-over the metal-edge-flashing on the board-boxed roof-deck. The scrapyard also loaded on several free large pieces, which were shared-locally with other natural builders and proved useful on other projects at the ranch. The cabin roof-roll of HDPE was lashed to the Blue Ox, pulled up the gulch-road, and down the logging spur to the gulch-bottom.

As with the sill-logs, it was winched-up the fuel-break and then up the east-wall of the cabin-frame onto the board-deck. Luckily, a friend showed up to do the last push up-onto the roof-deck. This one-piece-membrane will solar degrade, one mil a year in full-Sun where it is exposed, but the roof-deck is at a shallow enough pitch (1:10) to hold a living-roof. The membrane is buried under succulents that interrupt Sun, protect the membrane, and keep the cabin climate-controlled. Once the membrane was wrapped-over the roof-box edges and fastened, side-boards were added to hold the living-roof soil. Hopefully, any burning-embers landing on this roof will not ignite the cabin, as the soil and succulents-garden might resist.

*Are you,
the reader,
following all this?*

PART-TIME WITH HAND-TOOLS: THE TIMELINE

It took a year to put in the foundation and sill-box. It took a year to raise the pole-frame and roofs. It took a year, or more, to put up the diagonal-bracing poles and the cribbing, and it took two years—at least—to fill the walls with brindles and plaster the brush inside-and-out. Having the porch as a workshop and having a roof-deck to keep the cabin-frame dry, were priorities in the sequence. The cabin-long pile of adobe stayed inside the pole-frame until it was gradually used-up as plaster (see below).

More than three quarters of this labor was done solo, the rest with volunteer work-crews. Twice, early on, there were visitors who could give advice, and they saved the cabin from disaster. This whole process was off-the-cuff. No one had ever built a cabin quite like this. We had to wing-it. The story of the Three Pigs and the Wolf kept coming up; this cabin covered all the bases: straw, sticks, and adobe (the bricks?). The real Wolves are now back in the neighborhood, but so far the cabin has stayed up.

AND THEN THE WALLS CLOSED IN

We began to fix the cross-diagonal cribbing onto the inside and outside of the post/beam/braces in the cabin pole-frame. These skinny Doug Fir poles/struts are screwed to the frame members at about a forty-five degree angle, except where interrupted by window and door board-boxes. They go one-way-slant on one side of the frame and cross-slant on the other. The two torsion-box lofts are sitting on wall beams, and have shelving hung from them, inside of the inside-cribbing.

Once the pole-frame had doors, windows, lofts, and shelves, but no flat floor, the whole-frame was wrapped with salvaged greenhouse-plastic from an organic farm. The south end of the adobe pile got moved to soaking and curing-buckets, opening up a small floor space between the mountain-of-adobe and the solar wall.

A light-weight sheep-herders stove (see above) was perched just inside the solar-wall, on the first bit of flat-ground excavated from the adobe-pile. A wall-mounted stove-pipe thimble allows a flue-pipe to zig-zag up, through the south-wall and then alongside the metal-flashed south edge of the roof-deck box.

A small-chair was set by the stove and a step-ladder was propped on top of the adobe-pile, so that the front-loft was accessible. At last, the pole-frame felt a bit like a cabin; one could sleep in the loft and cook on the tin-stove. More like camping in a barn.

THEN WE WENT FOR THE BRUSH

Next, the intersection between the cabin-construction and the forestry-work moved up-slope. The pole-frame was built with poles from the riparian zone in the gulch-bottom, and the restoration-effect was to move the riparian-forest towards old-growth characteristics by preserving the biggest trees and allowing the forest canopy to close more vigorously, shading-out excess-competition from any further conifer-reproduction.

The next fuel-reduction task was to coppice the Buckbrush and Deerbrush in the Oak/Pine savanna. The cabin was gaining a fuel-reduction perimeter, as it absorbed ecologically "surplus" materials. By skimming the overstocked-understory, the catastrophic wildfire-hazard was reduced, and the potential of reintroducing under-burning increased. As more ladder-fuel was converted into building-materials, the zone of fuel-reduction expanded in all directions.

The forestry camp had moved in, and one small part of the gulch started to look more open. Another, better, spring was found in the gulch-bottom, under a tangle of logs, just below where the bulldozer had pushed soil into the stream bed and where the pole-peeling tripod was set up. The spring-pool overflowed, and the flowing water sank quickly into the gulch-bottom gravels. Very sneaky. It was a small-pool sitting on lime-clay; with a little bit of shaping, an HDPE one-inch (2.5 cm) poly-pipe was inserted and laid down-gulch to a 120-gallon (<600-liter) water-trough.

This became a plunge-pool of cold-water on a hot-day, when the spring was running. This spring is seasonal: it only runs for a few months after a wet-winter and not at all in a dry year. The spring-tub/water-trough is far-enough from the upper-spring and base-camp, and from the cabin, that it is accessible to wildlife. Another eco-benefit emerged with our careful-settlement-disturbances.

The Buckbrush and Deerbrush that had grown back, after being wildfire-coppiced in 1987, was tall and lanky. The Buckbrush—with its thorn-like entangled-twigs (divicarate-branching)— was hard to pull-apart, but with loppers and gloves, the coppice-clumps disassembled and were sorted into size-classes. About three-to-six stems could be bundled together and lassoed with Sisal-twine. These twiggy brush-bundles were piled-up near the cabin and selected to fit into the cribbing, pounded down. Thus the cribbing filled-up with brindles.

Filling the crib-walls with brindles took a while. They are stickery to handle, slipped-in-between the cross-diagonal-cribbing, laid down horizontal, and pounded into position. It takes a lot of brindles to fill a crib-wall, but we had lots of fuel-load to convert. The first wall we filled was the north-wall.

Our initial brindles were too fat and gnarly. That later meant that the north-wall took more adobe to plaster—extra-mass that never sees sunlight and helps with the thermal-stability of the cabin. Eventually, we made custom-select-brindles with the best branches, and those stacked better and laid in easier.

Before we tried plastering, we tied the diagonal-cribbing-poles to suck them into the brindles through the wall. Long staple-shaped U-wires were wiggled though the brindles, to enfold cribbing-poles on either side of the wall, where the poles crossed. Once the U-wire protruded on the other side of the wall, we could twist the wire together and suck in the opposite cribbing-pole to tighten up the wall, so as to better hold the weight of the adobe plaster.

We set up a board-jig, to be able to sit in-the-shade and make a lot of U-wires with ten-gauge galvanized-form-wire. We had learned to set-up-jobs, from first tying lassos for the brindles, then making brindle-tongs to squeeze the brush and hold it while tying. After the cabin walls were stuffed, we found other, on-going uses for brindles. We also learned to wear heavy-cloth, leather-gloves, and hug the brindle, while lashing and pulling tight with the lassos.

Once a wall was stuffed to the top beams with brindles and sewed-together with the wire-loops, the brush-quilt then got a twig-trim. With hedge-shears, the twigs that stuck out of the cribbed-wall were cut-back. This helped make the first-coat of adobe-soaked-straw easier to apply.

PLASTER IT

Although the alkaline, weathered-in-place adobe is high quality "ready-mix," it still benefited from curing. The Japanese-potters learned long ago to "sour" clay, so any residual-carbon fermented-out. The clay-pots did not crack or burst in the kiln. Carbon-debris in clay keeps it from sticking-together and can lead to crumbling or cracking when *kiln-fired*.

Pottery is different; we need to mix in some straw or fibers with adobe to help it hang-together, brick-like, as reinforcement. The fibers mixed into plasters are long and stringy (straw) and do not rot, once the plaster is troweled-up and dried. The carbon-debris in local, dug-up, clay-heavy-soil, can be from root-fragments and leaf-litter, which is not a suitable reinforcement-fiber.

Clay that cures in a bucket for a long-time (months or years), also "develops" better, meaning the clay-soil-particles swell-up with attached-water and stick to each other better. A potter works-the-clay, on-the-wheel, that has been cured in buckets, to get it smooth-enough to throw a good pot. One of the potters' tricks is to put urine in the clay-bucket, to hasten the fermentation of the carbon-debris. The nitrogen in the pee feeds a biological-fermentation, and the carbon in the clay-bucket out-gasses.

The adobe at LWG sits in buckets for years, and we like to say "the choir is singing" as the clay matures: bubbles are rising in the soaking-water. The adobe that we keep-wet in buckets also changes color. Our adobe is magnesium-dominated and alkaline: If oxygen is depleted by the souring, the adobe turns blue. The anaerobic-reduction of magnesium is a lovely blue, but we do not get to keep it or use it: As soon as we mix in new oxygen, the iron in the adobe rusts (oxidizes) and turns adobe-red. Red is fine, but blue would be great, if we could preserve it! No such luck. Our walls are red-brown.

RECIPE FOR CURING CLAY

Yo! Chelsea,

I do not know what the name is for the fermentation, but I have done something similar here. I think the method is: Soak the clay in a plastic bucket until it absorbs as much water as possible, you should be able to stir it with a narrow wooden paddle or stick. Then when the clay is well softened, add the urine. Any biological activity in the clay has been awakened by your soaking and now, with the nitrogen from the urine—and perhaps some microbes from the air—fermentation will ensue and eat all the carbon debris in the clay, outgassing it as nitrogen-oxides, carbon-dioxide, and methane.

I know that fresh-urine on some garden-crops is diluted 10:1—water to urine. So since you already have some water standing above the settled-clay, I would start with a quart of urine. Estimate three-gallons water total in the five-gallon bucket-of-clay, with one-gallon of water above the clay: one quart in three gallons is 12:1?

Do not seal the lid. But put it in the shade, so it does not dry out. You might see algae growing on the clay and in the water? I pour off biological buildup, such as algae, and scrape the top of the clay before adding more water and stirring up the clay again. You might have to go through some skimming and new water/urine twice? My buckets have sat for years, and when I am ready to mix new plaster, I pour off the top-water, skim the green off the clay-mass, and then dump the cured-clay in the wheelbarrow for mixing.

I also have lime-putty that has sat for a decade, at least, and gets stirred maybe once a year. That is how one "slacks" lime to make whitewash. I add the thick whitewash to plaster for the final lime/adobe top coat. Lime putty has no organic-matter when made from dry lime, fresh out of the bag.

Add the dry-lime slowly to water,
not water into dry lime:
prevent the caustic-alkalinity
from spitting in your face or eyes.

I am trying to point out that clay develops two ways: It soaks up a bunch of water and the clay particles swell up, and that can take some time depending on the type of clay; that long-soaking improves the stickiness. AND if there is any organic-content in the clay, it needs to be fermented-out, to allow the clay to stick-to-itself enough to have integrity in the kiln and not explode. Got that?

If you are cleaning up wild-clay that you dug, there is some sorting to do, by soaking and allowing the clay-soil to settle. As in a soil-shake-test, the sand and gravel will sink to the bottom, and the clay will be on top of the settled-soil under water. So you can skim off the clay to separate it and perhaps repeat this stirring, settling, and skimming to collect the finest-grained clay. Silt settles just below clay. Great for gardening but weakens clay-crafts.

All the best glue to you,
Hazel

Stick to it!
Don't fall apart!

Even as we have lime-sand in pockets of the gulch-bed, we still brought in builders-lime in a bag from the hardware store. Lime is an alkaline-clay derived from limestone, by baking in big-hot-ovens. One has to handle it carefully, with gloves, and not have it get into the eyes or even on the skin. It is caustic and can dry-things-out fast. Lime that gets soaked-in-water and stirred a couple of times a year, develops much like clay, discussed just above.

The wet-lime becomes "whitewash," the most basic of paints. We use this "slacked" lime, mixed with our mature-adobe, for the final plaster-coat and as a paint, over the red-adobe-plasters. The lime-plasters set hard and resist erosion where the simple-adobe can act as a sponge and soak up too much humidity, slumping or crumbling. However, we have noticed that the humidity meters in the plastered-cabin do not vary year-round. It seems the adobe-walls modify the internal-humidity by absorbing and releasing moisture; they buffer the humidity.

As no one could give us advice on plastering brush-walls, we had to figure out some way to make it stick. After trimming the brush walls, "light-clay" or "licht-loam" (clay coated straw) was stuffed into and onto the twiggy-brindles. This, when dry, gave a bumpy, straw-filled, base that allows adobe to be troweled-over. Light-clay is used in natural-building as a wall-filler, for insulation, and as a base for troweled-plasters. It is usually stuffed into the standard 2X4 or 2X6 inch dimensional-lumber-framed "balloon" construction with slip-forms (moveable flat-walls) on two sides.

After the light-straw/clay-mix has set, the forms can be moved-up or over and used again and again until the wall is full. Light-clay is made by dampening straw with water, in the mixing-trough or wheelbarrow, and pouring *clay-slip*—a thick-soup—onto the straw, tossing it until all the straw is coated with clay. Adobe-plaster is sloppy-clay with rough-sand (*grapple*), mixed-well in a trough, and then straw mixed into the adobe. Light-clay is clay mixed into straw, and adobe-plaster is straw mixed into clay.

We call our light-clay plaster-stuffing technique "papering-the-brush." By working light-clay onto the brindles, we create a rough, bumpy, but continuous layer of straw. When this dries out, we can mist-it with hand-held water-sprayers and slather-on adobe with trowels and spreaders. After that first adobe-layer dries, it takes a couple more layers of adobe-plaster to bury all the cribbing and framing, making a somewhat-flat wall. These subsequent layers can be added over the next year or more, as once there is wall closure with the first adobe wall plaster, we can take our time in adding more layers. When the plaster layers have added up to significant mass, the wall can then be whitewashed or lime-plastered. The accumulated inside-layers of plaster and the outside-layers are about two inches (five cm) thick apiece, which is actually quite a lot.

WHAT THE WALLS WROUGHT

Traditional desert mass-houses have very-thick-walls, commonly a meter thick, and they go through a yearly mass/heat cycle. Thick walls are a heat-battery (climate-battery), that slowly cools-off over-winter and slowly warms-up in summer, giving the impression of a warm-house in winter and a cool-house in summer.

The cabin has a daily heat-battery cycle; it can be cooled-off overnight and kept-cool all-day in the summer. In the winter, it can be warmed-up with solar-income or the wood stove to keep-warm all night, without an overnight-coal-bed. This is sometimes called the "thermal-lag-effect," or "thermal-flywheel."

The outside walls have only ever had test-patches of lime-plaster, as the adobe-plaster has held up so well. The adobe-plaster tends to slough-off more dust than lime-plaster, as it ages and expands-and-contracts with temperature and humidity-swings, contributing to dustiness in the cabin. More on that below, under "The Floor, At Last."

Remember that pile-of-adobe inside the cabin? After much mixing of adobe-plaster and filling-of-buckets in the "choir," it was gone. Little by little, a flat-floor appeared inside the cabin. Many ladders, and scaffolding, and ropes, and muddy-gloves sealed up the walls.

The best way to trowel plaster is from the bottom-of-the-wall, up in short-strokes. Every layer of plaster is started at the bottom-of-the-wall and extended all the way to the top, before the next layer is applied. It is best if there are many-layers that add-up to the desired thickness, as this creates a laminated-effect and it is easier to end up with a flat wall than trying to put up thick, heavy layers. The traditional, thick-seasonal-flywheel-walls mentioned above are built with sun-dried adobe-bricks, which are stacked up massively and then plastered, in and out, with a layer of lime-plaster.

It is better for the stability of the architecture if the wall is thicker at its base and thinner at the top, especially if the walls are massive. The thicker bottom-wall-plaster carries the higher-up wall-weight and adds to the integrity of the whole-wall. The thicker-adobe at the bottom-of-the-wall also resists any splashing water, in spite of the wide overhanging-eaves of the roof-deck. It also insulates the foundation-sills and rubble-fill, and adds to the geothermal-stability of the bedrock-perched and tamped-adobe floor.

The brindle-filled walls offer some trapped-air spaces. The insulation value is helpful. Keeping the outside adobe-mass separate from the inside adobe-mass is more significant. The two plaster-loads have different heat-storage fluctuations. The inner-mass wall heats up faster than solid-wall cob and adobe, stays warm longer, and cools off faster during summer nights. The outer mass-wall is a buffer, lessening the effect of extreme heat and cold.

Air is able to penetrate these walls, slowly. The walls breathe. Keeps them alive and competent. Keeps the small, captured cabin-atmosphere healthier. We can feel cool air coming through the walls in the evening of smoky wildfire-days. The walls filter out the smoke. They also absorb indoor pollution. Clay is a hungry crystalline-complexity, open bonds available for attachment.

THE FLOOR, AT LAST

The floor started as a big, long mound-of-adobe. The long-pile of adobe slowly levitated, with some muscular-assist, and stuck to the walls. The shelves and the built-in table got embedded in adobe-plaster walls. The cabin got cozy. A bit dusty, though: dirt floor, dry cabin. As the pile retreated, the exposed floor was leveled, filled, and tamped. Only the northeast-corner was down on soft-bedrock and needed chiseling.

The tamped-adobe floor was dampened with sprayers to get it to settle down tight, sprinkled with dry adobe, and tamped again. A layer of newspapers and magazines was laid down, with carpet-scraps and patio-runners (consumer detritus) on top (see tent-floor discussion, above). This layering needed no moisture-barrier. It is a bit of insulation on the geothermal floor mass, and hard to sweep. We use a roller-sweeper: a "floor sweeper" that throws up dust. Dust falling off the fresh-plastered walls accumulated in the carpets. Not too bad. Take out and beat the carpets periodically. Later we learned to mist the floor and sweeper before sweeping; this keeps the dust down.

15 Natural Building

When we found that crawlers—such as Mice, Snakes, Scorpions, and Mole Crickets—were finding their way in, we had a plastering-party to seal-off the exposed-rubble below the sill-box. First, the outside of the rubble-wall got chinked-with-small-rocks and plastered. Then, because the Mice were moving into the cave-system between the foundation-rocks, those cracks and entrances were sealed off. Live and learn! Natural-buildings with adobe-walls demand continuing-maintenance. At least the materials are right at hand? Now, whenever we build something, we ask, "Who is going to move in?"

We were told that tiles could not be set into an adobe earthen-floor (too loose, too fast, and too soft beneath to prevent cracking). None-the-less it seemed like the best-idea. We did not want to do the wax and Linseed oil natural-plastic treatment. Slowly, wheelbarrow by wheelbarrow-load, a pile of random scrap-tiles accumulated. After a death in the family, as a mourning-distraction/honoring, the adobe-floor got tiled anyway.

We tamped tight and re-leveled the floor, starting from the solar-wall. Then we laid down an inch-and-a-half (4 cm) of stiff adobe-plaster, heavy-with-straw. We sprinkled dry

CABIN FOOTPRINT

Cabin oriented 15° east of south

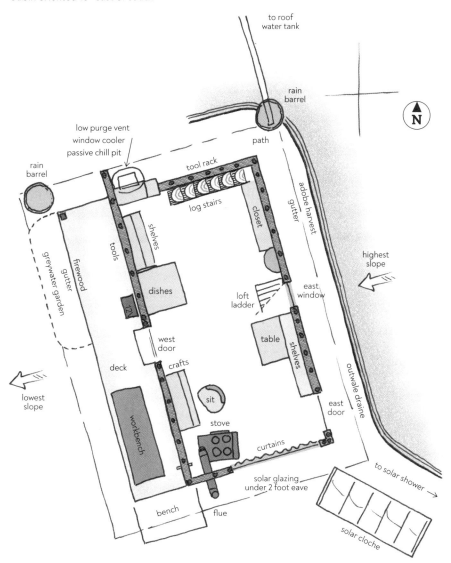

adobe and tamped, while it was still slightly damp. Next, we troweled on a sloppier adobe-plaster with a-bit-less-straw and shorter-pieces. We laid wet-tiles in the mush and wiggled-in-the-tiles with a thumb-wide *grout-spacing*. We found that broken or loose-tiles are easily reset and we learned to scrape the almost-dry, extruded-adobe off the tile-tops. After the *grout* (the adobe between the tiles) dries and stiffens—while still dark and damp—smooth and polish the adobe-grout with a burnishing tool (waxed stiff rawhide, salvaged HDPE scraps, smooth-bone, wood, or shell).

This tiled-floor was a big improvement. A lot less dust. The loose tile is easily reset and the grout re-shaped. We learned that traditional polished-adobe floors were swept, using a soft broom and sprinkled water, so we used a hand-held spray-bottle with our push/pull floor-sweeper or a moistened-broom. Perhaps we will try to apply a Beeswax and Linseed oil-seal on the grout.

It was good to graduate from the rugs-and-papers floor. We get compliments on the fancy-tile-patterns, necessitated by the scavenged and collected odd-assortment of tiles. Starting at the solar-wall, we laid down the black-tiles to absorb and hold maximum-heat from the wood stove and the Sun. We laid a strip of lighter tiles down the middle of the floor for low-light guidance.

The mix of sizes fed patterns that eventually filled the whole floor. We discovered that the traditional one-inch thick (25 mm), low-fired, red-adobe ceramic-tiles, with the wide-grooves on-the-bottom, worked better than the fancy-colored, but thinner, high-fired porcelain/ceramic tiles.

The paper-and-carpet floor did not have a lot of insulation, but the tiled-floor immediately had a buffering-effect. The exposed mass-floor keeps the cabin way cooler in the summer and does not seem too-cool in the winter, at least with slippers. Without earth-berms, or a dug-in-the-hillside-cave, the cabin still exhibits the stabilizing influence of deep-Earth. The floor stays about 55°F (12°C), year round.

The northeast-quadrant of the floor is on-bedrock, which is even more thermal-stable than the tamped-mass of adobe-fill underlying the rest of the tiled-floor. With ongoing plaster-repairs and some Lichen-chinking of cracks, the walls stopped leaking-drafts in big-winds.

Similar to the Tipi floor plan and flow (see above), the cabin layout mimics the landscape flows, especially the daily wind and breezes that sweep the gulch. Whereas the Tipi is a semi-nomadic thin skin dissipative structure, the Ranger cabin is fixed in a location pattern. The cabin is built and operated in alignment with the landscape. Both are similar to sailing a boat. Adjustments are made as conditions change. If the structure is operated in sync with air flows, very little extra energy is required to cool or heat, or to provide high quality air to breath.

16 Airflows

INSULATING THE CEILING AND CLOSING UP TIGHTER

The uncontrolled-drafts were coming from where the membrane sat on the cross-layered (board-deck on skip sheathing) roof-deck on-rafters. The HDPE-liner, even with all the weight of the living-roof, did not sit perfectly-flat on the top board-deck. These boards were single-sawed, not planed to one thickness, and spaced from each other by one quarter inch (7 mm) to allow for shrink-swell under changing humidity.

The adobe-walls buffer humidity, so there has been no evidence of board-swelling. For a *quick fix*, we mud-chinked the board-gaps at the top of the outside-walls, stuffed lichens up into the gaps between the deck-boards from inside, and came up with a plan to insulate-the-ceiling. A slight eco-spasm of small Miller Moths undoubtedly bred in dry-lichens, enlivening the reading-lights at night.

From the lofts, under the roof, long-slices of foam-board were snaked up into the wide-gaps (5-8 inches, 12-21 mm) between skip-sheathing-boards, chinked with insulation-scraps, and taped with painter's-masking. The ceiling was now a criss-cross pie-crust upside-down. With silver insulation-board patches.

A production cutting-area was set up in the Palapa. A reflective mylar-backed/recycled blue-jean cotton (off-the-shelf, big-box-store) insulation-quilt, was cut into four foot long (one meter plus) strips to match the width between the rafters and cover the first layer of in-fill. The rafters are not evenly spaced (see about poles above). The flexible insert-panels are stapled to the rafters-sides, under the skip-board/foam-board junctions, and stapled up into the skip-boards. This leaves the pole rafters visible. A peek into the roof structure reveals it is filled with reflective-foil batts. Almost everything we build during Transition will be Retrofitted from salvaged and local materials, hybrids of before and after. Ultimately, we will learn what the drainage-basin wants and how to do it.

Now the cabin is easy to heat in winter and stays cool in summer, what with the double mass-walls, tiled-floor, and insulated-ceiling. We are getting used to the reflective-silver, and we can still see the Doug Fir pole-rafters. The solar-wall has three layers-of-curtains, and with good airflow-management, the cabin is often 30° F (14° C) cooler at the end of the day—during heat-waves—than the outside temperature in the shade.

BRUSH-WALL SHELTERS

The Ultimate shelter, used by many Pacific Coast First Nations, is a brush-wall. These are made from on-hand materials and can be very-temporary or last for years. Melt-down for sure. When the brush-wall ages out, turning brittle and crumbling when pushed, snow-load crushed, or wind-flattened, the mulch can be a planting-spot, or the wall or hurdle can be pile-burned for charcoal (see chapter 8).

A brush-wall shelter fits into the airflows. If the wall is tight and vertical and set perpendicular to the wind-direction, the wind might spill over the wall and crash-down on the "sheltered" side. Better angles and slants will ease and turn the wind around the brush-wall, leaving the small hearth and seating out-of-the-wind. A leaky brush-wall will let some breezes through and give a down-wind shadow, with more protected space.

13 Windbreaks

A *hurdle* can be assembled in a hurdle-vice, until it can be moved as one unit. The hurdle-vice (two logs, bolted together) holds a two meter reach of *staves* (*shores or sails*), spaced close enough to weave in the chosen materials. Whippy, flexible Willow, Hazel or Dogwood may only need 20 cm (8 inches) gaps, while Maple, Ash, Mock-Orange, or fatter *rods*, may need 30 cm (1 ft) or more between the vertical staves. We may use brush, twigs and all, or long-skinny brindles to fill the hurdle. In the end, the bolts are loosened and the hurdle is lifted out of the hurdle-vice. A portable-wall or fence-section.

The same can be done with sharpened staves, driven into the ground to hold the filler-materials. The wall could be curved some, or run down-slope and tapered (a side-wing wind-break). Good, solid, brush-walls can be built-to-site. Hurdles can be wrapped around a tripod to make a solar-bath shelter or a council pit-Fire-ring back-stop.

Hurdle-walls can be plastered and make a lasting panel, if under a roof. That would qualify as wattle-and-daub. Perhaps to fill in post-and-beam frames? Hurdles can be custom-made as walls, securing the staves to the pole-frame and weaving in filler-rods.

Brush-shelters can be made much easier, just by putting up some poles and piling brush against the windward side. A brush wall allows some light airflow through but slows down wind nicely. The heat from the hearth is reflected back, and if the wall is a lean-to, the loss of heat to the sky is diminished. A tarp or bark (on a strong-enough frame) can give some rain-shelter. All with local on-site materials and carried in hand-tools, cloths, and ropes. Can be fun and last for a while if used daily, kept up, and smoked out. After we move on, other locals will move in and the next time we come back, we will have a Pack Rat nest or a Bald Faced Hornet paper-ball.

TRANSITION-TIME REVIEW

The zoning-codes of the colonialist-government and the older forestry-based principles of law, in this timber-extraction federal-state, allow for "access to economic opportunity." There is a concept of "temporary housing for seasonal workers" that allows lower standards of construction. There is a universal-building-code maximum non-permitted building-size that is ten by twenty feet, and not taller than eleven feet (3 x 6 m footprint, and 3.5 meters tall).

A forestry-camp can have tent-sites, work-sheds, and small-cabins, as long as there is no internal-plumbing and the camp is not occupied full-time. This implies composting-toilets, dish-pans, and water-jugs, off-the-grid solar electricity, and permanent addresses elsewhere. This suggests semi-itinerant hoop-culture and cottager-coops (see subsequent chapters).

Once a ranch or farm has a main house with several bedrooms and bathrooms, there is allowance for "remote bedrooms," as long as they do not have kitchens or bathrooms. Non-permitted buildings are considered temporary-improvements and are not added to the tax base; they are not high priority for protection in case of a wildfire. The government may not tax, but they may not protect either. Usually, the phone-company will not install land-lines to remote-buildings without asking the county if these are permitted-structures. Areas without cell-phone service may be served by satellite companies, but they are not affordable.

Most organic-farms in the Siskiyous have untended-woodlots. They will benefit from forestry-camps. If they follow the rules, principles, and ethics, there may be a way to do this as a *retrofit* during Transition. There is a need to do this for long-term landscape stewardship. See elsewhere and everywhere in this book.

LIVING IN PLACE—VERNACULAR LIVING

So it has always been for Horticultural-Human-Clans worldwide. The work and the Place of the work dictates or suggests the resonant-response of ornamental-decoration. Small but sufficient-shelters that vary with seasonal-work and travels have concentrated/consecrated-meaning, and are respected and reinforced by the dense TEK wisdom, carried by traditions. Everyone is held-safe and known-at-home, with one's own special worth, names, and skills.

A tipi, wigwam, yurt, or brush-shelter, all have the immediate advantage of aural-continuity. The world is close by—we can hear it. Continuous-observation with open ears, is sound through thin or porous walls. Wind and rain or snow, comings and goings on the landscape, the music of the cluster-of-shelters, birds high in the sky, and the rustle of small-rodents looking for scraps and treats; all are available for aural-perusal.

The vernacular palette-of-styles and materials tends to develop on the village or local-drainage level. The roofs are all the same? The plasters are seen in these hues? Local artists and builders know the styles, and the materials, and the special-details. Barn and lodge-raising is collective and social, and we all look forward to the occasional frame-and-roof-raising; so much seems to be accomplished in a short-time.

Another example of local solutions is the Japanese vernacular-architecture in snow-country that can involve paper-screen houses. These have wrap-around screened-porches that act as double air-locks and contain small-rooms with sliding paper-screen doors, all under a steep, snow-shedding-roof. The wind does not penetrate. Everyone wears warm, quilted-clothes. Folks sit around low, wide-tables, with insulated table-skirts. Their legs are under the table, with a small charcoal-burner.

The charcoal is venting very-few toxic-gasses, if fed enough oxygen, with a good air-flow (duct from outside the building or under the floor? Peak vents?). The whole-building is airy but not drafty. Folks sleep in cool-air under warm-quilts and can quickly relight charcoal-stoves in the morning for spot-heating and Tea. As the whole-landscape is being carefully-tended by specialized guild-workers, most folks are outdoors or working in heated crafts shops during the day.

Similarly, semi-nomadic Horticulturalists use temporary or portable-structures that breathe, block-the-winds, and provide a quick on-the-spot warm-up. A brush-shelter can have a small-flame and modified airflow, so that the air-quality is good, depending on where-we-sit. A can stove for charcoal can be brought into a tent, lean-to, or brush-dome (with bark-roof) to keep several folks warm, in a tight circle. The key is ventilation and smoke exit. Block the drafts and rain, wear warm clothes, and be social, and all-is-well, or at-least-better.

ADAPTATION TO PLACE AND TIME

There are limits to all ecological-roles and evolutionary-trends. Living and working within the limits of the drainage-basin, and within the limits of our Human-participation, has always been an art of balance-and-persistence. Climate-weirding has forced Human-migrations and re-inventions many times in the last quarter of a million years. Although individual-ambition and appetite may seem evolved for opportunistic-expansion, fitness-evolution has its problems, which are mitigated by social-evolution.

The scale-of-community which co-evolves with a local Horticulturally intensified bio-diversity is limited (Dunbar's Numbers). The relationship-web between drainage-basin-councils is not limited. The web is horizontal, with the relationship-maintenance delegated to travelers and regionalized through fairs and gatherings.

Social-forestry-practices will also vary within the limits-of-the-landscape. In the Siskiyous we have small farms and ranches, on mostly valley bottom-land, with a few small-farms on upland-benches. There has been almost two centuries of industrial resource-extraction. The Indigenous-management was compromised before that, by two more centuries of epidemics and tribal-displacement that reduced the TEK memory-bank and epic-story-trove. Many elders were lost and clans reduced by diseases brought by traders and water animals. We only have tattered-remnants of that previous Indigneous-participation in the whole-landscape, but we can still see some evidence of what they did, especially on the lands that no one wanted for resource-extraction.

Many riparian-exuberances were eliminated by Beaver-removal, then gold-digging, then overgrazing, then road-building, then saw-log-mania, then ticky-tacky McMansions in the woods, with no-one tending-the-land. Now we are building the cottager and hoop-walker-cultures that could People the lonely, private, manor-house-estates and return vitality-and-fecundity, working with what-we've-got and where-we-are.

PART III

Toward Cultures of Place

The Principles of Invisible Structures include

48.
Invisible-structures are hard to see. We can understand them through naming and mapping. Natural patterns shape invisible-structures, as well as Life.

49.
Social arrangements are learned. We share these ghosts, they haunt us. The stories and ways we relate help us navigate challenges. Imagine a guided change.

50.
Home ways need to fit Place. And Time. The celebration of our livelihood is adorned by our fitness in Time and Place. What we do now is connected to the past and the future. Live for the many generations to come; appreciate the ancestors.

CHAPTER 11
Starting from Here

MEANWHILE, BACK AT THE TRANSITION

*Change is inevitable,
adjustment is optional.
We choose to live on.*

The situation is dire on a global scale. Previous Cultures of Place have experienced genocide and eco-destruction. Some have carried the story forward. We are blessed to have the reports and suggestions.

Starting from Here is about assessing reality during a time of chaos and confusion. We will do well to discern our opportunities without succumbing to despair and depression. The law-books of the empire are old and deep. They hold some ways to finesse collapse while avoiding too much attention.

We would do well to engage the shifting policies, finding restorative and generative actions to move forward. Staying connected to eco-restoration efforts world-wide holds the promise of real change through information-exchange that fosters cooperation across continental-scale regenerative work.

Scenario explorations (imagined futures) will help us see the alternatives, inevitabilities, and challenges. The political tumult of our time may be a late-emerging reaction to ongoing systems changes. The rise of capitalism, and industrialism, was catalyzed by discoveries of energy-supplies undreamed of by our ancestors. Steady-state peasant farming-cultures, muscle, wind, and wood-powered conveyance-and-machinery was eclipsed by oil, natural-gas, coal, and eventually nuclear-power. The application of these powerful-energies blew up the whole-Earth balance. This experience of POWER is addictive. We struggle to imagine living without it.

Recovering from modern civilization is going to take some doing. We are addicts, hanging on the spigots.

The Social Forestry scenario of recovery offers psychological and ecological benefits to the largest diversity of participants, with the highest level of average health, at the edge of chaos (Kauffman 1995). We will surf the waves of healthiness and dodge the newest pushers that aim to distract us. We will need to be clear, tell the real-story, know the back-story, and hold the visions of recovery and re-enchantment.

ENERGY DESCENT SCENARIOS

The book that this drawing is adapted from is an exploration of four future scenarios (Holmgren, 2012). If climate-change is fast and petrol-fuel and atomic energy are failing fast, we are in free-fall. Civilization Triage suggests a future of radical re-localization under harsh circumstances, with random pockets of survival and persistence, scattered about the globe. This scenario sends us to the woods, just to get by. Building local eco-resilience will take a long time.

Feudal Fascism follows fast climate-change with slower-weaning off the petrol-spigot. This scenario is not pretty. Some technological-civilization will persist with pollution and destruction, in a desperate-authoritarian-consolidation of power-elites. The push to escape Earth using every scrap of technology left will be disastrous for the whole planet. The top-down state-power effect will be brutal. Mob-reactions, and secured and gated-enclaves for the rich, will lead to constant-awfulness. Seems we are already there in the early 2020s.

The remaining two quadrant-scenarios are based on slow climate-change. The softest-landing gives us the Eco-Deal, with political-adjustments, but continuing industrial-burdens to the eco-sphere. Fast energy-decline with slow climate-change offers a sobering reality check, while big-energy and connectivity-dependent global systems-collapse. This gives us our best opening to becoming Earth-Stewards, making our adjustments, while we let go of the consumer-madness and material-exuberance (party-party).

We are assuming (and hoping for) this Earth-repair scenario with our hopes for becoming People of Place. Again, we find it is not popular to talk about decline—folks who were happy with the accustomed-addiction will complain: "Someone needs to fix this!"

We are getting fixed by the laws of physics: there are limits and we must adjust.

Energy Descent Scenarios

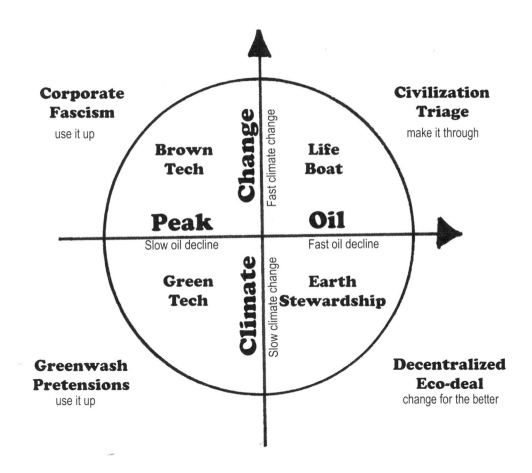

TH/KT 2022
From Future Scenarios
by David Holmgren

GLIMPSES OF POSSIBILITIES FROM THE PAST

In Europe there's a traditional hut system that supports graziers (Herders) and woodworkers (Bodgers, Charcoaliers, Rangers), going back thousands of years. Interestingly, we had a facsimile of that in Southern Oregon and Northern California before the Second World War. Many families and extractive workers were living on mining claims and leases–dependent on hunting, gardening, and scrounging to survive–surrounded by federal and railroad lands. Since then, the state and federal bureaucracies have burnt down and destroyed most of this housing, which was built by Native Americans, people on mining-claims, sheep-herders, and herb-gatherers.

That trail and cabin infrastructure was intentionally destroyed to move people off of the landscape, into the industrial-economy, and into the military (with conscription). There's a problem here, a political challenge. Local drainage-basin-councils want to re-open those trails, re-establish hut-systems, and support people in the forest, for the work of tending.

Ecotopian syncretic culture-creators, very recent settlers in the Pacific Northwest, are brave in trying to cobble together a new, Place-appropriate culture, fast. We are culture-Bodgers, fitting pieces of borrowed-ways together to build a chair-of-wisdom that might support the new People of Place and the integrity of drainage-basins. We are cobbling culture, by hook and by crook, through scenario-exercises based on the best of council and science, spiced by world travel; travelers who learned from distant-cultures and came Home to become locals, thanks to the kerosene-fueled air-bus-rush.

FINDING A NICHE

In Oregon law, Social Forestry has these little-known niches. One is called "access to economic opportunity." It's usually twisted nasty in its application. With a homestead and a piece of forestry-land up-drainage, the logging-company can put a road through the front yard and log the 500 acres behind. The homesteaders can't do anything about it. In Oregon law, the loggers have access-to-economic-opportunity: If there's timber standing, they can cut a road through any blocking-property. But what they usually do is arrive with their equipment, freak people out, and say, "Well, we'll sell you that land," and they make more money with less work by extorting a high price (you have to buy the timber too). Instead, we can use that law to justify eco-restoration camps and forestry-camps on private-property.

In the late 1980s Guy Baldwin (of Bangor, California) and I rough-mapped a seasonal-work-pattern, starting from limited-access—but cheap—parcels at lower elevations, moving higher in summer and back down for fall and winter-camp. The circuit started in the inner Northern California coast-range, circled up to Mount Shasta or Mount Lassen, and looped-back to the northern Sierras for a good south-slope winter-camp, above the fogs, where we process our material-bundles cache into craft-items.

Oregon statutes allow "seasonal workers housing" for forestry, which are supposed to be occupied for only 11 out of 12 months. I had a friend whose sister was married to the forestry tax-officer of the state of Oregon, and I interviewed him over dinner. He said there are many tax-relief opportunities for alternative-forest-products. This is also vernacular-knowledge (redneck TEK), picked-up while talking to small-time loggers ("Gyppoes"). One has to show ongoing-work-activity to justify the camps. Bill Burrell, a Kalapulya First Nations member, gave a workshop at a *Permaculture* convergence in the 1990s near Eugene, which confirmed the rumors.

First Nations folks know imperial-law better than citizen-consumers; they acquired skill under duress.

ONE DAY I WAS GLAD NOT TO BE HOME

We built our forestry-camp here at Wolf Gulch Ranch in the early new century. One day I was working a couple of mountain ranges east and my partner, Elizabeth, was at the camp. Every other property in this neighborhood, when visited by the Jackson County or the Sheriff, had to tear down their unpermitted buildings. I had briefed Elizabeth, so when the Sheriff showed up and asked her all these questions, she gave the right answers, and they never came back to red-tag us.

She was very-clear: this is a forestry-camp; we work for Wolf Gulch Ranch; we are doing fuel-hazard-reduction. They said OK and walked off, but the story begins before this, when BLM burnt 500 acres next to me in 2008. I welcomed them and worked closely with them, so they knew I was a Syracuse University, Forestry College graduate. I had established my credentials and found the right legal-language to use, and we passed. They never wrote a letter to the ranch.

During the remnants of global-consumerism, here's the marketing challenge: multiple products. Good *Social Forestry* has multiple-products, diverse-skills, and varied-crafts— economically, this is not "efficient." It doesn't work in yesterday's economy. It's very difficult to market mixed-seasonal-quantities of multiple-products. We are supposed to have one product, externalize all costs, and deliver quantity to market at a globally-competitive-price.

Folks, that's not Social Forestry; that's industrial-colonialism, and it's killing the planet. Instead, we create all these value-added products and make our money teaching courses. These prototype (actually traditional) "products" are examples of what-we-could-be-doing in a local gifting, barter, or trade-economy. Neighbors in the Little Applegate Valley have sampled Manzanita charcoal and think it's great, so we sell some charcoal. Baskets do not sell. Jewelry does.

*The problem is
we are not thinking
household economy;
we are glamoured by markets!*

It's a slow, long-range-process, trying to pay off your college debt this way. In the 1990s I talked to people who were using mill-ends (salvaged waste) to make birdhouses. "Only pay people $10 an hour, cut costs here and there, only have one product, market it online."

Perhaps we can learn something from corporations too? An eco-accounting-path that uses business-like decision-trails, scenario-practices (imagining and mapping alternative futures), guild-contracts with job-descriptions: this might hurry the *retrofit* paste-up/hybrid, while we surf and dodge.

Written-language has long been evidence of vertical integration and bureaucracies (Shlain 1998, Illich 1988). Documentation is confused with transparency. Laws can be extended to far reaches of colonization. Check out this book! Do you notice the computers behind it all?

*Any energy-intensive technologies
that we use
should be used
to the greatest good,
while they still work.*

STEWARDSHIP CONTRACTS

The desired, principle-and-pattern-compatible, horizontal-common-culture-contract (local culture rooted in Place) is more transparent and cooperative, while still interfacing with the courts. There may be a *Transition-retrofit* that includes stewardship-contracts for management of whole-drainage-basins. This might suggest negotiating long-term-tenure with social-arrangement-tolerance and local-decision-making.

Might remote-powers be willing to settle for clean-water, clean-air, and some value-added forestry-products? New-settlers, contracting for stewardship, need emergency health-care backup and may be willing to recognize a monopoly on military-power in exchange for regional-political-stability.

*Retrofits need
a lot of flexibility—
we should talk.*

We propose Ecotopian-reservations: ecological-opportunity-zones. In these ecological-opportunity-zones (also known as stewardship-contracts), barter, local-currency, tax-free-operations, tolerance for social-arrangements, and health-care services are all in the contract. There's a lot of work to be done, so this is what we need. The promised technological-global-economic-future produced lots of poor (deep-in-debt) college-educated people without employment options.

*Some of us
will want
to get real.*

In Oregon, 52% of the state is government-owned-land. It's not even in the tax-base to support local-government, which leaves commons-rich communities strapped for funds to keep up extraction support-services in one of the most economically-depressed rural-areas in the federation. What any "pro-growth" administration is doing is privatization-and-liquidation: no-tree-left-behind. There needs to be bioregional and drainage-basin-governance, interacting with remote-powers and stranded-investors. Relocalization, within shells.

Discrete drainage-basins ought to be the boundaries of the stewardship-contracts in the ecological-opportunity-zones. The existing road-systems desperately need to be reconfigured, and most of the individual roads themselves put-to-bed, to decrease erosion and sedimentation-loads to help the Salmon. These *retrofitted*-road-systems (fuel-breaks and detention-swales?) need to be gated and controlled; limited-access will have to be negotiated.

That suggests a gatehouse or a hamlet; that means there's an interface between the contract-lands and visitors. That's part of the vision: a river-valley spokes-council with drainage-basin-villagers and craft/work-guilds that are under contract (with attached eco-covenants) with a planetary-climate-council. We're talking reservations, or better, monuments-to-restoration (eco-tourism? pilgrimages?).

We may benefit, during Transition, from solar-powered, portable up-down satellite-link-communications. We don't need to be Luddites, more like use-it-or-lose-it (a very short-sighted time sense). In order to do a complex-business-plan, with multiple-products that are only available on a seasonal-basis, we need a sort of brokerage, an information-rich business-plan. When approached, in the last century, by lawyers and CEOs, they wanted to know one thing right-off-the-top, "Where is the economic-product? Don't talk to us about multiple-products. How do we make money, fast? We'll get to complex-business-plans after we make some money."

*Our motto
after all that,
is "No More Gold Rushes."*

That's already been overdone here, and most everywhere. First it was the Beaver furs, then the gold and other metals, next Cattle, then the timber. The first roads were built on ridge-lines, where they *broadfell* (clear-cut) all the watershed (drainage-basin-boundary) fog-brooms: drought followed. Some of this we can repair.

Environmentalists fought hard, through litigation in the courts and direct-action), to protect some old-growth. Now we have small-diameter, second and third-growth trees and a whole new industry, and that industry is very-low-margin, high-volume, high-subsidy, and automated, with few workers. Do we want to recommend another gold-rush to anyone at this point? I don't think so. Watch out for charcoal! May be the next boom, as we lose easy-petroleum. (More on that in Chapter 8.)

During these Times of Transition, we could use a sophisticated business-plan. We need layers and layers of maps to support information-rich stewardship. We need a council-sponsored brokerage, where come a field of Yarrow, "We've got some prime #1 Yarrow here, it's about 300 lbs. Is there a buyer?" The amount to harvest is limited by the site, but the type of product could be handled for special-use.

This is similar to organic-farming market-research before planting. In the headwaters, we would be able to make seasonal small-quantities of high-quality products with some value-added-processing and deliver to mostly local-markets through not-too-many middle-people. Perhaps, only seeds and extracts or essences can be traded afar?

We need a culture
of tending for the long-term,
to keep up the investments
in ecosystem functions.

How many of us have anything at Home made of any local-materials? What global-industrial-items do we consume that local-wood products from-the-forest can replace? That's what we're competing with. We're competing with plastic, which comes from petroleum. I'm hoping that we bring-to-market the idea-of Social Forestry, by showing people what real, Local-Life looks-like.

"Get a local-life with natural-style,"
could be a slogan.

THE CHALLENGES OF HUMAN SOCIETY

Human relations to other Humans, in close-quarters and working-relations, can be complicated, incredibly-challenging, and down-right-awful, depending on the camp-conditions, the level-of-trust, and rapport built-up by the cooperators. People have been finding ways to get along since time immemorial.

They still get along during emergencies when they have to. Start-up community-process with newly-dedicated Nature-nauts seems to take-up too-much-time, especially if not well-facilitated. The expected promise is comfort-and-familiarity, releasing our anxiety, so we can be present. It takes patience, while we slowly get well-practiced. Experienced-elders are crucial. And lots of love-and-forgiveness.

"And a sense of self and belonging so fierce . . . "
it can never be shaken,
only momentarily forgotten,
until the Clown comes by to remind us.

There are documented ways, oral-traditions, and process-tools to help us; we *can* grow into a Community in Place, as we muster the visionary-intentions collectively, practice-patiently, and get our hearts-into-it. This is where being attached to and enamored by Place is a balm. One can go-for a slow-walk in-the-forest (Plevin 2019). Becoming, and maintaining, a cooperative People-of-Place is serious relationship-work (Brown 2017). Becoming naturally-Human with ourselves, each other, and All Beings, is the Work.

Human groups in the past, such as the Anishinabe and Sioux of North America (Turtle Island), and the Hebraic peoples, originally of North Africa, have managed to stay-together for hundreds and thousands of years, through climate-weirding, countless natural-disasters/shifts, and forced-migrations. A collection of deeply-compelling-stories helps, especially if accompanied by a full-set-of-icons and rituals.

Actual Place, right-here, inspires and teaches us. We learn the protocols of carrying-capacity and the protocols for how-to-handle challenging-times and interactions, through acknowledging real-limits and tending inherited-TEK, while co-evolving *local* IEK.

Always know
the wildfire escape-routes.

Let us remind readers that we are silly-Humans and make lots-of-messes, but we still have useful-contributions to make. The many trickster-stories, featuring Coyote and Raven, help reiterate-the-reminders of our essential Human condition, ignorance-and-foolishness. The useful-way to show respect and learn-from-others is to observe-carefully, without too-much intervention, and certainly no staring. Predators do that!

Good-manners include asking questions instead of making accusations. Better not to jump-to-conclusions. Perhaps you're really hungry, but you may be missing-something-important, such as leaving-enough for regeneration-and-sharing.

Coyote
is endlessly searching
for all the parts.

WHAT WE LEARN FROM NATURE ABOUT OUR WE

One of the hardest problems to solve, for modern science and math, is to find what-is-missing. This is why Aldo Leopold says "first, keep all the parts" (Leopold 1968). Old-stories sometimes remind us of what-is-missing. Better not to lose too-many parts in an ecosystem-mosaic, on a complex-drainage-basin or mountain-range, when you do not, and cannot, know the whole-relationship-web.

14 Village Homes*

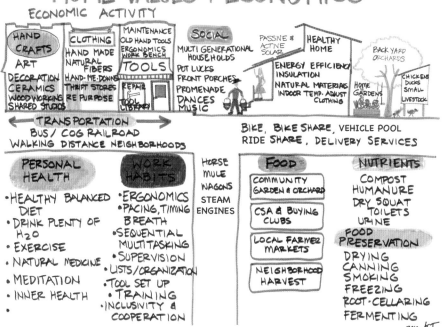

15 Natural Building

*Editor's note: the community-generated posters in *Social Forestry* retain their original spelling to keep the integrity of the original hand-drawn artwork.

16 Airflows

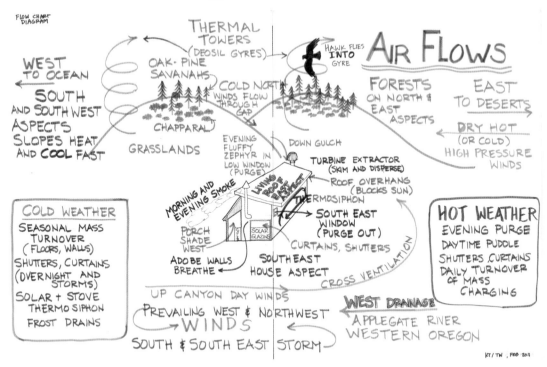

17 Now Next Forever

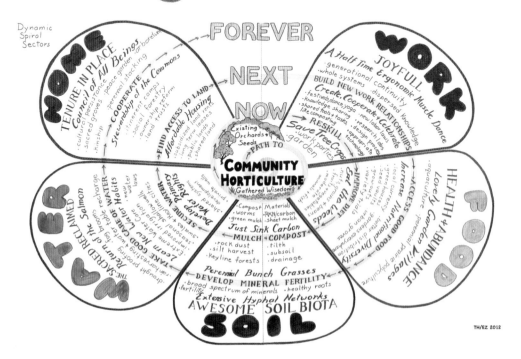

18 Inventory

19 Wagner Councils

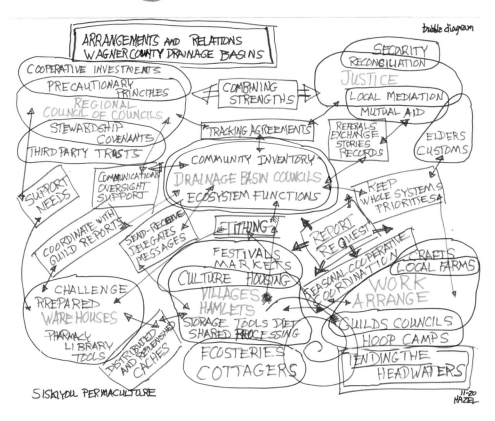

20 Assumptions Matrix

MATRIX CHART

SOCIAL FORESTRY ASSUMPTIONS MATRIX

	LOCAL COUNCIL	WAREHOUSES	WORK	SHERRIF	VILLAGES	SUBSIDIES	MOU	SALMON
LOCAL COUNCIL		X	X	?	X	X	X	X
WAREHOUSES	X	#	X	/	X	?	X	?
WORK	X	X	#	X	X	X	X	X
SHERRIF	?	/	?	#	?	?	X	?	!
VILLAGES	X	X	X	?	#	X	X	X
SUBSIDIES	X	?	X	?	X	#	X	?	*
MOU	X	X	X	X	X	X	#	X
SALMON	X	?	X	?	X	?	X	#
				!			*	

#	Common Category
/	Negative Potential (Conflict?)
X	Positive Mutuals
?	Perhaps Trouble (Needs monitoring)
!	Questionable Row
*	Positive Row

21 Just Sink Carbon

22 Nutrient Cycles

NUTRIENT CYCLES

NITROGEN — N
- air loop
- soil loop
- water loop

- compost
- urine
- night soil
- lightning
- fixating/bacteria

- nitrates—NO_4
- nitrites—NO_3
- nitrogen oxide—N

PHOSPHORUS — P
- water loop
- rock loop
- plant loop

- insects
- ramial tissue
- seeds, buds, flowers
- guano
- volcanic rocks

↕ SALMON

CARBON — C
- air loop
- rock loop
- soil loop

- peat
- sinks
- chlorophyll
- sequestration
- compost
- entrophication

- CO_2 — carbon dioxide
- CH_4 — methane
- CO_3 — carbonate
- CO — carbon monoxide

OXYGEN — O
- air loop
- rock loop
- water loop

- fire
- anaerobic
- aerobic
- Gaea

- O_2 — oxygen
- O_3 — ozone

POTASSIUM — K
- soil loop
- vegetation loop
- ash loop

- fire
- alkaline
- compost
- parks on clay
- insoluble

{ MUSHROOMS }

23 Geo-General Places

Geophysical Generalities

Boreal — forests, bogs, lakes, rivers; acid soils = surplus males = armies; food preservation

Deserts — dry, temperature swings, oases; mineral fertile soils, seasonal rain + flushblooms, special shelter + clothes; alkaline water = surplus females = harems; soil erosion, stocking, brittle

Jungles — rapid carbon cycle, stacking, thin soils; delicate (brittle), local weather cycles, treehouses

Islands — dry + wet sides, rampant exotics, tight ecosystems, boats, fishing, trade, travel, global warming = rising ocean

Arctic/polar — vulnerable to global warming; permafrost, long dark, long light, whole animal diet, tight family culture, special clothes

Subtropical — typhoon/monsoon, aquaculture, origins of agriculture + civilization, tree crops, year round production, casual shelter and clothes

24 Soil Water Rock

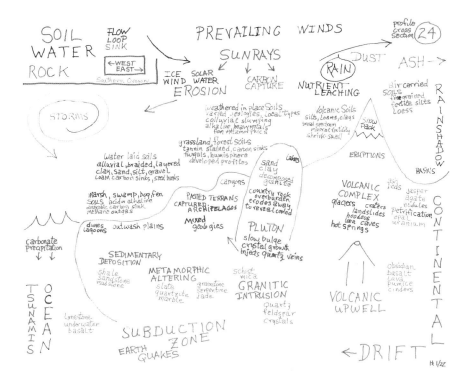

25 Water Use

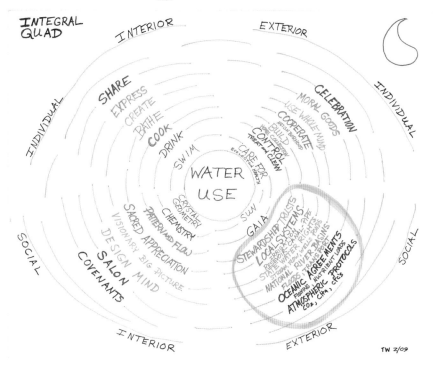

26 Water Cycle

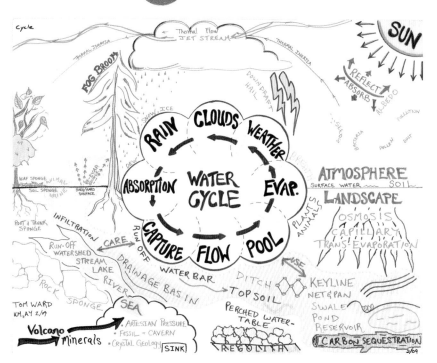

27 Water Care

28 Water Stats

CHART

WATER CATCHMENT FORMULAS

1 cubic foot (ft³) = 7.48 gallons
= 62.4 pounds

1 cubic foot wet soil = 172 pounds

1 cubic meter (M³) = 35.31 cubic feet

1 acre foot = 325×10^3 gallons
= 325,000 gallons

50 feet of vertical head = 21.6 p.s.i.

horizontal roof plot × net rain × 7.48
(square feet)   (feet)
= potential gallons of rain water

2 inch pipe carries ~ 50 gallons per minute

* a water pipe <u>less than</u> 1 inch diameter restricts water flow

BA. 10/09

29 Loop Law

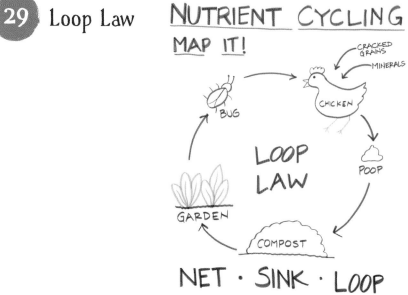

30 Poster of Posters

There are more relationships in the implications of Oak/Pine savannah-cohorts than we will ever be able to fully describe. We can appreciate the unknowability. A high-degree-of-interdependency is the outcome of long co-evolutions, and the greatest resilience-and-fecundity is achieved through increasing-biodiversity, even in brittle-systems.

23 Geo-General Places

As we strive to learn the descriptive-language of relationships, while remembering who-we-are as Human-cultures and practicing good-social-order (see Chapter 14), we seek to support this entanglement and coherence, roiling at the edge of chaos. The right sorts of complexity, such as genetic-diversity, support *meta-stability* and resilience. Over-specialization is usually brittle, and poorly-woven-complexity can collapse as well (Tainter 1988).

Sometimes, we Humans amplify and intensify Nature (*bio-magnification*) by flinging the right seeds; or flipping the right rock; or lighting the appropriate-cool-burn; or singing the resonant-song—and *emergent-properties-of-complexity* unfold. Most often, we only know we did something useful when an edge-effect clues us in. We mean well, but take a lot of risks while practicing our *precautionary-principles*.

4 Ecological Analogs

2 Indigenate

See our selections of Sociology and Anthropology references, in the Bibliography to get a glimpse into Indigenous holistic thinking. We will return to community-in-council again, further on, in Chapter 14.

19 Wagner Councils

5 Animals in Ecosystems

THE SENSIBLE DRAMA DIET

Some drama *is* sensible. There are real and deep Human reactions to stress, angst, and joy. We grow and heal through Sensible Drama. We have a lot to learn, and re-member, as cultures. There has been a divisive (*alienation-and-fragmentation*), hyper-individualized, self-improvement colonial-franchise, but it is the local-culture that holds Wisdom, where we can find it when we need it. It would be great to get help from local Human Elders and All Beings.

Composure only **looks** easy in Humans. Our feelings can be overwhelming. Practicing detachment, in order to open up compassion and reach *enlightenment*, is much talked about in spiritual-communities. This Seeking-the-Light is a grasp-urge at understanding reality. Humans, especially on their own, have a difficult time accessing this capability. We, as a species, have evolved many reactions that are automatic and often act-out through feelings. Some of these feelings, preferably the sensible-ones, can be amplified by social-activities and ritual-repetition.

Not all of these dramatic-capabilities are always appropriate or useful. Feelings can be manipulated by clever-predators, who most often appear composed—perhaps they are focused on their prey. The prey finds composure admirable and is thus distracted (hypnotized?).

All organic-Life, everywhere, is experimenting through evolution, even the *Trilobite*, who seems to have stayed the same for eons. That conservative-evolutionary-strategy is: "Don't change, and stay deep; hide in the muck." Trilobites have Ultimate-composure.

Drama seems to be inevitable for Humans. If we sort out Drama into categories and notice relationships, perhaps we can find Drama that is nutritious and avoid Drama that reduces our resilience.

Melodrama clogs up your wavelengths. It is sometimes useful for distraction or diversion, but is always laden-with-judgment. Melodrama is a crude way of reinforcing common-perceptions and social-mores. Melodrama can be addictive, as a distraction from complex-reality.

Angst Drama is tweaky. It makes anyone who is susceptible twitch with discomfort. This is a potent feeling-full reaction. Paranoia, frustration, projection, shaky-breakdown, and gut-wrench, can ensue. Triggering these dense-reactions is not appropriate without a lot of shamanistic-preparation and psychological-competence. Not recommended: very-risky. Angst can break a person, or angst can push a person through to new-understanding. Competent-practitioners and excellent facilitation necessary.

Growth Drama is the culturally-preferred-experience. This category allows wide-participation and transparent-gains in skills, knowledge, and community-integrity. Storytelling with background, exposition, and moral-lessons, keeps previous-lessons learnt and holds complex, symbolic, ecological-memory in community.

In a Culture built in Place, our phraseologies refer to shared-stories for growth-drama, to circle around. Lots of repetition and practiced-formats are appreciated. This type of drama can blossom to staged theater-productions, but the hearth-side story still holds primal-fascination. Long-epics of a peoples' narrative, kept over many generations and migrations, have proven to be lasting-glue for holding language-skills and cultural-values. Taboos, etiquettes, and social-processes (such as courtship or Eldering) are taught best through dramatic-storytelling.

 ## Sensible Drama Diet

There are creative possibilities of Growth Drama, played-out through envisioning-exercises, problem-solving salons, play for the sake of play, Council of All Beings, and trance-dance rituals. Drama is the natural-method of child-play. This natural-impulse can be facilitated, with a bit-of-support from the Household, but flexible latitude is likely to nurture natural-learning.

These creative-play events can also be practice-grounds for personal-identity and are vulnerable-to-domination by children suffering traumatic-reactions (see Angst Drama, just above). Bullies, scapegoats, gender-stretchers, alter-abled participant contributions, leaders, nurturers, and cooperators are all possible-behaviors, emergent from dramatic-play. Some adult-observation, perhaps by a Grandparent, is advised. Elders are invaluable in community. Sometimes we need an Elder to cue the right-story. Perfect timing of simple-interventions can be very effective in a cohesive-culture, held-together by shared-narrative.

> *"Queen Salmon"*
> *is a traveling play*
> *about the Mattole River basin*
> *and Salmon recovery.*

A community carrying displacement-trauma will have Angst Drama emerge. Experienced-facilitation can finesse much reactionary-behavior into Growth Drama. Trauma-healing work is delicate. We are dealing with deep-seated feelings, powerful-imprints. A Culture of Place, which has accumulated-wisdom, will have protocols for healing. Naming-the-trauma and group-work, through the Council-on-Truth-and-Reconciliation (as in South Africa, in the late last century), is one path toward restorative-justice.

There is much to learn from cultures who have practiced restorative-justice. During Transition, we will benefit from conferences (a council of councils) with mutual-aid-agreements between a diversity of cultural-experience. In Ecotopia, this is culture-building through a syncretic-process with careful-studies of options. Non-violent-communication can be learned. Healing cultural-trauma is necessary for a People of Place to feel and hear landscape-trauma, and dream the local dances-of-renewal.

DEEP IN THE SENSATE PSYCHIC WEB

Drama is real Human-experience. Not necessarily the Truth, but Real, as fiction can be "true." We have much to learn, and there are many sorts of people, and peoples, with their own narratives and traumas. Our Human-species works best, in regenerative-Horticulture, when it is not so distracted as to miss clues-and-opportunities. We should be working as passionately on our community-building as we are in the ecologies. Much study is warranted, but the land-itself—the Place— will be telling us what-to-do and how-to-be. We have to live within ecological-limits and we have to behave within cultural-limits. Learning all-the-rules ("Where is that Rule Book?") and becoming "smart" is not the same as being intelligent-and-observant.

Continuing-revelations
arrive endlessly.

Sadly, cultures who are displaced or oppressed, who are having a hard-time-surviving, do not have a lot of time for Growth Drama. Dealing with emergencies requires a different set of protocols, such as *triage*, than the patient-deliberations of a settled People of Place, with just enough comfort.

While we do the Quick Fixes,
and discuss the Retrofits,
do not forget the Ultimate.

Global ecological-decline and climate-weirding are aggravating all cultures-and-people in the modern-industrial-age, the *Anthropocene*. We are all under stress and most of the other species are at least as put-upon.

Even the Wild
has nature-deficit-disorder!

Adjustments are available, but they are stopgaps. Until we get Human-cultures back-in-balance with-the-Wild, progress on renewed co-evolution will be delayed. Without the talk-back and gifts that Nature offers People of Place, we are anxious, confused, and desperate to settle on a convenient, passable, narrative to calm-ourselves. We can be misled by demagogues. Oh, and then there is substance-abuse, although some medicines are called for to relieve the symptoms-of-aggravation. Acknowledging the situation helps some too.

Obsessive, compulsive-anxiety is demanded by *industrial-consumerism*. "Lean in!" now and check-out-later, are artificially scheduled. Pills are handed-out to ease the wrenching-experience. The ecological-problem is not how to adjust to the modern-economy, it is how to change the economy so that Humans and Nature are healthier and resilient-enough to allow elegant-Horticulture.

The peak-everything and climate-weirding scenarios (see above) both demand re-localization and de-petrolization. That implies more physical-labor and simpler-house-holding. The sequence of Transition suggests *Ecosteries* and *voluntary-simplicity* (Drengson, 1993). That will be brave-and-lonely until we have drainage-basin-councils meeting again and the local, community-discussion is inclusive-and-polite.

Meanwhile, densely-populated peoples in South Asia and East Asia have learned social and individual methods of adjustment-and-transmutation to allow paths to wide-perspective. Yoga is great for discovering tension and breathing into release. Zen Buddhism is great for quieting-the-mind, reaching past cognitive-dissonance and inherent-bias (to the "natural-mind" or "original-mind"). Quakers (Sheeran, 1996) and the Ruckus Society (Brown, 2017) have developed social procedures that help bring people together for community-efforts to transcend-the-aggravation and facilitate some healing social-movements.

Ancient cultural wisdom and advice is derived from real Human toils and joys, and guides whole-communities through to good-social-order, while holding the fallen In-The-Light. It is these cultural-opportunities for social trauma-release that take the longest to build into Place-based protocols. We have some therapies, developed by Moderns to guide the individual forward; some of these are just band-aids or distractions, and some are personal-growth-accelerators, but all individual quick-fixes lack the deep-healing in-community that Humans really-need to achieve collective-focus and enthusiasm for the Great Work.

A traditional village-clown might be following you; everyone is laughing, and you don't know why, until you turn around.

One of the hardest things for Humans to do individually is to admit-mistakes. Deep wells-of-shame can be so debilitating that cognitive-dissonance is triggered and attention is directed anywhere-else (*Mistakes were Made, But Not By Me*, Aronson & Tavris, 2007). Learned-bias, somewhat useful for tribal-bonding, is a set-of-blinders that keeps us from seeing wider-reality. However, there is a path-of-learning-drama that takes us through shame and chagrin, to apology and restorative-justice.

To realize a mistake is naturally a bodily, sensate, experience—not a-lot-of-fun. The actual, real, feelings feed memory-reinforcement, which limits repetition of the mistake. Or, an identity-crisis may trigger denial (cognitive-dissonance).

An experienced-culture allows the guilty to process the trespass-and-harm, but then ushers the embarrassed-member back into society, perhaps even with a form of congratulations-for-lessons-learned. For example, learning the difference between fierce and angry: fierce is purposeful and intentional; anger is emotional and can be triggered.

Individuals and groups can behave angrily and become a mob. An elite can organize in reactionary ways, which leads to authoritarianism. Or a culture can discipline itself to resist non-violently and reach Peace. Sometimes fierceness can be mistaken for anger.

Ask questions:
avoid accusations,
remain curious.

CRESTFALLEN

After being humbled,
embarrassed and red-faced,
now searching to be contrite.
After having been taught my lessons,
the nauseous feeling, butterflies
in my nervous stomach.

Discomfited to discover
a mistaken assumption,
and immensely chagrined that only
regret can drive apology.
Thus chastened: SORRY!

> If one has to change, to become one's True Self,
> no quick support from family and old friends,
> they defer to the familiar, worry for your risk.
> However, pay attention:
> One may learn something.
> Or at least compose a silly song?
> Look behind for the clown.
> Then, try negotiations.
> Community is waiting with an old story that may be familiar.

IRONY, SARCASM, CYNICISM, AND HUMOR

Perhaps we can learn to avoid sarcasm and cynicism by learning to appreciate irony. With irony, we admit involvement and include ourselves in the lesson. Complaints that seem to suggest that "someone should fix this" avoid inclusion and are rude in company that has memory of long-sufferings.

*Irony
is feigned ignorance.
We make fun of
our own ineptitude.*

Some cultures enjoy arguing-with-vigor. This can be scary to visitors from cultures with staid-mores (such as "waxy-oatmeal Quakers"). The structure of any spoken-language constrains the sorts of conversations available.

In English, we are forced into authoritative-pronouncements. We do not have a lot of subtlety of gender-pronouns, verb-declensions, sentence-structure, or subjunctive-phrases. This is the convenient-failing of most global business-languages—keeps the sales rolling through enthusiastic-simplification.

Local languages-of-Place have lots of reference-phrases, and these phrases are shorthand for long-stories. Having a polite-conversation in English is difficult. Especially since the speakers are, most often, not aware of their language-constraints. See Chapter 14.

> *Getting-the-joke,*
> *between cultures,*
> *is fun to play with.*
> *Sometimes just repeating back,*
> *with a straight-face,*
> *the blunt-reality,*
> *can be hilarious.*
> *"Did I really just say that?"*

THE SHAPES OF CULTURAL LIMITS

The language-behaviors that take the most air out of the conversation are accusation, argument, and micro-aggression. Modern norms of constant-call-out-culture and anonymous-abuse (trolls) reduce the options and can channel public-discourse into exaggerated-showmanship for entertainment-consumption and advertising-exposure. There is a scale of functional Human communities called Dunbar's Number, which suggests that we only get to know 150 people by face-and-name. That would be the size of a very-small-village, a hamlet. Or a walled pocket-neighborhood in urban-sprawl.

All the distraction of modern-digital-media is taking us away from Place and neighborhood. The proper social/cognitive scale allows us some comfort. These small-scale, radical-solutions are invisible to global corporatism and get replicated only by shared-experience. Big systems are slow to change (*institutional-inertia*) but are *brittle* for the same complex challenges; they can fall fast.

Traditional-cultures, in the widest-definitions, account for less than ten percent of the North American Human population. Moderns—secular-individualists who believe in Science and technological-progress (techno-optimists)—are the majority of citizens. Then there are up to a third of North American citizens who are social-change-activists in many small or ambitious ways. Through this oversimplified social-census-scenario, we can emphasize that there are very few traditional-cultures and lots of Moderns. The deep-misunderstanding of the Nature-of-the-universe, exhibited by beliefs in Growth, Technology, and Globalism, is a spiritual-failing at-heart. Empire has no soul. Nor much sense-of-humor.

Coming back-into Place-based relationships is the miracle-option. We can relearn how to be co-evolvers with Nature, but we will get scant-help from the Empire. *Ecostery*[1] restoration-work (Drengson 1995) and other new-monasticism options—anchored with the "lifeboat" nodes recommended by the energy-decline-modelers, Transition Townes, and Peak Everything activists—seem to be obvious-strategies. However, at the cusp of phase-changes in systems-logic, nobody knows what-happens-next. These times are the chaos, the aggravation at the edge-of-change.

---

1 Religious and meditative land trusts that protect ecosystem functions; guarded nature retreats.

In Social Forestry camp, we enjoy repeating the phrase, "everything to do, nothing to get done," as we are multi-tasking and sequencing complex-Horticulture. *Synergistic-productivity* is difficult to quantify, especially when there are multiple-values and diverse-products.

With our English language restraints and the industrial-mechanistic-priorities of Globalism, we are pushed to "rationalize" our work, "externalize" our costs, and limit wages-and-benefits. These demands come-across, over-the-media and from-our-bosses, as: insist, demand, cajole, and accuse.

Very difficult to respond to in a cooperative-way. Yet, back-at-camp, there is a balance that shifts-with-season, between introspection and polite-discussion, then getting-something-done with elegant and perfect timing. All with some rewarding physical-labor in the theater of Nature.

AND NOW, A DISCUSSION OF COLONIST EXPECTATIONS

We can prepare for this best-case outcome by indulging in neighborhood-inventories and disaster-preparedness, before the Changes. Transitional Human/ecosystem treaty-opportunities and stewardship-contracts are crippled by the economic-pressure of growth-demands from capitalist corporate-extractors (colonizers and their investment-class backers).

Nonetheless, some enlightened estate-holders are finding room, outside-and-alongside the constantly-changing county regulations and the protests of environmental-defense struggles, to assemble *life-boats* and nonprofit cooperative-incubators. Some of what we do now will be useful after the phase-shift (Orlov 2013).

Generally, mortgage and deed-keepers distrust caretakers (cottagers). The conceit is that poor-country-folks, who are offered a footprint and shelter, tend to collect junk and have funky-guests. Visiting estate-speculators and part-time (vacation only?) residents feel-and-act more lordly and try security-cameras, instead of caretakers, to avoid messy Human-relations and local realities-and-compromises.

When landless-workers and crafters do manage to attach to a farm, mansion, manor-house, or retreat-center, they are often more knowledgeable about local-ecologies than the busy market-and-investment entrepreneurs, who are "employing" them. The speculative-price of rural-land is way-in-excess of its sustainably-managed production-capacity. A young-farmer can only lease, rent, or be subsidized, during the real-estate-bubbles of imperial global-finance.

The first systems to fail
are the big, overly-complex,
global-arrangements.
They are brittle.
We can be ready to be flexible,
if we have discernment-through-knowledge.

A FEW OBSERVATIONS OF CULTURAL PREFERENCES

Looking someone "in-the-eye" is considered to be very-impolite in many cultures but is mandatory amongst-the-English. The direct-stare is a challenge. Do you measure-up? Are you ready-to-do-business? Predators do this to capture the soul-of-the-prey. Hustlers and salesmen want to seize your focused-attention.

Any cultural-practice can have natural-roots, mimicking-Nature, but some *etiquette* can become divisive, keeping class-structures rigid or making overly clear-distinctions from outside-groups.

Humans are specifically known for having an interest-in-caring for any, or all, other Humans, unlike within many other species. Although, what do we really-know about how this plays-out for *others*? It is common for both settled and nomadic folks to isolate-themselves culturally, to keep their distinctions, but to offer kind-welcome to strangers as long as they keep moving-on. This insularity comes with benefits-and-restrictions.

If the group of Humans is properly-scaled and is small-enough, internal-cohesion is necessary for successful-cooperation and persistence. If the group is too-exclusive, there are problems with genetic-inbreeding, knowledge-restraints, and good-relations with neighboring-groups. Prevention of continuous-war has often rested on intermarriage and fluency in several-tongues. Geography is the crucible of conflict-resolution (Kaplan 2012). Big, open-plains demand mobility and complex mountainous-regions resist invaders.

There are still semi-nomadic groups with tight social-cohesion in Modern-times. There are still edges-and-connections that enable sub-cultures to persist. In the midst of the industrial-grid-format, these groups usually have distinctive-clothes, by which they can identify each-other out-on-the-road. They also have internal-languages and phraseologies that allow polite, convivial community-experiences, while excluding strangers and other-cultures. This builds tribalism and serves the tribe-at-large with extensive relationship-webs, helpful-referrals, and shared-opportunities. But impolite-behavior is reinforced and propagated by hostility and mis-trust towards strangers.

Roma people,
have been in circulation a long-time.
Once Eurasian,
they are now global.

Indigenous People, still living in ancestral ways, have complex-ways that visitors cannot appreciate. First Nations have to inform Modern visitors that what they hear and see is not for sale. The way-thing-are-done has a wholeness that cannot be taken apart or mined for tricks.

There has been a long-story of expropriation that is colonial in essence. Some Indigenous ways are nonetheless familiar to other traditional peoples. We can see that there is a palette of cultural-ways that is vast, yet patterned. We can talk about the principles and observe patterns.

Settled, isolated-villages might be friendly-to-travelers, but are probably terse-and-indirect about village-affairs when interrogated. Quakers, in traditional farming-communities, send messengers to visit-widely among related communities, carrying a *letter-of-introduction* with the *traveling-minute* (message-of-greeting, and concerns).

> *A traveling-minute,*
> *also called an epistle,*
> *carries a community message*
> *of welcome and concern*
> *to other communities,*
> *in order that all come*
> *into common-understandings*
> *through the exchange of ideas.*

Visiting Cosmopolitan-Moderns do not appreciate much of this Place-ness, as askance-locals limit their access and opportunities-for-exploitation. There is a large set of interesting anthropological-observations to study (see Bibliography). The Human social/local arrangements that get reinforced by insular-cultures also take away some individuality and replace it with warm-belonging. Members of closed-communities risk being "*written out of meeting,*" excommunicated, or externalized-by-shunning if they break-taboos or disrupt Convivance (getting-along-with).

The experience of losing the belonging that is woven into everybody's psyche in these communities is devastating to the one-sent-away (shunned). This is an existential-plight; with no-help-from-Home, one has been externalized. Most people shrivel-up and collapse. For a person to be shunned, the social-situation has to be really-stuck. There are always cultural-ways of dealing with conflict or difficult-people who do-not-fit-in, among long-experienced Peoples of Place.

> *And the options*
> *in hyper-individualistic*
> *consumer-culture*
> *do not replace*
> *that lost sense of belonging.*

Cultural-discipline, through constructive-shaming, is powerful in traditional-societies. Good-social-order is taught in stories to the young-child, and reinforced through shared-rituals and *Eldering*: the focused-experience of being talked-to, in a formal-way, by a respected-older-member of your people. A mature-community has ways to bring dissenters and wayward members back into-the-fold.

> *Quakers*
> *value the dissenter*
> *in Meeting-for-Business,*
> *while reaching for*
> *unanimous-consent.*

Using strict-discipline should not be over-done; perfect timing and restorative-promise is the art of Eldering. Being shunned or feeling-shame, in anyone, triggers deep-seated natural-reactions. The Human emotional-repertoire has co-evolved with group-dynamics and can be triggered, both for useful discipline-and-learning and for oppression-and-control. Empathic-care is the glue that keeps a subsistence Culture of Place resilient.

> *The "peculiar-people,"*
> *Quakers,*
> *are said to be*
> *terminally-codependent.*

Some First Nations peoples have Clowns (another guild, or ceremonial-society) that playfully mimic non-approved behaviors in ways that exaggerate the behavior, shaming the target and entertaining the by-standers. The celebratory-flavor in the assembled-community is in the moment-of-opportunity to share cultural-norms. The "target" is actually a participant and recovers quickly, hopefully a bit wiser, if in on the joke ("Whoops!").

A *Clown* might start following one person who is acting haughty or angry and, without the target knowing what is going on behind them, everyone else starts laughing. The effect is sobering and built into the cultural-expectations of the community. The pageant reminds All-present of the shared-stories of how-to-live, who and where they-be.

> *Certain behaviors*
> *are considered dangerous.*
> *Pride, Greed, Tyranny, and Hypocrisy*
> *are famously cited*

> *as ultimate Human misbehaviors.*
> *"Abiezer Coppe did away with Sin!*
> *My body is my Church, said he,*
> *God's Spirit lives within"*
> *As by Abiezer Coppe, of the Ranters (Hill 1991).*

Let us repeat here how important the size of a village or community is. Once a group exceeds 150 people, there will be some members who do not know others. Decision-making becomes more difficult, for lack of familiar trusts, and stretches the oversight-work of the Elders too thin. If a large-group can divide up in a friendly-manner, spokes-councils (see Chapter 13) with delegates can facilitate inter-group cooperation. The new manageable-sized sub-group can now reach consensus.

> *At least eight,*
> *gathered to do council.*
> *Best thirteen,*
> *to have everyone heard.*
> *Twenty-three is still good,*
> *and thirty is getting unwieldy.*

Different from the manipulated-urges of hyper-individualistic consumerism, a great-comfort can be found in Place and subsistence-based cultural-conformity. Communal-friendliness and support relieves much *existential-angst* and *cognitive-dissonance*. The great majority of Amish youth come back for confirmation and join-the-church after their teenage "devil year" out amongst "The English" (Amish-category for Modern-societies). One gives up much autonomy to the group-think and hierarchy-of-Elders, in exchange for a simple-life, while held lovingly-and-tight, in the ways of Amish-culture. Good farming-decisions do seem to be made by the Amish. These folks work hard. They are hired to raise-barns for "The English."

Keeping People of Place in community and doing the Work should avoid *shame* (only used sparingly, as we have discussed above). Constructive uses of momentary-shame support the whole-epic-story-memory. There is a difference to learn, between cajole and consent, between rules and advice. Shame, argument, and strict-rules can be used to trigger reactions, to force conformity.

When the social-process is not transparent, and a person or clan feels dispossessed, there is shadow-resentment. Only mature Cultures of Place, with practiced spiritual-ethics can avoid many of the pitfalls that Humans have, built into their psyche, and juggle the collected-symbolic-referents to restore a sense of justice-and-wholeness.

Bringing new people (adoptees?) into the cohesion of Convivance can be facilitated by processing their fears-and-reactions, through rituals and rites-of-passage, while teaching the wisdom of good-social-order. Elders might make weighty-comments, and thus move discussions, but at best Elders refer to the held-in-common epic-narrative and the collection of First Instructions, inherited from mythic-origins.

QUAKER CONSENSUS WISDOM

By moving decisions through unanimous-consent procedures, the whole-community can come to agreement and move-forward, unified, All-in. Consensus has ways that steer-the-gathered, but with practice, we all know good-order (transparency). With the goal of bringing-everyone-in, the member who feels they must object, is closely-listened-to; they may have something important we forgot to consider.

Before the group-attention shifts to the one *being-truly-moved*, that member, rising to speak, feels a twinge of care-filled shame at bringing a problem-to-consideration. One member does not object to a nearly-unanimous decision—or *speak-against-the-minute*, during deliberations—lightly or impulsively. Fierceness is not necessary here. Be present and in alignment with Spirit: Speak softly your truth, and it may come to be seen as a gift.

*On occasion
a Friend might shudder,
as they are Being Truly Moved.
"Friends," they might gently say,
"I must stand aside, at this late moment.
I am not finding the Light of my objection clear,
and cannot yet Speak Against the Minute.
Still, I must say, I am troubled."*

In the end, if unconvinced by the gathering, a dissenter can *stand-aside* to allow the decision, while sorting out the roots of the dissent.

*Be real,
with some good healthy drama,
and appropriate laughter,
reflective of our Human silliness.*

HOUSEHOLDING WITH SIMPLE LIVING

Coming Home is a gift of re-localization. Our commuting may be on foot or simple wheels, our work-areas near Home or near camp. The gear we take on hoop-work, in the headwaters, is necessarily light and simple. Our choices of tools and mechanisms, are appropriate. There is *no away*: all our materials are in cycles, all our inputs and outputs are compatible with organic-Life. The simplicity is elegant; graceful lines move through Nature smoothly. The shapes and colors are from the local-palette, the icons of the long-story are tucked-in, and perched. Beauty is a gift we can reciprocate joyfully.

*Home ways
need to fit Place.
And Time.
The celebration of our livelihood
is adorned by our fitness
in Time and Place.
What we do now
is connected to the past and the future.
Live for the many generations to come,
appreciate the ancestors.*

This is the principle from the top of the chapter. The details of our Home-ways, our householding, and art, are local, vernacular, and appropriate. The promise is that elegant simple-living in good-ways will free up our creativity for art and culture. Our daily-tasks are purposeful and stacked with observations-and-opportunities. As we go about the sequence of early morning chores, we are greeting the world, adoring the beauty, accepting the greetings of the neighbors and Wild, and opening up to surprise and insight.

Learn to love your chores!

The tools of our nurturance, tending, and maintenance sing with us. We are attached to our artifacts. The value of elegant-simplicity to culture and soul is a deep comfort, a trust in continuance. We celebrate the simple-beauty of the local-made crafts. The reason we do-things-the-way-we-do are multiplex, but there are stories that explain many aspects. As we have been repeating, *precautionary-principles* remind us of the dangers of untested innovation. Some of these principles can be structural, a sort of checklist for carefully-reviewed science: four principles derived from the Natural Step can guide businesses and corporations.

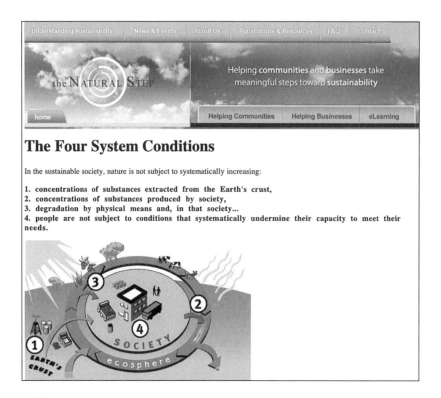

Household economy is built from the skills and materials available from the drainage-basin surpluses, with a few valued tools from distant crafts-cultures. As we are not putting our labor off on global labor-exploitation, we are making much of what we use. This looks like constant busy-ness around the hearth and in the sheds. Handwork such as knitting and fiber-crafts, food processing and medicine-compounding, caretaking and storytelling, are convivial—we sit around and work while we listen or talk or sing. Children see all of this happening and are taught beginning skills to participate and contribute.

14 Village Homes

A village or hamlet, manor-house cluster or hoop-camp, is a Home of households. A household includes the common tasks necessary to shelter a group-of-cooperators, not necessarily a kin-family. These are extended-families, or guild-collectives: they are sharing support systems. So everyone has a series of tasks, centered on the hearth, that keep the household happy. A cluster of households is doing a collective village-holding, as well. We have resources in common that we share.

The seasonal work-cycle, on the ecological-calendar, circles around our sense of belonging to a Place, and our Homes hold a special heart and health center that feeds and heals our community. We share a lot of knowledge on how-to-do-Home, which is very local. The social-ways that we bring this together vary by Place as well.

2 Indigenate

COMMUNITY INVENTORY, A STORY

In the spring of 2014, a couple of neighbors met to plan a series of community-meetings to map our resources. We came up with a facilitation-recipe. Thirty folks showed up to our first meeting, from various neighborhoods and farms in the valley. Hazel used some storytelling skills and posters to set the stage, talking the group through community-asset-mapping (explained below). We emphasized that none of these lists and compilations was to be photographed or posted online, to reduce fears that local-specifics would go viral, exposing sensitive information to whomever.

18 Inventory

In a second meeting, we reviewed the project and brainstormed what-was-missing, identified what actionable-projects emerged out of the information, and where we would go-from-here. We discussed external-challenges to our valley. We began to think like a drainage-basin-council. Committees formed on delicate-questions, new inventory-categories, or to move forward with the obvious-projects: a bulletin board at Buncom, a ride-waiting shelter at the Crump mailboxes, and a serve-yourself farm-store at Yale Creek got attention. Several small-businesses talked about sharing bulk-buying and recycling by-products between operations.

There was a lot of enthusiasm for this community-inventory, and some of the projects have seen progress. By our third meeting in June 2017, we had assembled our bundle-of-poster and lists divided into three categories: mapping, big ideas, and public services. We reviewed our story-thread with 22 folks, again—half of them new. Systems-posters hung about to keep us oriented.

At this meeting, we realized how little we know of folks living in the valley. The Mapping-group suggested we assemble a map-bundle, all maps to the same-scale, to overlay and see the coincidences and think strategically. This bundle would include geology, timber, mines, trails, wildlife, water, and wildfires. Collect, organize, and interpret. Big Ideas group reported lack-of-housing, need for family-support (daycare and education), interest in cooperative-projects (barn raisings and cleanups), and concern about food-insecurity, even with all these small-farms. The public-services group focused on communications and connectivity-trees, the well-being of our community, and the organizing of emergency-response.

We now celebrate a community calendar, creek-drainage neighborhoods, connection-trees, the ride/wait shelter, and ongoing interest in sharing. There is energy for three council-meetings a year. All-age educational-evenings are popular; we have had two "report nights" hosted in local homes, where a queue of presenters get five minutes each. Much fun is had. This whole-storyline is deepening our sense of Place.

COMMUNITY-SCALE PLANNING

Scenario practices that set the stage for long-term plans, through brainstorming and visioning, orient and educate local-folks to see Ultimate-optimums, beyond imagined-constraints. Community-inventory and assessment, on a drainage-basin scale, is the foundation. As knowledge is gathered and mapped, tabulated and modeled, we see potentials by peering through the layers of information, alert to *quick-fixes*, retrofit-challenges, and opportunities emergent-from-complexity.

Making speculative community investments during a phase-shift, when everything changes, is risky.

Community-scale planning is massively imaginative and ambitious, but keeps us thinking. Conservative, precautionary-planning is the lesson to follow during Transition. Releasing the resistance to hopeful-visions is the first-step. Elise Boulding facilitated community-gatherings that dared-to-dream in the late 20th century (Davis, CA, 1980s). Getting real, but still moving forward, is challenging.

Now, further down the cascading-effects of failing-globalism, back to Earth, let us imagine side-creek neighborhoods, sending delegates to drainage-basin spokes-councils. This process makes clear who collects-and-stores the tools, books, and hard-copy maps, to facilitate community-inventory. We should be collecting hard-copies, even as the satellites are still up.

During scenario-imagining, the stress of community-inexperience can be relieved by reminding folks this is play-and-practice. Are we limited to being true-believers, extroverts, and introverts with performance-anxiety? Best get real-local-folks, who may already know each other; family-delegates; and spokespeople who are from this farm, that lane, or hidden gulch to show up for scenario-practices.

RECIPE FOR COMMUNITY INVENTORY, REVIEWED WITH SYSTEMS JARGON

For your entertainment and elucidation.

When gathered and grouped by neighborhoods, after orientation and songs of gratitude-and-praise, the ten groups each get a large piece of paper with a category-name at the top (see Inventory poster, above). Each group brainstorms the details-and-information within their category and adds the details to their poster. Then the posters are in circulation. Thus ten neighborhoods, so as to have every group with one of the posters, at every rotation. This is the accumulating-inventory of the drainage basin-wide, the hyper-local aspects and resources, listed in ten categories.

Rotate the posters every few minutes. After maybe an hour, there will be ten posters, full of lists. Re-sort folks from their neighborhood-groups into affinity-clusters that focus on each of the category-posters. The new-groups can then clump-the-notations from each poster, sorting out the brainstormed-lists into categories within the category and drawing a diagram-of-bubbles on a new poster. Lines can then be drawn to show relationships, actions, or bridges between bubbles. Perhaps a Venn diagram is appropriate with a space shared among radiating and overlapping bubbles? Or some other system type? The type of systems-poster chosen matches the nature of the information.

Out of a two-to-three-hour meeting comes at least 20 posters: the original lists and ten posters that reorganized the lists. The overall cache of the community-inventory gallery will suggest a meta-poster (see "Community Resources" poster, above), implying a multi-dimensional map/model of categories, in relation to each other. The afternoon's work is rolled up and handed to the Bundle-Keeper.

The next re-convening of community-delegates for council walks the poster-gallery gauntlet, on parallel string-lines (with clothes pins), so that participants can see, update, and review the mapped information and process. There can be a one-half turnover of delegates from one council to the next, and everyone can still get up-to-speed with this review. The meta-poster will help set the stage for the second meeting's agenda. The first exercise is to brainstorm the hopes, and then the fears, in order to clear the air for visioning. We can hear a review of the first meeting. Then we brainstorm emergent-possibilities suggested by the posters.

By the end of this second session, obvious take-aways and to-dos emerge from the Light-of-the-discussion and action-groups, or committees, coalesce. For example, if we start with brainstormed outside-threats, perhaps a committee-of-the-wise can deliberate more deeply and report at the next council-meeting on internal threats (a more delicate topic). Work parties need organizing. More posters, modeling committee-work, can also be generated. The list-of-maps and categories of information-collected—the meta-poster—can be shared, to help other inventories elsewhere. The local bundle stays Home.

At the third council-meeting for community-inventory, the conversations get practical. We can now get going on clearly-important tasks. With the reports from our previous community-efforts and the poster-mapping, we are now holding a complex-set of ideas about our valley, and the priorities-and-sequences of regenerative-tending.

SYSTEMS POSTERS

Posters in this work cite the systems-tool type in one corner.

There are several types of systems-posters that help display the context and bubble-clumps of categories in a chart, cartoon, storyboard, or nest of lines-and-relationships. The Ken Wilbur quads, dynamic-spirals, zones-and-sectors, bubble-webs, and Venn-diagrams can grace the walls or clotheslines of the council-lodge, as icons. These maps-of-knowledge surround the deliberation-space, and remind us, while we dance through the conversations. As in visiting Stations-of-the-Cross, or touring the Gaelic Sacred-Trees-Grove, these posters-as-icons serve as mnemonic-devices. The scaffolding of Traditional Ecological Knowledge, arranged around us, in council.

25 Water Use

We can see on this Wilber Quad systems map of Water Use (25) that there are many aspects of Water Use to be considered, in four quadrants. However, we also find that most council conversations are concentrated in the lower right, the exterior/social basin. We would do well to consider water in more contexts. This systems map type can help widen an agenda or discussion so that more bases get covered.

Another very helpful multi-dimensional mapping approach is to collect as many map-type layers as possible: topographic, geologic, hydrologic, wells, mining-claims, ownership-boundaries, area-use-plans, infrastructures-and-utilities, wildfire-history, ethnographic-archaeology, layers-of-colonization, and extractive land-use of the past. If all these can be reproduced in the same common map-scale, on see-through plastic or thin paper, then taped-up on a picture-window that works as a light-box, sequences of selected layers-of-information can be super-imposed and scryed-through (read-for-clues). This is referred to as *palimpsest*, as in seeing shimmering-depth in layers-of-paint, or seeing meaning in chaos.

Three-dimensional (relief-projection) maps, sand-box mock-ups, and constructed-models (paper-scissors-glue) gather endless attention. We can hover over them and imagine our whole-drainage-basins. Models can be hand-made, with layers of cardboard and cut-to-match topographic-map elevation-lines, then glue-stacked and elaborated with representative stick-ons. Be creative! Add texture and color. A sand-box model can offer a seldom-experienced perspective; we hover above, as if an Eagle. We can practice, using our mind's-eye, seeing dimensions that are hard-to-model, such as Time. The mysterious (to Humans) dimensions are the air-we-breathe, the earth-beneath-us, and the surrounding-horizon (Abrams 1996).

The Principles of Transition include

51.
Beware taking from Nature as if it is owed us. Avoid overfishing, overhunting, and overharvesting, of any sort. It is not virtuous to exploit Nature just because your ancestors were removed, placeless, and disenfranchised. Sit down and think.

52.
All forms of social arrangements might best be modeled on natural patterns and built within the limits of Place. The local biological carrying-capacity determines our Culture. Celebrate what we have. Build a new tomorrow. Go Deep-Local.

CHAPTER 12
Cultures in Transition

PEOPLES HAVE OFTEN RE-ENVISIONED EXISTENCE

Scenario-practices (see Chapter 11) are visionary and create a new-story. Propaganda can also use new mimetic-ideas (contagious word-viruses), as does consumer-advertising.

Vision-councils have kept some Indigenous-peoples oriented, during displacement-and-migration, by holding the True-story dear, while considering options. Mixed-provenance, hunting and horticulture groups, have found ways to settle into a new hoop and tend a changed landscape, following climate-change.

Colonial displacement moved the Sioux and the Anishinabe, or Chippewa, peoples west in the 18th and 19th centuries (Farb 1968, Martin 1978). The Sioux were pushed out of riverside-Horticulture, onto the northern-plains. A strong foundation of origin-myths and traditional-instructions on behavior and the nature of being, persisted, and a council-evolved way was found. Many new-instructions could be revealed and received. New tools came into the bundle: Horses, steel, and Hemp allowed a wider, faster, further hoop-year. Buffalo tending with Fire and Horses allowed a powerful new/old epic to blossom and sustain Culture.

The Anishinabe sat through a series of council-fires (conferences), as they moved from the east-Atlantic coast to the wild-rice lakes south of Lake Superior. They carried a set-of-pictographs (bark-paintings) that helped them remember who-they-were, and how-they-moved and changed. These language-in-common groups kept essential-TEK while they moved and changed—the base of symbolic-wisdom that fostered continuing co-evolution and nurtured cultural-continuity.

European-migrants, who had been removed-forcibly from their cottages and shipped to some-other continent, with broken-traditions and scattered-relations, were more susceptible to becoming Moderns.

The carried-record is commonly in symbolic story-paintings and ideograms, on bark or skin or sticks, with notches, paint-marks, shells, or quills (Martien 1996). These mnemonic icon-strings cue the epic-story in memorization, and the carried-record can be displayed during council, to hold-the-circle with the ancestors and epic story-bridges represented.

Pre-colonization social change among semi-nomadic Horticulture-based hoop-folks seems to have been incremental. The language gains words and phrases. The list of taboos rotates with the cultural-priorities and with ecological-conditions. Sets of patterns are widely adopted (etiquette). These clan-based peoples have been in-Place (however large) for a while, and have learned from Place by making-mistakes and then being wise-enough to celebrate-mistakes. It is the held-memory of long-times, and hard-times gotten-through, that builds the cultural-ways of a People of Place.

*Traditional Peoples,
entangled in Horticulture
and Hunting,
hold the keys
to our post-Modern repairs.*

When Indigenous-languages expire, we all lose huge-nets of TEK, fine-tuned to Place. When new dialects emerge as pidgin-languages (mixes), trading-languages (Swahili and Chinook Waugh-wa), and slangs, they are seldom ecologically rich. Traditional storytelling is often in older-language, and the storyteller training happens within mnemonic-architecture (natural-cathedrals), to preserve exact-repetitions of inherited-epics (Illich, 1988). These trainings can be fierce, as the People hold their heritage.

Both taboos and etiquettes can become entrenched and lose their ecological purchase, especially when Cultures of Place are force-marched into industrial-extraction, armies-of-empire, and mercantilism. The original meanings of the cultural practices were tied to living-on-the-land, and when the story leaves the origin-Place, basic-wisdom may be kept, but the ways-of-being become abstract.

This holding to the storied old-ways can be a glue-of-necessity, but the whole-story will remind everyone what the work is now. When Bio-dynamic Horticulture was transplanted to the Pacific Northwest, my students told me about a German native who came over to teach and wrote a critique (Storl, 1977) suggesting we-here need to learn new-ways, appropriate to our new Place.

*OK then,
let's get at it.*

Traditional social order is a multi-dimensional dance-ground. Some social order is so ingrained as to be sub-conscious. Often only visitors notice critical-practices, or shadow-work. To our benefit, reports have come from village-Horticulturalists, semi-nomadic hoop-camps, and Herders. At best, these cultures facilitated high-biodiversity and high nutritional-quality, from diverse, selective-harvests.

Humans make messes—disturbance-regimes—yet the disturbers can be enhancers. The increased-productivity, from ecological-interventions in semi-Wild ecosystems, is called *intensification* or *magnification*. This usually refers to Human-benefits of harvest; there are other-benefits that can be called *eco-system functions maintenance*.

Perhaps some of the most-diverse and complex, tightly interwoven species-densities have been on Human-affected landscapes? Such societies, living in ameliorative-relations with their Place, always exhibit intact, inherited-TEK and an attitude of gratitude-and-humility. The useful-lesson here is that Human-cultures, embedded in abundant-Nature, have co-evolved in-parallel-with their species-cohorts, constantly-learning and keeping Wisdom.

EMERGENT ECOTOPIAN CULTURE

Ecotopian syncretic culture-creators, very-recent settlers in the Pacific Northwest, are brave to try to cobble together a new, Place-appropriate culture, fast. They are culture-Bodgers: they fit pieces of borrowed-ways together to build a chair-of-wisdom that might support the new-tribes and the integrity of drainage-basins.

Meanwhile most moderns are lost. They need to get-rooted and become Place-wise in order to survive the mess they find themselves in. Big global-systems have unfolded without social consideration-at-large and without anyone actually being-in-charge. This is an explosion of mechanics and numbers that has no heart-or-ears. We-All, on this small-planet, are along-for-the-ride. Dig-in somePlace, and grow-culture through council and maps of social-relations.

*"We've got
to get ourselves
back to the garden."*

There is a common-set of universal-patterns (see "The Common Core Model," Mollison 1988). Consequently, there appears to be a common-set of Human-social-arrangements that owe their universality to direct-observation of natural-systems over multiple-generations. Even the promise of this tendency is enough to spark-the-interest of bold social-change artists.

1. Observe

2. Indigenate

If a language-group of villages, and/or hoop-walkers persists long-enough, some predictable social-arrangements emerge, relative to their habitat and diet. There are Place-based exceptions, but generally women's-council is the coordinator of extensive and intensive-Horticulture. The men's-lodge is crucial with semi-nomadic Hunters. Mixed-providence peoples exhibit social-balance. What-we-eat teaches us how-to-relate. There are practical-roots to these emergent social-relations.

"You can have some of my purple berries . . ."

When an appropriately-scaled group of Humans (best 13 in a clan, and then 13 clans in council?) achieves a cultural-body of use-patterns, which are appropriate to regional co-evolution (with All Species) and intensification of edges and their yields, social ranking-systems emerge. The abundance-blessed coastal-cultures of the Pacific Northwest developed a sharing-system called Potlatch. A village social-ranking (and guests) sit within circles of circles and pass a circle-of-gifts, and feasting, that flows-through-them (as in *dissipative-structures*) on a seasonal calendar.

The order-of-service (sequence of eating) at regional ceremonial-feasts seems to display a linear-ranking. In Potlatch culture, the poorest or lowest (captured strangers, who are becoming adopted?) can gain-fitness through gifting, making the line into-a-circle. The awesomely-appointed (adorned) host becomes poor, through giveaway for a feast, or a festival day. The European Gaels had Fool-to-King stories and ceremonial-regicides following eco-collapse or crop-disasters.

Collecting a set-of-guidelines for Culture is a fool's-errand, as we fumble with peak-everything. The promise is that we might get-something-right, and persist, and see progress, if we make the commitment to Place and People. We can change the old-stories to make sense where-we-are.

DODGING AND CULTURAL REFUGIA

We are in the aggravation-phase of the decline-of-empire, and our job is to dodge. Even though there may be continuing urban street-fighting, we still have to get working in the headwaters. Our job now is to make the new happen, and it's going to be completely helter-skelter. We need to have complete-empathy for people's-choices. Nobody knows what the perfect personal-choice is next, except clean-up-your-act, and a couple-other-things (see Ten Steps to Becoming Local). We are going to all get-closer to each-other as surviving-cultures, in the near-future.

Making a prediction—we're going to work closer-together. That's the theme of today's story: working more-closely, with the landscape, listening to what-the-landscape-wants. The landscape is coming-into-council with us; here comes landscape: it's at-the-table with us. That's *indigination*, finding ourselves in the Place where we feel-we-belong, where we don't feel like-a-visitor or a tourist, where-we-belong, and the land wants us. Our Siskiyou experience of the persistent Wildlands—mostly the mid-elevation fragmented-Wildlands, where we see the persistent-Ethnobotany—is that the land is lonely-for-us.

*We can be useful
and entertaining,
all the Time.*

While no one can predict, by the principles of *Chaos Theory*, how exactly the empire crumbles or where the best-Place is to stand, or which-way to dodge, we do know the working-principles of productive-Horticulture on a landscape-scale. We might as well explore scenarios, just to keep our group-mind in practice, while still paying-attention and offering-aid to the casualties.

During the period-of-aggravation-and-change, we locals are faced with shell-shocked-survivors, refugees-and-seekers searching for community, or pillage. Local rural-neighborhoods and small-hamlets may not be prepared to offer triage, decompression, or even re-skilling (Orlov 2000). A staged re-localization-process would be more survivable, but who is going to facilitate that? That would be you-all and us.

As financial-globalization devolves to regional-sovereignties, with their salvaged-militaries, we may want to organize-and-negotiate, as local-citizens.

A *life-boat-strategy* of caches-and-caves may carry a few survivors long-enough to re-emerge in survivable-*refugia*, and local-wisdom could begin to re-accumulate. Holding rural-culture, during the last-gasps of imperial-greed, is a matter of knowing the local carrying-capacity, carrying a vision-bundle, and having enough-isolation, with perhaps no precious-resources still-evident.

TRANSITION CHALLENGES

The cooperative-anarchism of Deep Ecology and Forest Defense activism is consistent with ecological-principles and TEK, when it emphasizes restoration-strategies. Relation-web-maps suggest inter-dependence, in balance with self-reliance. When the systems-tangle of social-challenges, resistance to climate-shift, racist-cultural-inertia, institutional-obsolescence, and psychic-storms of confusion all seems to demand re-localization, we may not really-need to directly-attack or protest symbols of global-corporatism, unless we are threatened in our local drainage-basin, and then we might need to muster mass-resistance and call-up mutual-support by treaty.

If you think that sounds wonky, try this: The magical skill-to-practice may be *dodging*—trying to be in the right-Place, at the right-Time, without a lot of predictive-confidence. Thus we as social change-artists have a lot-of-compassion for multiple, potential action-paths, as appropriate wide-experimentation. If enough-of-us are lucky to find the conditions we are exploring-for, or can adjust to, in this dystopian-unfolding, we bring great good-fortune to All.

If none of the scenarios envisioned in our community make sense in times of chaos, we need more council time to mix-up a new batch of strategies. Our goals, beyond survival, still rest on the necessary Truths held in our principles. The story we are telling here is restorative. We still have the long-time vision to lean on and the ecological-goals: return of Salmon, Beaver, Fire, and Keeping-It-Living, with biodiversity-enhancement. This strongly-suggests post-petroleum, locally-self-reliant cultures that are working-with Nature.

This art-installation type culture-implementation impulse may go-through a variety of Transition-stages. We may find ourselves contracting with landholders as co-operative societies (Drainage-Basin-Council franchises?) for seasonal-occupancy (caretaker-cooperatives), while walking through the hoop-of-seasons and work. So far, Ecotopian Anarchists (anti-authoritarian by protoplasmic-reaction) tend to spend a lot-of-time in-council, just to cut one tree.

*Be sure to drink plenty water
while in council-lodge.
The Mind is thirsty.*

Many hyper-individualistic neurotics (Moderns) who intellectually have Place-based re-integration in-mind and in-sight will have a challenge with discipline-and-protocols, or just trusting-the-potential of cooperation. Prototype Hoop-Cultures cannot serve as mental-hospitals, although they may serve well as Spirit-maintenance clinics. We will need a miraculous and dedicated cadre. Lots of *holding Friends in the light*. And Elders.

Thus, we think about multi-dimensional social-relations. Communities of new-settlers might want to practice the ritualistic Council of All Beings (Macy 1988) and try to channel (act out, receive messages from outside of language, dance) the spirit (*light*), representing All Sentient Beings of a bio-geography/drainage-basin by giving-voice and expression, in parade-and-council. Excellent social-orientation.

Systems-assessment and mapping, along with discussions and plans for potential ameliorations, then emerge out of a better informed mass-consideration, with folks knowing-their-Place, their position, and their responsibilities: *good social order*.

After we-All get enough practice with council and culture, we will be able to turn our attention outside of our Human concerns and see the communities that surround us. Not only do we learn the restoration-work of enhancing ecosystem functions, we learn the social-relations of community-to-community within whole regions, across drainage-divides and within collections of drainage-basins. See below in "Guiding Principles for Culture" for mapping extensive-relations.

*We have a whole planet
to consider.
Bigger work than
Human-cultures
have done.
This is eco-globalism.*

Once we really get rolling on Earth Repair, local council-consensuses can lead to negotiations (*epistles* and *treaties*) with other cultures. In the Pacific Northwest, this first means the reinstitution of First Nation treaty-rights, and the establishment of mutual-support agreements. There is a wonderful net of information-sharing that everyone, everywhere will appreciate. This can be done sensitively, so that Human cultures do not feel colonized or exploited, as the industrial-globalists have done. Mutual-respect and careful-diplomacy is the natural-way (Akwesasne Notes 1978).

Good foundations, for useful taboos and etiquette, can be built by story-sharing. Storytellers from many cultures tell familiar stories, reinforcing common-wisdom. Some early First Nations treaties were poetic, and thus more whole, as they laid out the foundations-for-stewardship and mutual-aid.

Many traditional-educations tell stories of the receipt of Original Instructions. These core-values suggest job descriptions for a wide range of diverse participants, on a well-known landscape. All Beings possess categories of honor—their stories are told over-and-over, in the referential-epics of relationship.

TEK seems to always insist on the *outside-category* (*Other*) relationship to Spirit: a mysterious, beyond-dimensional aspect of the universe. Spirit is due respect—no name is appropriate. We are encouraged to refer to All Entities indirectly: nicknames, vague-pronouns, or gestures.

*Pay attention
without staring.*

Any Social Forestry culture is almost a living-entity itself. We can speak of Social Forestry as gifted from the Deep Local, a precious-being played-out with communal-cooperation, inclusive-and-wise.

Modern individualistic consumers seem to focus exclusively on, "Where do I fit in?" They have identity-concerns. As if the best-choices are found in accumulations of personal opinions and decisions, or the solution-to-anxiety is to self-reinvent, over-and-over. The inclusive-discipline of traditional Indigenous-peoples' practices compensates for the insufficiently co-evolved urges of rootless-individuals. The community fills in the gaps with *good social order*. The common village-culture approach is the use of shunning and shaming. Or, with more finesse, clowns and perfect comments by Elders: find the wayward-passion and direct it to benefit All.

The rewards of joining social order, wide-relations, taboos, and etiquette are not obvious to Moderns. Convivance has implicate-value, in that one has a comfort of knowing-support and cooperation. One becomes known to others, in many ways. The elaborate Place-ness (seasonal festivals) and good-order can be comforting. One becomes we, and we can concentrate on approved-and-supported tasks with focus-and-enthusiasm.

In order to find ways to fit-everyone-in, the Elders are watching the children. using the sort of bond that skips generations. Careful observation of skills-and-propensities leads to mentoring and, many times, personal-naming. Children might be apprenticed at ten years old or earlier. Special honors come with being who-you-are and playing along: hats, sashes and ceremonial inductions? Rites of Passage! The stage is set with stories that explain why we do this, this way.

If all participants have usefulness, honored-at-large, perhaps individuals and we-as-a-community can find contentment-enough to keep going. In some traditions, healing is seen as a process of re-integration. One is visited with ceremony and reminded of family/clan/society/Place/Spirit/name/song and is thus encouraged to heal. The opposite of shunning! An isolated-individual attitude, in youth, fades fast when they accumulate the belonging-experience from birth.

If you are run out of community, where do you go? On the other hand, when solidly a member of a community, folks at passage-junctures are sent-out-on-quests that are just as disorienting as shunning and return wiser. The accumulated-ability of Elders is to facilitate the integration of some personalities that otherwise might be trouble if insufficiently engaged. Some advice comes with TEK, stories that tell of consequences-of-behavior. This sort of social-relations-knowledge is key to persistent ecological-thrivance.

*In matriarchies,
the rambunctious lad
is given options.
You can be chief,
if you learn oratory.*

The overweening experience of Moderns is fragmentation-and-alienation. They feel lost and needy: the perfect set-up for consumer-addictions. As Humans, we have the biochemical template shared by all Earth Life and special-adaptations on our evolutionary fitness-peak. We have dwindling-options for such an opportunistic-species.

A species being selected out, as conditions change, collects random code-glitches on our genetic-memory, through stress and chance. This is what-we-are, but who-we-are is a cultural-evolution, a social-opportunity that requires a different code-memory than DNA. Our Human evolutionary adjustment is TEK or deep-wisdom, and the expression of this mimetic-coding (reinforced by repetition) is appropriate good-social-order.

Our cultural memory is modulated by our language, and symbolic codes (ideograms, icons, pictographs), and learned mental processes. Thus we are theoretically capable of compensating for our genetic impulses, with social-discipline. Taboos, etiquette, and storied social order are the traditional-arrangements.

The tools for a common-discipline are found in phraseologies, procedural-protocols, and celebrated-gratitude. The storied natural-principles modulate and constrain the perceivable-options. The whole kit is derived from Nature. We can only try to better-our-ways. Culture is practice.

Perhaps an interesting question in our discussion of Social Order is constraints. Personal self-assessment is very popular, which seems to take place out-in-public but is not processed-by-public (except superficially, the gaze). In socially evolved culture, we now know that observing variant-learning-styles, skills, and physical-propensities can help focus mentors on feeding passions and thus facilitate integrative-orientations. This is a public-process of assessment and appreciation. Fragmentation leads to alienation, just as shunning can. The modern, and inHuman experience, has similar disintegrative-effects. Healing is available.

RETRO-FEUDALISM WANTS A TURN

Feudalism is an old authoritarian arrangement. The monopoly of violence first sequestered by chiefs, in villages, graduates to an inherited order of power. Many Indigenous peoples have allowed monopolies of violence to consolidate after too-many generations of brutal revenge killings and local-warfare (Diamond 2012). European feudal-systems are built around Warrior castes, with serf-built fortifications against invaders.

The other castes were the Clergy (who keep the divine right to rule reinforced), the Crafts-guilds (which later gave rise to cities), Servants (many sorts, professionals included), and Peasants (people of the land). These castes were held together by threat-of-force and by mutual-needs. Feudal-arrangements were stable for long centuries, except during war, plague, and famine. Cultures evolved within the feudal-system—festivals and decoration bloomed in good-Times.

Wild people
have always lived
outside
the feudal boundaries.

After the Roman Empire faded, it was monasteries and nunneries that kept libraries and learning stashed away. The dispersed Royal caste claimed divine-right to rule based on their direct descent from James, brother of Jesus. Pilgrimage routes that crisscross Eurasia, lead west to Fin de Tierra and the St. James monument.

Thus early Christians took power and used convenient translations and interpretations of scripture (Clergy) to justify oppression. The Peasants often kept many land-based natural-practices, which the Clergy incorporated into their liturgy out of necessity: How else do we get the crops to thrive, so that we can feed the warriors and cover the overhead?

*Retro-feudalism
wants to celebrate
the useful-flows
of an old
and never-vanquished pattern.*

Capitalism and industrialism grew directly out of feudalism. The Christian gloss (easily adjusted to fit imperial goals) enabled abstract ideas that justified the destruction of Nature to become a planetary-ecospasm of European culture. Protestant rebellions, based on new translations of scripture, further muddied the deep-connections of all peoples to care-for-the-land. Hidden in the massive propaganda of imperialism is bits and pieces of Earth-wisdom. We need to suss those out, and shine Light on Truth.

RETRO-FEUDALISM SLOUCHES TOWARD THE HEADWATERS

*Retro-feudal visioning
tentatively unfolds possibilities,
based on principles, ethics,
and patterns of persistence.*

Perhaps there are still some private-parks and public-lands allowing some stewardship-tending, but if any land-owner class persists through *Transition*, they will need help to stabilize the rundown ranches, farms, and woodlands that they intend to lord-over. The *retrofit* for social-Transition and de-colonization might include a phase of small-hamlets, clustered around manor-houses, as long as they keep making sense!

Can manors (McMansions!) be retrofitted for low-energy-use or do they become factories and warehouses? And how will they do in wildfires? If the managerial-caste (owners, for how long?) needs the manor-space, they probably ought to share it with Servants (record keeping, libraries) and offer public-event access in season.

Just imagine this as an emergent-pattern. We are doing scenario-visioning here.

Perhaps local on-the-ground folks will be appreciated-enough to be given status-and-tenure? The Servants-quarters may be near the manor-and-fields. They may wear many hats, showing their craft-guilds or ceremonial-societies as appropriate. The woodlands need Cottagers to see real-progress in drought-proofing, wildfire-allowance, and fecundity. The Cottagers may have relations-with Hoop cultures and keep a forestry-camp ready.

Rangers (Sheriffs?) walk the connective-trails between hamlet/manors, villages, and watershed-ridges, as wide-scale reporters. They watch the passes. They wear a big-hat, fluid-as social-intercessors, bringers of news-and-discussions, announcers of opportunities-and-necessities, and predictors of perfect timing opportunities. Drainage-basin councils draw a big circle that includes Human-cultures and All Beings.

Retro-feudalism is a scenario of near-future decentralization, obviously imminent. The *Chatelaine* is traditionally the "woman who watches."

She manages the affairs of the manor and holds its hamlet, while the globalist-men are out trotting around to wars and mines. This woman knows how everything is done. She carries the ring-of-keys. She takes notes. She manages the purchasing-and-sales. Her partner (or cohort) is away on business (or civil service) a lot. This is a special skill. A very-competent, locally savvy manager.

If the Sheriff has a concern about the estate-Forester, she (the Sheriff) presents the case first to the Chatelaine, who settles the matter or calls for the person. This political-courtesy supports the Chatelaine and her community. The Sheriff is elected locally and serves everyone.

20 Assumptions Matrix

An eco-savvy Cottager is a guild-member with celebrated-expertise. After tenure is established (Life Lease), they may oversee side-drainages associated with the Manor, as well as other local reserves-and-commons, in consultation-with the local drainage-basin-council. Manor Servants also participate in land-work, as all-residents pitch-in for seasonal work-parties and celebrations.

CALLED TO TEA BY THE CHATELAINE

The ranch Cottager/Forester, Hazel, in their associated role as Ranger, had been visiting over the mountain range in the next big drainage, Bear Creek. In public-presentations and salons, said-Ranger had been referring to Maud, the Woman-Who-Watches at the ranch, as the Chatelaine.

As gossip does, the word got back to Maud that Hazel had been suggesting the ranch was being run as a medieval feudal-estate. The Ranger had been trying to talk to folks about how the collapse-of-globalism might find older-forms of social and economic arrangement, appropriate to organizing ecological-restoration. The term the Ranger kept using was NEO-FEUDALISM.

After a few months of the Ranger talking this *neo-feudalism* up with posters and panels, Maud invited Hazel to late-morning tea. "You are not my serf!" she avowed, "And what's with this *neo-feudalism* bit? Sounds bad!"

"Ah," said Hazel, "I meant no harm. I appreciate the support and allowance that the ranch has facilitated so that we can do Social Forestry research and provide the ranch with ecosystem services. What would you suggest?"

"*Neo-feudalism* sounds like *neo-liberalism* or *neo-conservatism*," said Maud, "And we know where those have gone: global-ecological-catastrophe! Surely we can do better? *Neo* implies bringing back failed-ideologies, merely for continuing exploitation-and-profit."

"How about *retro-feudalism*?" said the Ranger. "Could be more like *retro-hipster*, with the aspect of art and clever-appropriation? We are trying to rearrange, not fall-back to authoritarianism."

"Sure," said the Chatelaine, "But please be a bit more careful when you are using the ranch as a canvas to sketch-on and color-in this speculative-visioning. We are blessed with family and community support, in our endeavors during this time of Transition and experiment. Perhaps we can work out the forms locally and avoid the political-global megaphone? We still have to get through the phase-change and its chaotic-unknowns. I am feeling protective of what we have got."

Absolutely-appropriate for the Chatelaine to have this conversation with the Cottager. The Ranger, with their base-camp lease-exchange, is finding ways to

(continued)

balance the big drainage-basin picture with the infrastructure of settler homestead/farms, considering carrying-capacity, social-change, evolution, and multi-dimensional eco-relationships. Perhaps this *retro-feudalism* can birth-and-rediscover some vocabulary-and-understandings that facilitate discussions in local-council and even clarify the wider inter-basin mutual-support treaties.

In Retro-feudalism, the Chatelaine is the Woman-with-the-Keys. She needs to know everyone's relationships and arrangements. She knows everyone's jobs and basic needs. She keeps the ranch whole and manages the budgeting (on many levels), while other roles, such as Farmer, Forester, Craft-Cottager, and even Traveler, can concentrate on tasks. The home-grown Traveler may have to be away a lot on business and connectivity-work. They may seem, to the neighboring-regions, to be the primary ranch-representative, but the Chatelaine is the decision-maker and continuity-conductor at Home.

RETRO-FEUDALISM DURING TRANSITION

This scenario concerns the re-inhabitation of the countryside, investment in small-farms, and the rise of local-culture. There is plenty of work to do on-the-landscape and fewer-and-fewer people have been working-outdoors. Meanwhile things are changing-fast; we might have refugees referred to us, and we may need a lot-of-labor.

The Transition Times continue to hold private-ownership of land, but the Ultimate drought-proofing of the valley will involve the whole-community, in cooperation across-caste-lines. Similarly, the relationship between managers, workers, families, and syncretic culture-change artistry needs to get communicated in an organized fashion. We could all have a good-time, if we get that right!

The value-and-honor of roles-and-expertise can translate into things-getting-done, or at least in motion toward regeneration. All of us will want to understand systems and should be able to help-out-with almost every task. Work-blitzes can manifest from this redundancy of skills and collective-cooperation.

A small-village of small cabins; a barn; a large manor-house with storage and guest spaces; lots of small-sheds infrastructure and some animal-support; a campground for seasonal-workers and Hoop-guilds; and pretty soon the small-valley or farm is looking rather medieval? Not so, what with remnant Modern-technology still helping and great-advances in science and medicine.

Nonetheless, perhaps we can learn-a-lot from the previous dark-ages. In our imagined Transition, we see forestry-camps on undeveloped-land; we see a few Life Leases for eco-services and ecosteries (meditative care-takers); and we see some winter-camps for seasonal-forestry. These Time-bridges can go somewhere, with some council-Time, and favorable fates.

Humans are profoundly interconnected with the whole-biosphere. The industrial-scale messes, left by colonial-industrial capitalism, could doom our species. Up river and down river, over the mountain range and across the ocean, folks need to learn-again to indulge in negotiated and re-negotiated treaties that open some value-added trade, cultural-exchange, methods for conflict-resolution, and friendships. We have some landscape-wide coordination to discuss. A new story-line will celebrate a rich-epic of ecological, ethical, and plain-sense principles that strongly urge us to get-along.

Negotiating social and work-trade arrangements 'twixt the manager/owner caste and the manor Servant-folks, with the local drainage-basin-council in oversight, is a challenge of cross-cultural translation and storytelling. Settler-folks need a lot of education—takes-Time.

Meanwhile, most old-homesteads and small-farms have been taken out-of-production, save hay and Cows, to keep irrigation-rights. Here is that classic-conflict again. Absentee-landlords do not allow counseled local-investment in persistence, they just want to preserve their privileges. The Retro-feudal alternative is, at least during Transition Times, that the manor-house-folks need to show-up and repair their broken-lands and their broken class-relations. Most of the mid-elevation forests of the Siskiyous are in private-ownership and awful eco-repair.

Excess roads,
a legacy of petroleum,
have fragmented woodlands.

Retro-feudalism is a quick-fix scenario; it gets us through early Transition. Eco-restoration-camps can be organized by dodging the capitalist social-patterns that still hold momentum. Social Forestry is a retrofit; the vision of craft-guilds and a drainage-council is an inventory and planning task. To see the bridge from quick-fix to retrofit, think of the Chatelaine function as the Mothers-guild, managing the warehousing, guiding the Horticulture, and tending to the wellness of the community.

Perhaps the Servants of retro-feudalism are supporting the Mothers; they would be the Hoop-guilds, as they pass through on assignment. That would consist of many different guilds. The Elders-council would be the equivalent of the Cleric-caste: they hold Wisdom and keep us-All in-the-loop. The craft-guilds have access (and then gain usufruct-rights) to workshops and sorting-yards, spread out over several hamlets. The retro-feudal castes quick-fix melts into the retrofit-redefinitions of Guilds in spokescouncil—different pattern.

HOOP CULTURES

Working back from how we might live sustainably on this changing-planet to what-to-do just-now, brings out the but, but, but questions. The practice of envisioning and scenario-playhouse, in reviewing Retro-feudalism, allows us to explore the path to ecologically integrated enhancements, as First Nations have done before, on most of this continent.

On this planet, we are still looking-over and living-near co-evolved ecosystems that are bereft of song, rhythm, mob-grazing, and cultural-fire. With TEK wisdom, eco-science mapping, and monitoring, there is a great-promise of useful-employment for Modern Humans during Transition.

Now, it is put-back time, give-back time, time for reciprocation. Too much bureaucratic skimming and commercial concentration has outlawed real Human lives, discouraged enchanted-children, and erased deep cultural-memory.

*Hoop-culture folks
burst into song,
as they reach
the first long-view
of the headwaters,
fresh out of
sedentary winter-camp.*

Ultimate-indigination will nurture re-emergent Hoop-cultures. Hoop-culture is a semi-nomadic gardening-circus. When they come to a new camp, there is celebration, connection, and gathering. The local landscape gets a work-over by Hoop-blitz. The seasonal-series of camps are at different elevations and aspects, in sequence.

The disturbance regimes and regeneration of harvest and thinning are ameliorative and appropriate for increasing local biodiversity and closing nutrient-loops. The neighborhood loves the music. A hoop-camp entertains and provides for local craft-specialist guilds and provision-processors.

*The inter-visitation
of the cooperating guilds
mimics a flock of small birds
of different species,
working different layers of a savannah
as a Wild chirping swarm,
moving through.
There is a sequence
to who works which layer first,
as that can free up or add value
for the next niche-crew.*

Traders, packers, pilgrims, travelers, messengers, and poet/elder wanderers with their storytelling are the inter-connective pollinators, traveling between manors and camps. The songs, decorations, personal-ornaments, and pattern adornments around-and-about all artifacts of work and life, surround the telling of the long-epics at the hearthside. These stories are a colorful-basket of interwoven concepts, deep-Time memories, skills, taboos-and-etiquette, Place-names, trickster/creator lessons, and original-instructions. Deep TEK.

The holographic-dances of imagination in the local culture approximate the cosmic-flows (night-sky processions) and personify the many-beings that keep us in-the-complexity. All to better our social-collectivity and cross-pollination, to reveal the shape-of-our-work. We are imitative-participants in a real-world, where we all stand-in-awe. Stories are a collection of universal sets-of-patterns, celebrated in various Places and Peoples, who are experiencing eco-stability or Home-center.

See how easy it is to wax poetic about Deep-patterns of Human-cultures. Especially with forests, savannas, travel-stories, and legends? We do not need to travel very far to have these icons-of-culture appear, in some way. When we recognize the forms and practice the steps, the journey seems magical.

Meanwhile, we have some intentions, as semi-nomadic Hoop-folks, to foster the abundance and fecundity we need to support the work and the story. We step into sacred-tasks. How we Place our feet is reflective of many dimensions of consideration. We need the help of many layers of culture, some local and settled-folks, visitors from other drainages, guild-workers on assignment, and Rangers facilitating the flows.

In the last century Hoop-culture persisted in tree-planting co-ops, rainbow-gatherings, communes, fairs, and emergent Wild-tending vagabonds, working with remnant First Nations.

Can we imagine the village-scene in mid-winter, when many participants in this dance are in the guild-halls, crafts-sheds, market-places, and council-lodges? We can remember these old patterns from Deep Time. We can intentionally reconstruct key pieces of the tapestry.

FESTIVALS AND THE SEASONAL ROUND

One vision that feeds becoming People of Place is about a series of cultural-festivals, spread through the year. In early February, in the Siskiyous, it's the Manzanita Festival. The Manzanita is in bloom, and there are all kinds of gifts that are associated:

Manzanita-flower wine, Manzanita-berry powder (smoothies and bickies), the carved-wood of the Manzanita. Seasonal festivals are market-opportunities for the guild-workers, near gathering-meadows or villages. At these seasonal festivals, many things transpire that help people remember who-they-are, embedded in a forest-landscape. Read Chapter 16, "A Year in Wagner County."

This is a vision of re-emerging culture.

Our Social Forestry has roots in the village of South Glens Falls, in the remnants of Quaker farming-culture. Every Yuletide in snowy upstate New York, we'd go out and collect two kinds of wreath materials—two species of *Lycopodium*, Wolf Moss was the local-name, both are now threatened-and-endangered. Perhaps from loss of tending? We would thin a patch, not strip it. We'd also gather other forest-understory plants that would be the wreath, hung on the front door. You'd walk up to our house and see, "We are people of this forest; we live here; here's our wreath, and look what it's made of." We cut our own Balsam Fir Yule-tree. We celebrated the forest, by bringing the forest into the creche and the warmth of our house at Yuletide.

Another seasonal example is May baskets. The first of May, we'd make May-baskets out of milk-cartons, crepe-paper, and candy. A generation before, these were woven of forest-materials and filled with the first-flowers: Anemones, Violets, early emergents. May-flowers, they were called. We'd sneak up to the front doors, ring the bell, and run and hide in the hedge. We'd wait for the door and watch our friend's surprise when they saw the May basket. Through the neighborhood, the baskets announce, "It's spring. Come out of the winter-house; here's the tidings-basket."

The village had lots of gardens, hedges, fruit trees, and even backyard Cows.

Celebratory seasonal-activities help us remember who-we-are, how-we-live, and why-we-are-here. Village families supplemented their diet with foods gathered from the Wild-lands near the village. That included summer-fishing, fall-hunting, berries, spring greens: the remnants of a local Social Forestry. Factory-workers made root-beer in garages with Wild-crafted ingredients. There was tending of special-patches and building gardens at home, of Wild-flowers. Let's imagine a new participatory-culture, for the Places we-are-now-in.

SCENARIO PRACTICE AND LOOKING TO THE FUTURE

Scenario practices set the stage for long-term plans, through brainstorming and visioning. They orient local-folks to see Ultimate-optimums, beyond imagined-constraints. Community-inventory and assessment on a drainage-basin scale is the foundation of our scenario-scaffolding. As knowledge is gathered and mapped, tabulated and modeled, we see potentials by peering through the layers of information, alert to *quick-fixes*, *retrofit*-opportunities, and *Ultimate*-vision, emergent-from-complexity.

> *OK, there is that review,*
> *now for some speculation*
> *built on our conversation so far.*

    Let us assume some remnants of corporate-capitalism persist for a bit, especially the information and communications networks. Perhaps these have merged with regional-councils (multi-county pacts) and devolved federal bureaucracies? Land-ownership covenants (ecological-constraints, attached to tenancy-rights) with local-oversight stewardship-agreements are crucial to continued food and energy-sharing on the drought-prone western landscape.

    The irrigation ditch-systems are long, extensive and need huge maintenance yearly. They are ripe for *retrofit*. The trading-routes are long and go over a-lot-of mountain-passes. The village-clusters, tied to drainage-basin-council, are far separated from city-state supported restoration-programs. Regional-scale cooperation is essential, to inform local decision-making and warehousing. Headwaters eco-restoration is a layered consensus priority-process.

> *First Nations*
> *in Northern California*
> *commonly had five years of acorns*
> *cached in granaries.*
> *Must be from experience.*

ASSUMPTIONS LIST THE FIRST

This exercise does not name the guilds from previous visioning. Think functions, not players.

Now let us slip into *retrofit*-exploration mode. We here brainstorm a list of assumptions for the most-hopeful, but grounded, vision. These assumptions are displayed in a bubble diagram (see below, "Bubbles of Assumptions") and relationships are shown through functional-words on the lines between the assumptions. We can also build a reality-check matrix with the list on both axes and look for contradictory assumptions (see Poster 20: Assumptions Matrix). Party-down; salon-time!

Here is a proposed list of principle-based *retrofit*-ideas for Cultures in Transition:

1. Local drainage-basin-councils will manage community resource-assessment-and-mapping and establish inclusive decision-making processes based on precautionary-principles.
2. Emergency-preparedness teams will warehouse enough—perhaps barely-enough—locally-produced staples, an herbal-pharmacy, sets-of-tools, and information.
3. Work arrangements: local organic-farms carry a significant population of workers. Semi-itinerant gatherers and council-observers monitor work and travel on the commons-that-remain. Winter-camps support restoration-forestry, charcoal/crafts, and are often associated with manors and small-farms. Temporary-occupancy field-camps use walking or Human-powered wheels, such as bicycles, to get to nearby-work. Cottagers (caretaker-cooperatives) work with land stewards and manor-houses through co-op contracts with ecological-covenants and third-party oversight. All this is coordinated by local drainage-basin-councils.
4. Reestablished (after failure of central powers) monopolies-of-violence (Sheriffs with militia?) restrict-and-process chaotic-movements of refugees and secure the main corridors of travel and access (local sovereignty). There are no violent local-conflicts or wars. Mediation-services are centered in local-councils. Tribute or taxes may be requested and required for security-services. Control-of-weapons and emergency-incarceration will be arranged through mutual-aid-arrangements.

   Reconciliation and social-justice would start with local-councils and move to regional-justice, through cooperative treaties-of-support among collections of drainage-basin-councils. This allows multi-party conferences with mutual observation-and-monitoring (transparency) by councils-of-councils, organized as regional spokes-councils.

19 Wagner Councils

5. Drainage-basin village-clusters and hamlets set-up a central-consolidated-council, a spokes-council of neighborhoods and enterprises. An evolution of subsequent relocalization, as Place-identified Cultures, allows for negotiations-and-exchanges with persistent First Nations and nearby emergent drainage-basin-councils. Stewardship contracts and occupancy-covenant-treaties can justify whole-systems drainage-basin local oversight.

    Settled-villages, with landscape-monitoring skills, enhance and support subsistence-flows, while coordinating ecosystem functions management, such as wildfire fuels-management, road removals, access-pattern *retrofits*, erosion control, water table recharge, late-season creek and river-flows (preserving low-temperature holding pools), air-shed amelioration, eco-tourism, and cultural-exchange opportunities. And all-this, with net-positive carbon retention.

    In exchange for ecosystem functions maintenance, regional councils-of-councils provide satellite eco-monitoring and communication-links (with the least signal and power-pollution), emergency medical and dental back-up (special-skills expertise, coordinated to support local-clinics), and ongoing basic-monitoring and mutual-care. Everyone exhibits tolerance for social-arrangements and vernacular-solutions.

    The villages stage, or support, celebratory seasonal-festivals and markets where local value-added goods, derived from the streams of incidental-materials and associated with eco-system management, would be available for regional-trade and cultural-exchange (see Chapter 9).

6. These cooperative new societies negotiate stewardship-contracts and long-term tenure with any remnants-of-empire to ensure local long-term-investments are better protected. During Transition, any demands for commodities-extraction (exported-raw-materials) must conform to advanced eco-system-management principles and monitoring must show net-biodiversity-gain and renewable-natural-resources recovery with increased resilience and fecundity. Precautionary-approaches will apply to materials-and-tools used. Subsidy-support will be provided, through trade-and-contract, to local rural-populations, First Nations, worker/settler homesteads, and gateway-villages.

7. The stewardship-MOUs (memorandums-of-understanding), along with eco-covenants, are negotiated to be soundly grounded in Place, with a process-for-amendment so that any imported-support is balanced with exported-support in the form

of improved or stabilized ecosystem-functions. Besides structuring flows, these understandings contribute to the continuity of the drainage-basin-councils and craft-guilds spokes council, with the goal of greater regional eco-social-resilience. Eco-systems-management, contracted with regional councils-of-councils and private-landholders (stewards), uses appropriate-technologies to build a strong-sense of drainage-basin-culture.

And so, *Ultimately*:

8. Generations later, with accumulated TEK and social-cultural-skills, regular broad-scale-burning is back and Salmon has returned. Relations between city remnants-of-civilization and upriver Hoop-cultures, with their complex, vernacular-emergence, have evolved to become the exchange-systems of mutual-benefits-and-appreciation.

Reciprocation and awe,
allows extended-persistence.

RIGHT! NOW HOW ABOUT THAT *TRANSITION*? ASSUMPTIONS LIST THE SECOND

Back to systems-theory: It is difficult to imagine really-big-changes and phase-shifts. The emergent forms of new organization, on the other-side of a phase-change, are unforeseeable. Some of what we are doing now is creating a set-of-prototypes, of new paradigms. Remember old and practice new-and-old—but appropriate—*retrofits*. Our visionary-work, just maybe, births seed-like innovations/renovations that take root and grow in the unimaginable but *Ultimate*-beyond.

Here are some guiding-assumptions, and suggestions for cultural-values during *Transition*, based on universal patterns-and-principles and interpretations (reverse-expropriations) of academic anthropological, sociological, and biological sciences. May we find the graceful acceptance-of-limits, live within-natural-laws, and nurture Deep Spirit-full appreciation.

1. Some sort of spiritual-practice has always proven necessary to re-mind Humans of non-duality, participatory-service, endless-Light, and humility. Place will teach us.
2. Cooperation, empathy, gratitude, patience, and nurturance need reinforcement and cultural-imprinting, through rites-of-passage, careful plain-language, full-sets of cognitive-symbols, and endless storytelling.

3. Specialized-Humans of different genders, learning-styles, physiologies, and passions prove useful with enough recognition, support, guidance, and facilitation-without-compulsion. Honor, grief, and praise hold-us-together.
4. Wildlife-corridors, trade routes, intermarriage, and cross-fertilization of methods and materials are essential to promote diversity, which supports fecundity, resilience, and complexity.
5. Appropriate-management-choices rise out of a council-process that includes the representation of all-voiceless-sentient-beings and incorporates a view-shed that encompasses Deep-dimensions-of-Time.
6. Our inherent humanitarianism, or philanthropy, when we all sit in the Council of All Beings, comes into perspective after plenty-practice: Human-culture thrives best when the biosphere is allowing fecundity, resilience, and diversity to persist at the local-level. We find our-utility on-the-ground and we find-ourselves on one Earth.
7. Earth-life is completely interpenetrated: the Human-genome contains contributions from four billion years of sharing. Lots of virus-codes and mysteries. The Human "self" is mostly composed of commensalistic, cooperative, fellow-travelers: the gut-biome, skin-ecosystems, and the eukaryotic-organelles inside every cell. Without complex soil, water, and air-born microbiologies that are largely invisible to Humans, we are helpless. Thus reciprocation, sharing, living-within-limits, and humility, are appropriate.
8. Long-term, multi-generational space-travel, over light-years to distant, very-foreign planets, is probably unlikely (Robinson, 2015). The biological complexity necessary to support co-evolved-immune-systems would degrade in the relative-isolation of space-travel. Protection against cosmic-rays is difficult; DNA damage is likely. We need to make-it-here.
9. The long-evolved complexity of life-on-Earth challenges our imagination,and our world-views as scientists and as cultures. Winning or losing is not the question-before-council; we need to find balance. The Human species may be a temporary-strategy of Gaea's (see everywhere here).
10. Complex-benefits to Humans derive from, and are gifted from, the general-thrivance, that fulfills Boolean-logic: the most resilient, and persistent, complex-living-systems in local-situations operate at the edge of chaos. Chaos has form but is not predictable. Life, as an anti-entropic organization, makes the-most-of available energy-inputs and when the inputs change drastically, such as too-much energy to handle, the organization explodes, or implodes.

11. Life does best on a fairly-level evolutionary-landscape, with not too many species sitting on hyper-evolved peaks with inflexible-fitness (brittle-evolutionary position). The greatest-diversity of participants with the highest-average-health, fully-engaged with the energy-flows available, offers the best dynamic-persistence for complex-whole-systems.

*Holistic Management
is not easy,
just natural.*

The principles of Social Categories include

53.

Different genders, sexualities, learning-styles, physiologies, and passions prove useful to Social Forestry, with support, guidance, and facilitation-without-compulsion. The wide variety of Human interests and abilities allows complex cooperation. Many of these tendencies may have co-evolved with Horticulture, gifting us the social structures that support tending. Honor, grief, and praise hold-us-together. *Diversity is beautiful.*

54.

Common traditional patterns of social relations seem to fit eco-systems, as we have ancestors that lived those ways. Even where social categories prove useful to holistic strategies, keeping perspective on unexpected troubles and shining Light, with advice and questions about rules and rigidity, will be our never-ending delight. *Pay attention, but don't buy it.*

55.

Prejudice based on superficial reading of perceived markers is shallow and inappropriate. Pre-judging limits our relationships. We do it all the time for convenience, taking short-cuts, but do not forget how easy it is to make mistakes and how much we can grow through learning the stories that nourish our compassion. *Did we notice?*

CHAPTER 13
A Place for Humans

FINDING THE RIGHT FIT

In *Social Forestry*, everyone in community has job skills, everyone has something-to-do, everyone is honored, everyone is included. Got hands or feet, there is something to do. Everybody is kept-on-board as limits allow; there are ways that we can value a wide range of contributions. That doesn't happen in a consumer-society. In Nature, a great diversity is on-board—we want this, we need the challenge, we want to celebrate. This is the ample-concept: *Social Forestry* contains all these levels.

This does not mean that tending cultures are refugee camps. We cooperators do have some goals-in-common to reinforce and sustain. Becoming a member of a clan, guild, ceremonial society, crew, or hamlet is only by adoption. Disrupters are not welcome. We change by process, not provocation. *Friendly persuasion* is slow but inclusive.

 ## Sensible Drama Diet

While working together in emergent cultures and restoring ecosystem functions, people's interests, passions, skills, and cultural experiences are gifts brought together to magnify our collective efforts. Community processes that may relieve forms of disconnection and harm will bring in trauma healing, anxiety relief, and inclusion reinforcement. Especially for those whose lives passed through the hell of rejection and resistance from systems, laws, and institutions that have locked in privileges for elite classes and colonizer cultures, perpetuating racist behavior and white privilege. This includes many targets of different sorts of prejudice; we all carry some confusion—belonging and mutual support is healing. We can call out the injustice and we can learn from the long stories.

Start at home and in local cultures,
Take it to salon.
Don't ask the sufferers
To carry the load!

Moderns have been taught to live within narratives, definitions, and categories. Wild ecologies are not discretely-organized; they exist with continuums-of-variation. We here, in this chapter, paint a bark-scroll for an ecologically-appropriate Culture of Place, with ideas that are gathered-in-the-field, through watching-Nature and living-with-Humans. These are not the academically agreed-on-categories used by approved-science. These are our terms, to support these discussions and explorations.

CLANS AND TOTEMS

Clans are groups of extended families that share a totem. The bloodline-relations may be complicated, with marriages and adoptions. Place is woven into Clan-names with iconic-totems. Totems recall the symbolic-essences of origin-stories or identification with natural-forms, a relationship to Other. The totem tells a story about a People and their Place. It is a shorthand reference for an epic association-with-wholeness. Raven Clan. Orca Clan. Three Trees Clan.

Totems
are symbolic
representations
of felt relations
with other species
in our Home Place.

A clan in-the-mountains inhabits a drainage-basin. They travel/observe (while working-the-land) boundary-ridges and burn/maintain corridors for eco-continuity and travel. Mountains and a complex-of-valleys hold local-dialects of a regional language-group. The local-stories refer to local-landmarks. Vernacular styles, colors, and patterns abound so that affiliation is evident immediately-on-meeting within larger regions. Then we start talking and the accent tells-the-person.

Clans follow social-practices that organize marriages and relationships within the clan and between clans. Genetic diversity, to dodge in-breeding, is facilitated by rules against marrying within the clan and suggestions on which clan would be good to marry-with. Matriarchy keeps lineage through mothers, the obvious parent. Council is flexible and practical. Patriarchy traces fathers, the questionable parent. Patriarchy uses rules and laws to keep order. Clans carry special skills and responsibilities, and their ceremonial-societies are honored, recognizing the Spirit that they carry-and-share.

Trudy Fish,
village of South Glens Falls,
the Fish Clan of Quakers,
farmers and traders,
married Tom Ward,
of Glens Falls (across the river),
the Ward clan of Lutherans,
engineers and warriors.

Tribes are sets-of-villages that share a common language. The local-dialects are mutually understandable. A village or traveling-band carries a selection or collection-of-clans, that work-with each other and share stories. Seasonal gatherings bring whole-tribes together, with enough language-similarities to hold council-of-councils. A tribe can embody a landscape, a set of drainages-and-ecosystems.

Diet
within carrying-capacity
dictates or suggests
different gender-social arrangements.
Nomadic hunters
follow the men's lead,
and sedentary Horticulturalists
hold women's council.
Mixed provender societies are complex.

Guilds are skill and craft-societies. A childhood-passion, with some facilitation by Elders, gets sent to apprenticeship, passes into membership, and dons work-related costume (and perhaps tattoos). Costume can combine gender, clan, Place, guild, and ceremonial duties. Each of these categories may have a totem association, symbols that remind everyone of everything.

Anyone, from any intersection-of-categories, can be invited to join a *ceremonial-society*. Community-rituals are facilitated and choreographed at councils, festivals, work, and markets. The calendar-of-the-year, with the cycles-of-cycles, is celebrated and re-marked (emphasized) by these collections of Spirit-called-folks. Holding-traditions and acting-out stories, with symbols, costumes, songs, and dances brings us All-together in the epic-stories. Rites of passage, naming, and joining/partings benefit from ceremony.

The genders are special-mixes of Human-characteristics that appear to be biological in origin. They are social-sexual types. Mysterious but observable. We have some clues about origins and reinforcements (see below). Once a culture has named a repertoire of gender-categories, these types seem to become associated with sets-of-skills, so that our sets-of-sets fit together socially, for Horticulture and eco-restoration. Genders enliven our communities. All sorts find comfort.

Totems are multi-layered-symbols, with complex-meanings, talked about in Creation Stories and Trickster Stories. A clan presents a totem that is useful as an introduction in seeking mates at festivals. One does not marry the same totem. This keeps things mixed-up and vibrant. Cross-cultural marriages allow (teach) clans to adopt speakers of other languages.

A child's personal-totem (or *name*) is recognized through an emergent series-of-attributes (growing-up). Perhaps the personal-totem is evident from observation, or something-special happens at birth? Then the child-in-Nature has adventures that are seen a-certain-way, and thus receives a *name* in ceremony (with perhaps a *secret-name*, held by an Elder). Other *reference-names* come with trade and experience, as we mature. Some nicknames (or indirect-references) are used at market, others at council, and others still in ceremony and healing.

There are a few ways to find proper relations-on-the-ground for these categories and aspirations. We have found that organizing broadscale-burns takes the whole community coming together to safely and effectively do something useful. The process of coordinating the guilds through council gives voice to the gathered interests. The burn necessities socially weld the whole cultural practice in a rewarding and cognitive-way.

Guilds and ceremonial-societies
are gender-focused
but not gender-exclusive.

When a tribe (or bio-regional hub) has brought delegates from a landscape-of-clans into council so as to establish peace-and-justice and shorten the revenge-cycles, the language-group becomes a trading-cooperative with local-specialties prepared (value-added) for markets in other regions. When the bio-region has come into combined-considerations and cooperative-restoration brings back Salmon, guild skill sets wade into story-depth and tool specialization evolves. Guild-members emerge from apprenticeships within tribes to travel to apprenticeships in other regions and improve knowledge. The conversation within-guilds rebounds, via delegates, to the drainage-basin-council, with careful and skilled-consideration based on shared-reports of experience.

Guilds sport costumes appropriate to the work that celebrate skill and natural-elegance. Special hats or vests are common as markers. Sets-of-tools, in baskets or on belts, are an obvious-clue. Guilds hold traditions and initiation-rituals with stories-of-origins, or miracles (and disasters). Guilds cooperate through the guilds-spokes-council that sends delegates to the drainage-basin-councils and organizes markets at seasonal-festivals. Guilds are conservative in resource-use (materials for crafts) and protective of their members-and-traditions. They foster concepts of precaution, insurance-warehousing, and planning-for-persistence.

19 Wagner Councils

GENDERS AND SOCIAL ROLES

There are more than ten genetic sexual-characteristic presentation-types. We do not have our categories worked out; there are more than have been named. All Humans have their intersection of social and biological influences that help them find the settled-identity that inspires creativity.

Are these DNA-variations sexes or genders? Genders, we think of as social-roles: How does one present and what work do they do? Genetic-variability is common, but often covered-up to keep-things-simple.

Social-training (acculturation) is mostly through traditional-cultural-practices; the range of behavioral choices is socially-constrained. We get fit into boxes—we should expand the list of categories. A deep-feminist science peers over, and emerges from, the evidence. All Humans are female-first. The type-gender for the species is a woman. Men are specialized-women.

The *mitochondria* in all-our-cells are inherited from our mothers. Dad just delivers a DNA-package. Those organelles (the mitochondria) have RNA libraries that can be compared, and thus the claim we Humans-now-here are all descended from one woman in North Africa, 300,000 years ago. Basically, our flesh-is-female, and *estrogen* is the primary hormone-type.

A Human, with XY chromosomes but hormone-variance that suppresses *testosterone* will present as a woman. There are some very complicated genetic-and-ontological expressions of development here, sometimes evident at birth and sometimes later. Many Humans know they are different but do not have a way to understand.

Even in the late 20th century, complicated genital-presentations at birth were often surgically altered without parental-permission or full-disclosure and the baby declared a girl. Trouble comes later, when the athletic "girl" does not grow breasts or start menstruation and gets kicked out of sports for testing XY. Modern science does not know what a woman is.

XXY and XXX are triple, instead of double, chromosome sets and lead to special-women, often very-capable Humans with observable gender-differences. Of course, common XX and XY double-sets produce a wide range of phenotypes on their own. The Y chromosome is notoriously-fragile, breaking-easy (brittle) and then duplicating, folding, or moving-over in novel-ways.

Displaced Y chromosomes deliver Humans expressing as *inter-sexual*, who are encouraged (directed) to socialize as-a-girl. Stuck in the binary-mode, Moderns are just catching-up to common traditional-knowledge of multiple-genders. This natural, reoccurring inherited-variability creates useful-Humans with storied-celebrations in TEK. Hormones can be adjusted, to some extent, and Moderns now have new-ways to reassign-gender (to one-or-the-other) by chemistry and surgery.

Colorblindness is a gender. It is associated almost-always with XY genetics. Men are the most likely expression. Colorblindness comes in different color-wheels: one does not see red/brown, or blue/green and more-complicated variants. The skill colorblind people all seem to be good-at is seeing-through-camouflage. They can see pattern-and-outline better because they are not distracted (glamoured) by color. A hunting-party good-size is eight and one-out-of-eight (or so) males are born with some sort of color-blindness (Ridley 1996).

The coincidence of crew-scale (one-in-eight) and inheritance is intriguing. Is this a set-of co-evolved talents? Very useful on hunting-expeditions: someone who can see clearly, not distracted! There may be a set of personality-traits associated with this too: These are imaginative-folks, nurturers and listeners, who get-used-to others not believing what they are telling them, which breeds a category-of-utility. If we understand the whole-person, we can celebrate the special-gifts and honor-them.

Social conditioning is more pertinent than genetics. Superficial physical characteristics do not preclude successful adaptation.

HAZEL'S PREFERRED PRONOUNS

I want to try to clarify my pronoun preferences and gender status. I am an in-between, an intersexual from birth. I have had many gender difficulties starting in kindergarten at Abraham Wing Elementary School in Glens Falls New York, 1952. The girls would not play with me, calling me a boy, and the reverse was true for the boys. I have had to make personal friends, as I was picked on in groups. I also tested as an introvert, which helped with the social isolation. By sixth grade, at South Glens Falls Elementary School in 1958, the teacher Mrs. Linke referred me to the school system's counselor, Dr. Cuchio, for gender non-conforming and anti-social behaviors. At the College of Forestry Summer Program in 1967, I was targeted for abuse and excluded from cooperation with the boys. The girls (there were very few, and they were also abused) were forced to accept me on their surveying crew. We got along fine without the mean boys.

All my life I have worked with women. I have participated on women's collectives and committees as an "acceptable other" all my life, much to my benefit. I would especially mention the Ma's Revolution Natural Foods Collective, Telegraph Avenue, Berkeley, California, 1971–1975. We were an early city-to-farm grocery start up. I feel honored to be referred to with the feminine pronouns.

My favorite pronoun is "Friend" from my Quaker culture. As in "Friend speaks my mind" or "Friend Hazel is present."

My preferred pronouns might be listed as Friend/They/She.

GENDER DIVERSITY AND ADAPTATIONS

This can be a tricky discussion. Moderns have been conditioned to react. Commercial efforts have driven many uncomfortable feelings and alienations. The essential flow is fragmentation and alienation. Identity becomes the existential necessity. This can be marketed.

Social Forestry is a way that Humans can learn from and integrate with natural systems. We have a lot of work to do to come into alignment with Place and tending ecosystem functions. If we can work together with social balances of gender diversity and visions of cultural wisdom, we can have whole landscape integration with Human tending as key to overall health and comfort.

Let's review complexity. This is the natural state of the Wild. We need not expect our complications, our adjustments to complexity, to rise to the sophistication of the Wild. We, as Humans, adapt as we learn from experience and remember from wisdom stories. Simplifications can be wise if we avoid dogma and stay open and observant.

The gender Quick Fixes during Transition are especially fraught as many of us Moderns are lonely and confused. We have been fed a lot of propaganda by social reactionaries looking to prevent uncomfortable change. We have a lot more good science and testimonies that open up the possibilities and options for gender diversity. Modern marvels have gifted some community members with adaptations that may include pharmaceutical therapies and gender affirming surgeries. These folks have gifts of value to return to our communities. There is an opportunity for an even wider range of options that support a greater diversity of gender adaptations. Most hormonal balance adjustments can be helped with local medicines, herbs, foods, and community support.

Personal identity options can be marketed as solutions to angst. Cultural belonging is way more comforting than self-reinvention. Cultures of Place have geophysical generalities that shape local cultures and physiologies. People of Place see the shape of their culture in the lay of the land. The chemistry of the soils and food, the mineral and climatic environment, and the inheritance from our ancestors all contribute to the expression of gender diversity.

In this time of aggravation during Transition, the reductive industrialism and colonialism has channeled Human relationships into a convenient but insufficient binary male/female lock down. Some cultural creatives are calling for a leveling, to ease the tension and inequalities. This can be called non-binary or gender non-conforming. By the principles of Social Forestry, we would rather elaborate the binary than flatten it. We need to think in more dimensions. As we discuss below, there are many natural gender variations and sexualities.

Only after cultural living and tending over longer periods of time and change do we become People of Place who know something of the range and relationships of gender and sexuality. The appropriate cultural adaptations to persistence and thrivance are discovered during the phases of Retrofit. This discussion is necessarily sketchy and complicated.

We are making this up as we go. Retrofit here means a lot of education, learning, and experiments with a wider range of social arrangements.

The Ultimate goal of integration with Nature and general health and comfort for the maximum diversity of participants at the edge of Chaos is just beyond our reach. We have a lot to learn about so many aspects of being fully Humans of Place. Our emotions, sexual interests, hormonal health, energetic health, and attention to the Other, all benefit with guidance from experience and community process. The land loves us. We can get support from a wider community of species and family shapes.

In Modern times we have at last noticed the intersexual and trans-sexual experiences. That a person in one type of body needs to be in a different body is known and reported. This is deeply mysterious, but compulsive for so many that attention is being paid. The lack of comfortable non-binary alternatives in social-market spaces leads many people to go through gender-reassignment, hormone-therapy, and surgical-reconstruction to allow relief for non-conforming gender-stress.

Some therapists suggest that if more of us can pass as legitimate-categories, perhaps we would attract less unwanted-attention. Modern medicine and sociology has created categories and pathways to normalize previously (in Modern-times) unimaginable identity-reconstruction (Brown 1996). But the science is still weak.

*We need to know
more about helping everyone
live a healthy and engaged life.*

Modern gender-nonconforming children who behave and express other than cis-gendered (born-as) are indulged by well-educated and advised parents; this is progress. Gender non-conformers in most colonized sub-cultures are brutalized and forced to conform, then externalized when they fail. When adolescence approaches, the queer ones are still children, but may be stressed with unanticipated hormonal-changes and the parents are not presented enough options.

The Modern legal-option is to use hormone-suppressing drugs to delay the onset of adolescence until the child reaches eighteen years old and can choose for themselves what they would be. Almost as if it were a consumer-choice. This protocol is really-invasive drug-therapy with unknown long-term consequences. Better we learn to support a wider range of gender expression and identity.

*It takes a culture
to raise the diversity
that gets the work done.*

The school-system may want in-betweens to choose and trans-gendered kids to act their birth-sex-role, so as to simplify the bathroom-quandary. The common medical-advice, on top of legal-advice, is to administer hormone-suppressing drugs to put off any choice until they are legally able to choose. But they have to choose? They get channeled into gender-reinforced service-roles? What if there were more choices?

SEXUALITY AND GENDER

Love manifests in many different ways. Family and clan-belonging nurtures a comfort and allegiance, affection-for familiar and supportive Home-culture. We love Nature. Life loves life. We come to appreciate that love is more than an emotion, it is a gift we share that reciprocates with a settled-happy, full-belonging feeling that thrills us naturally.

The ideal that we all deserve everlasting romantic-love, available to everyone, is a Modern concept, when it was special-and-fleeting for our ancestors. Certainly, there have been epic-loves and we celebrate them in song-and-story. The Modern nuclear-family (does that mean radioactive?) is a construct built for consumerism and social-conformity (makes industrial-production efficient). We get sold a template-of-junk, easily-reproduced wherever the globalists can manage. Nice economies-of-scale: a stove and refrigerator in every kitchen.

Traditional extended-families on Quaker-farms were in that Dunbar's Numbers sweet-spot, 20 to 30 folks. The marriages were socially-arranged, often by Elders, with many goals in mind. And marriage is promised support and back-up, from the community that marries them. But the marriage-partners did not have to spend a lot of time together exclusively.

Everyone's work holds communities-of-support and relationships outside-of bearing children. With grandparents, aunties, and uncles all-about, children were raised by the extended-family, while the parents were busy with other tasks.

SEXUAL ATTRACTION

Then there is sexual attraction. To wonder why we are driven this way is to dodge the reality: sex-happens. Is it mysterious in its passion or in its objectification? Projection of fantasy can be one-way and we can be oblivious to reality, if truth is inconvenient. There are traditional-ways, dance, song, and shamanistic-rituals to fulfill-or-distract the powerful-urges of reproduction and satisfaction.

Sex can be addictive when it is a displacement-activity where there is no other real-life to be had. This sex-mystery-show is a Wild dimension of culture and defies categorization, and declarations (such as these). Please excuse us while we make some broad-statements about sexuality, just to fit it into our visions. Certainly, if we can't dance-and-merge, we won't be useful to this work.

*The exuberance of Nature
we dream with
can be sympathetically mirrored
in our passionate-participation.*

The preferred urban-arrangements of Moderns sometimes implies any-which-way pick-up scenes. As the natural sexual and gender-role variability is uncovered through social-movements, anthropology, traditional-stories, and science, we are noticing the complexity-of-options for useful social-roles.

SIX GENDERS

When we look at all the coupling and social-role patterns available when sexual attraction, or partnering preference, is layered on skill-and-passion driven work, we can start talking about traditional options that have been useful in the past. If we call cis-women who are happy with Mother's-guild work and house-holding *first-gender*, we can call cis-men who are happy as Hunters and Sawyers *second-gender*.

Male-bodied children that get raised with women's-work, who may be hormonal in-betweens, can be *third-gender* and might marry second-gender. These men-women might be teachers to boys, so that they can learn about women. They are welcome in women's-council. To moderns, they might appear to be gay.

Female-bodied children that show fierce-ness and dominance, are skilled with men's-work, and tend to marry first-gender, are *fourth-gender*. *Fifth-gender* are inter-sexuals who, having been raised as women with women's skills, are still attracted to first-gender. These fifth-gender folks would be invisible to visitors, as they look, act, and work as women; to moderns they might appear to be lesbians.

Women oriented women, either both first-gender or a first-fourth-gender pair, are not attracted sexually to men and thus do not get pregnant unless by special means. Same-sex couples may also have children adopted from a previous marriage.

Women who live long and somewhat-apart also have a special-perspective on relationships and culture. The gender-categories amongst women-oriented-women are full of nuance-and-potency. Perhaps women who prefer feminine-roles but are in same-sex marriages are *sixth-gender*. These women are capable of multiple gender-identifications. Sixth-gender can use feminine-abilities in creative ways.

Traditional women's-work has long been Horticulture, fiber-crafts, care-and-nurturance, and lodge-ownership. Women who spend the great-bulk of their time amongst women-at-work gain experience and focus to offer special-advice and support (midwifery, doulas) and they hold the women's story-cycles and teachings. May they live long lives! Women's culture is essential to community-persistence (Barber, 1994).

All genders
Ultimately
serve women's interests
when a culture learns
to live-within-limits.

Men oriented men (second or third gender) might prefer non-procreative sex to limit liabilities, or they are amorously-focused on other men. Some men present-and-train as female from birth (third-gender) and marry cis-men as skilled-wives (no children, elaborated-sex) and as members of the women's-societies (Roscoe, 1991).

Some "men" (XY) are inter-sexuals with indeterminate sexual-characteristics and in ancient Greece and Rome they were raised as sacred-Humans (*hermaphrodites*), displayed as fixtures at shrines, worshiped (and perhaps abused). What do we know about the emotional interests of kept-boys? Cis-men have been guilty of vast sexual-abuse, variously-targeting all other genders.

Women
of all sorts
have suffered abuse and neglect
by European Patriarchy.
The witch-burning times
are still with us.

In many ancient royal-courts and harems, there are boys who for various reasons are surgically castrated very young. This can result in larger harem guards (who pose no sexual competition) or feminine-boys, kept as novelties and sex-slaves (short lives!). Some castrato are educated and recognized as courtesans and intellectuals. Boys can be castrated to preserve their pre-adolescent soprano voice-range and populate choirs of perpetual youth.

*Stories about
what has been bad or confusing
are challenging.
We still need to know
the types so as to be precautionary.
Most gender-types
have obvious value to culture.*

There is a clump of intersexual characteristics (in-betweens) including long-legs, long-arms, and short-torsos. These phenotypes are often athletic and long-lived. Sometimes they have heart problems or weight-gain challenges. They might have suppressed-hormones that do not fuel passionate sexual-attractions (they may be conservative of life-force). There is some evidence that androgynous-tendencies are associated with long-life, excellent-stamina, and long-distance running.

An in-between child might find themselves a Ranger, who reports to both men's and women's councils, who can cover a-lot-of ground, and who is comfortable with solitude in their work. As Rangers and in-betweens, these folks notice opportunities for perfect timing and report to the Mothers-guild, or women's-council.

The Berdache category recorded by early explorers and anthropologists in North America are third-gender men who are often derided as failed-men, but let's correct that reactionary-attitude. There have been male sex-workers (now a mostly-urban guild) who filled useful-roles in traditional societies, by training-up young-men in sex and by social-bridging with women's-work through skilled-crafts. They can serve steadily, without diversion by babies and women's cycles.

They dress as women and work-with women but also are also bridged to men's-council. These were visibly men-in-women's-clothing. They helped the women with the heavy-lifting and were great gardeners; European explorers-and-trappers saw them and wrote journals (Roscoe, 1998). Very likely, there were also intersexuals, who European-males would not have seen. These special-men, the Berdache-guild, were seen as special-women. Isn't that always the case, anyways?

Anthropology's collected samples of field-work and social-observations need some vetting, but we can glean some clues as to how-we-work-together as a Human-culture that persists, accumulating traditional-knowledge. For example, broadscale-burning and craft-materials fire-coppicing is traditionally women's-work, seemingly in every culture, worldwide. It has to do with perfect timing and the detailed plant-knowledge of the Mothers-guild.

The genetics, the physiology of it, only set the stage for what cultures do with categories. The key is to welcome, but be discerning. Our Social Forestry will benefit from a candid conversation about touchy subjects. Perhaps some nice cozy winter story circle?

> *Discernment
> is not discrimination,
> it is an opening to inclusion,
> through revelation.*

Opening up the gender-diversity basket doesn't mean that cis-men do not deserve honor and practice useful skills. Sawyers are a men-centered guild. Most of the other guilds are glad to have men's help and counsel. That is part of Social Forestry, learning what our collective skills are and learning to appreciate diversity.

LEARNING STYLES

The senses teach us as much as we can handle. When a Human is focused on one sense, at the lack of the others, a different world is perceived. Cultures of folks who have limited use of certain senses think differently. Bridges made between these vastly other world-views stretches all-involved.

Deaf youth are taught sign languages and can find each other and communicate in ways listeners do not see. The culture that emerges among deaf-folks is special, supportive, and insular: They cannot believe how the public is behaving! Blind artists deliver awesome gifts. Music, dance, emotional-perspective, and much more. Their senses of touch and smell may be amplified. We should be able to learn from everyone's experience and find ways to get along.

The learning-styles that get the most remedial attention are the ones that do not fit in the Modern consumer-industrial model. Efficiency (machines) demand simplified-categories and want to externalize anything that does not contribute to profits. So public-education has missed what folks that do not conform can contribute.

Folks with dyslexia, left-hand preference, and cross-dominance often do better at multi-dimensional tasks, and less so with linear, rote, technological education. The special-skills of folks with diverse abilities are always a contribution to the whole-work.

> *Three-dimensional skills
> in building and imagination
> are useful in Nature.
> Two-dimensional skills
> have been the demand of industry.*

There is useful work for everyone in Social Forestry. Humans display diverse potential-abilities and actual-behaviors. We can match-up their passion-and-attributes with the needs-and-outputs of community-in-Place. Subsistence with graceful-simplicity within carrying-capacity limits is how many intensive-care people can be accommodated. We-All wellness, our collective health, is among the endless-casualties of Modernity.

At D-Q University in the mid 1980s we trained for learning-disabilities assessment and treatment. What appeared to be problems turned out to be special-abilities. And when we found each person's passion and fed it with tools-and-materials, we saw their performance in academic-subjects improve. This experience led to testing for some learning-variations in classrooms that suggested to folks what they may be good at.

Fetal Alcohol Syndrome is a sort of brain-damage, and other organ drug-damage to infants, that can be tracked with therapies, but prevention is the cure. Brain-damaged folks can find praise-and-honor in work, but oversight-and-support of caretakers, is necessary.

There are many types of mental illness that have various-etiologies (origins). Some are Modern and result from the indignities-of-industrialism. Some traditional-cultures have found ways to honor speakers-in-tongues and ravers. Some types-of-brains exhibit very useful skills, while the person is less-able to take care of themselves. These special-brains have gifts to offer. Amazing math skills, clairvoyance, anticipation, comments with perfect timing.

The Modern industrial-Human is showing some evolutionary-adjustments to the stress. This means that there are experimental-brains showing up. Then there is the eco-spasm of Human-population, which through probability-and-chance is expressing mutations. Families with high scientific-intelligence often have a cluster-of-autism and other special-people.

There are many types-of-intelligence which are evolved specializations. And when we layer up the maps of Human-complexity with sexuality, gender, and social-training, we have lots to understand-and-facilitate. When we accumulate-knowledge and observation as cultures-of-Place, we can support-and-integrate all-people and the whole-Place prospers.

The two hemispheres and local-areas of the brain have somewhat-separated and specialized functions. When one part is impaired by injury or disease, others can reorient to pick-up lost abilities. This is wonderful-flexibility, but it has limits. All genders-and-sexualities have valuably different mind and brain abilities that contribute to the breadth of social perspectives.

As healthy-folks age, the *myelin-sheath* that surrounds neuron-bundles (*axons*) thickens and adds insulation, allowing faster-messaging. As this speed-of-communication picks-up, other parts of the physiology are slowing down (aging). Thus Elders make connections between remote-subjects and use metaphors readily. They remain-useful in council as they contribute-less to community hard-labor. This can be a handy compensation? Do they remain good at handcrafts? Is anyone, other than an Elder, able to pay-attention to subtleties? Do we watch-out for each-other and deliver praises-and-honor?

Cross-dominance means that one uses the opposite-eye for the usual hand-preference. So left-handed folks can be using the right-eye as their dominant. This can be tested by staring at a target, then covering one eye, to see if the target is still-visible. Then, covering the other-eye (or closing one then the other) we find which eye is dominant. With some practice, it is evident which eye is dominant, as the target will disappear when the dominant eye is covered.

Cross-dominant folks, in classrooms or intimate-conversation, may appear to not pay-attention and so they get in special sorts-of-trouble (non-participation) that can be avoided if we know-enough to check which eye they are actually-using.

Most teachers automatically talk-to and watch the right-eye; this is a force-of-habit. If the student seems to be spacing-out, they are unfairly punished or ignored. This leads cross-dominant folks to feel-unseen and unappreciated. The joke goes: Most-people in prisons are cross-dominant, but most cross-dominants are not in prison!

Try switching which eye you are talking-to, to see how the conversation changes? One can learn to assess-this, by direct-observation, without needing to test which is the active-eye with an exercise.

Left-handed folks are also misunderstood and tend to have variant-personalities. They, along with cross-dominant folks, have enlarged axon-bundles and enhanced three-dimensional abilities. When left-handed folks are cross-dominant, it gets real-interesting (especially with more-intersections).

Dyslexic-folks tell us that they cannot keep letters or numbers in order. Perhaps they are more-aware of universal multiple-dimensions than the rest of us? Perhaps written language is inappropriate to local Horticulture (Shlain 1998, Illich 1988). This is only a flaw for bureaucrats. Dyslexics also have the three (or more) dimensional-imaging skills. It is as if the Modern Human is diverting from the narrow-blinders of industrialism, in order to have the skills-needed to fix the mess it made.

We can predict that the child in-the-sand or on-the-floor with blocks, who can model-and-build better than most, may not-do-well with academic-subjects. They are, however, just who we need to do visioning-work. When we all learn to use more of our abilities, we will be better at modeling-the-work of restoration and re-enchantment. We need these folks, because they can swim in the models and 3D matrices that we use in visioning-council.

There are other learning-styles to notice. We can build a set-of-categories to use as a checklist, but the direct-observation of a child-in-community brings us their inexpressible-essence.

The child may remind-us of a totem-animal or a character in the long-story. The child may teach us something-new, that is actually very-old, from deep-in-our-inheritance, surfacing with sparkle, to our common-benefit.

The survivor of big life-changes or the newly arrived refugee is like a child, as they re-orient to the new-old world with our welcome. They will show-us what we can offer-them, so that they become Home.

COUNCIL

A spokescouncil, or syndical-pattern (delegates) of culture-relations is a wheel-of-clans, or guilds-councils, or council-of-councils, that brings-together Wisdom and report. Council is the cultural skill that connects. When societies (whatever category) have internal-consensus procedures and send delegates to the hub-council, the delegates come with council-skills. These designated-reporters and epistle-orators, do not make-decisions; the hub-and-spokes pattern allows Human-scale deliberation to be reinforced.

*When the wide variety of interests
in the drainage-basin
are gathered in council,
we feel the power
and strength of our people.*

The decisions are made in the spokes, and the process patiently goes back-and-forth until all spokes-in-council have reached *unanimous-consent*. This care is perhaps-protracted and yet practically-cautious and complete. Once everyone is on-board, great-things can happen in elegant-ways. The delegates keep going-Home, to the local consensus-discussion and then report-back to hub-council. The delegate is hearing all-the-other guilds' concerns, with a goal of cooperation, following the knowing of common-intention. The use of Precautionary Principles fits well with long-view cultural-processes.

*Never introduce anything-new,
unless it can be proven-safe.*

The principles of good-social-order include

56.
Good-social-order does not have to be class based or hierarchical.
It can be transparent and accessible. Good decisions get made
by communities when the process matures and all bases get covered.

57.
Shapes of social order become traditional, when they serve well.
Children learn the nuances early. Adopted members (*convinced-friends*)
can seem inflexible as they try on what seems to be a set of rules.
Experienced categories of participants carry *weight*.
Weighty Elders, Servant-clerk, Clown, Orator.

58.
Traditional Human social orders, at Home in Nature, are not
predominantly egalitarian or individualistic. Membership and
ceremonial positions carry intentions of service. Benefits accrue
to the community-at-large. Individuals sacrifice (gift) to the
whole. Social order carries the whole.

CHAPTER 14
Social Order

SOCIAL PROCESS PATTERNS

A culture of Place can use merit, wisdom, and work to weave a language of social-relations. The learned culture-pattern is ceremonially displayed in council and celebration. The council-pattern can be seen as circular, spokes-like, sequential, or multi-dimensional. The opportunity is to envision culture, appropriate to ecological-context, and food/fiber/shelter/Spirit-shed.

Here in Ecotopia, we have shamelessly borrowed (expropriated?) patterns from cultures-world-wide. This is an experiment in *quick-fix* culture-building, cobbling these patterns as syncretic-pastiche social-art. Wish us luck! Are we not needing to push the envelope in the face of climate-weirding, and peak-everything? We have learned how important it is to support First Nations and not exploit them any further. Very few *quick-fix* visionary-movements to build new culture have shown persistence enough to do a *retrofit* (Hill, 1991).

Indigenous-cultures,
with their Traditional Ecological Knowledge,
are assumed
to have built social order slowly,
through experience
gained in Place.

The Pacific Northwest Potlatch-tradition is a gifting-circle. Many Turtle Island cultures dance in circles, often around a tree. Quakers (Adirondack Hicksite traditions) are wisdom/Elder facilitated, with Elders on facing-benches, activists on the back-benches, and the room sometimes a grid of stoves with children and women on one side of the room, men and boys on the other. Many attenders are deep in mindless hand crafts, or breathing.

The Hicksite monthly meetings are local, the quarterly meetings are one-day's ride away, and the yearly or half-yearly meetings take some travel. The Industrial Workers of the World ("wobblies," or IWW) uses the syndicalist/spokes-council model. Different crafts-unions, from different work-sites, send delegates to regional-council, as we have discussed in other chapters.

Good-social-order
is a Quaker plain-speech phrase.
The details may be complex,
but the reference is curt:
All the Elder has to do
is mention the phrase;
everyone searches their behavior.

Societies working in Nature evolve multidimensional-relations for good-social-order that include All Sentient Beings, genders, guilds, and clans. These potentials for collaborative social-relations might apply-usefully in planning the sequences of disturbances, harvests, and celebrations in the seasonal-cycles of Social Forestry, dancing with every special Place.

HERE ARE SOME CLUES AND CONCEPTS FOR OUR DISCUSSION

Some source-materials for social order are writers including Calvin Luther Martin, Morris Berman, Ursula LeGuin, Matt Ridley, Robin Wall Kimmerer, Peter Farb, Stewart Kauffman, Leonard Shlain, Ivan Illich, Paul Shepard, Kim Stanley Robinson, and many others. See the Bibliography.

Many Ecotopian learnings have come with social-practice: collectives, cooperatives, communes, social-change events, political-action in the streets, and extended-families and friends. Some wonderful visits were to other continents and Indigenous peoples.

*The open opportunity
that we now have
on Turtle Island
is to listen carefully
to the original teachings.
Early settlers were offered cooperation,
but did not want to understand.
The essential philosophies
of First Nations
remains our reprieve.*

And so after wide-reading, local salon, and some visitation, this dense-essay, using peculiar vocabulary and syntax, arose from this wide-range of sources. Modern English and alphabetic-linearity are insufficient to the task. We go a-bit-poetic, hoping you end up wonderfully confused, delightfully overwhelmed, and usefully disoriented. We hope we can do our job with rhythm-and-rhyme!

The triad-portal to social-arrangements is good-social-order, taboos, and etiquette. Mapping the three as a Venn Diagram, the central-overlapping of the three bubbles—the inclusion—is Sacred Place. The overlap of taboos and etiquette is the commons, the ecology in which our social order has purchase.

The TEK stories that reinforce taboos and etiquette teach us why and how to do. We use stories to process in council and rise in celebration. The process-oriented social order bubble at the top holds the vision of Ultimate. Taboos can be quick fixes and etiquette facilitates retrofits. The Venn diagram shows how we use ecological-knowledge to facilitate inclusion and wholeness.

Human and animal-societies—and perhaps forests—exhibit these same themes when they gain enough social-development, within consistent-patterns of sustenance-and-regeneration, to exhibit genetic (or epigenetic) co-evolution with coexistent species and entities in their Place. When we hold TEK as language, stories, and other cultural-transmissions such as dance, art, ceremony, and pictographic-symbols over generations, we achieve resonant-inhabitation.

Animals, trees, rocks, and other entities are said to also have language. It may be slower or faster (songbirds) than Humans. Many animal-societies show cultural-development when they manage to hold multiple-aged groupings through a continuity of generational turn-over. We can learn to cooperate with all living societies so that they have every chance to mature.

SOCIAL ORDER

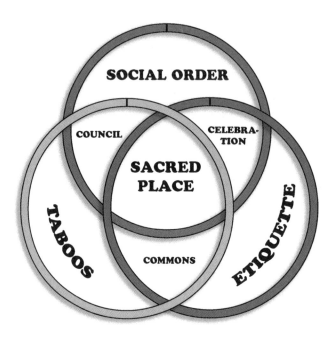

Enslaved Humans often do not have the benefits of cultural continuity when they are subjected to forced migration or separated from their families and practices. Slavery deserves to be or is now taboo.

TABOOS LIST LIMITS

Ecological-taboos are generated through close-observation of the loop-of-nutrients, predator/prey cycles, dietary-needs, and diseases. A taboo can outlive its usefulness when the culture has to move (expulsion from Home-land) and still carries the local-taboos, now anachronistic-and-inappropriate—so taboos have bad-reputations as superstition, though the stories may still hold general-Wisdom. Taboos serve a purpose: community awareness-of-limits. Indigenous-people collect taboos declared-by-specialists, such as Elders, Rangers, or Shamans.

New and timely taboos are processed-and-practiced communally, in Place, and thus people learn to change-or-adjust behaviors that could be ecologically inappropriate. Sometimes taboos arise and last for generations through renewed-observations, long-memory, and careful-councils. If the culture pays-attention to its environment, these behavioral-changes can have ecological-benefit and ease the challenge of abrupt eco-spams or collapses.

Hunter guilds collect taboos and celebrate with stories. As both the prey and the Hunters are in motion or have the opportunity to move, bringing food Home seems chancy-and-random. They feel they need to do everything right to propitiate the Spirit of the Game. The taboos are a checklist against making mistakes. There should be a taboo against trophy-taking. Hunters should help the Sacred-prey thrive. Other predators do this: take the slow, injured, tired, and vulnerable. Leave the herd (animal-society) intact and healthy.

Notice the gifts.

Indigenous Hunters prepare for the chase. They spend time in sweat-lodge clearing any ill-feelings and singing songs-and-prayers of alignment-with-Spirit, so that the animal seems to volunteer to the Hunter. If Hunters go out with power-in-mind and are not humble-and-patient, the ghost of the inappropriate-prey will haunt them. The stories Hunters tell themselves help keep a balance with taking-and-giving. It is taboo to not reciprocate with gifts-and-thanks when a generous-sacrifice is received.

Jared Diamond reports that a few human cultures have managed to modify and/or re-establish more persistence-oriented arrangements (Diamond 2005). The Emperor of Japan, many-centuries-ago, in danger of soil-loss from deforestation, declared taboos-and-repairs, such as limits on firewood cutting and reforestation, as a *retrofit* to turn-back disaster.

ETIQUETTE GRACES THE COMMONS

The *way* we do our tending should support the whole-drainage-basin. If we get our intentions and process right, we can avoid too-many taboos and our etiquette will be fine-tuned. If we just follow the rules, as if we did not want to know what we are doing, we will only have a list of things-to-avoid, as if gambling.

Etiquette accumulates through experience and can become iconic, celebrated-principles with associated-symbols, backed-up with stories of original-instructions. Etiquette, in an ecological-sense, is the accumulated-knowledge of how-to-do appropriate-living within the bounds-of-Place.

Proper, tested-and-true etiquette finds grace-in-relations, as we learn to care for others. The polite-sequence of Tea Ceremony in Japan, or Afternoon Tea in Britain, sets a frame where we can forget our troubles and find connections-and-insights. Ceremony uses etiquette to distract-and-remind.

Pay attention to how-we-do-this and we will show— you might notice— why.

A promising-vision, based as it is in old-epic-stories, is that cultures can change and prosper through Truth. Meanwhile, during this *Transition*, the reactionary-argument or reductionist-approach against suggesting such a high degree of cooperative social order, warns of inevitable channeling, class rigidity, authoritarian hierarchy, lack of personal choice, and loss-of-individuality. Good-social-order has been necessary to ensure cooperative-mutual-respect during previous times-of-Transition.

Modern-consumerism, supported by petrol-fuels and celebrating hyper-individualism, has seduced our fight-and-flight evolutionary-conditioning and re-trained our natural-tendencies. Advertising, repeated over-and-over, with lovely animated-graphics, flashes out, "You, too, can re-invent yourself!" Put on any costume, go somewhere-else and disappear, start-anew just by buying-something: "Live to your greatest-potential. Do not settle for anything less!" Whew! You got that? The wonderful-promise of Social Forestry will not be advertised by Modern-media.

Look at the mess consumer-advertising made! Anybody happy? How are your back-up-systems doing? Emergency-preparedness, anyone? Global-consumerism is *not* subsistence-within-natural-limits, with awe-at-the-immensity and praise-and-reciprocation to acknowledge our Place-ness.

The laws-of-physics still hold and Modernity has fallen-off-a-cliff of artificial-speculation-bubbles that burst with-nothing-underneath: free-fall for "freedom" with no-safety-net. Now, Moderns are really on-their-own, unless they accept some social-discipline, find that special-Place-to-be, and practice-service that brings honor. We can learn to enthusiastically step-up, to do the best-we-can, and accept the Grace-of-Love.

A TRADITIONAL SOCIAL STORY

Quaker farmers kept things simple, cared for the land (and all creation), and did community-business by applying deep-principles. They did not wear black-cloth, as the mercury-dye drove the cloth-workers mad. "Quaker gray" (undyed wool) became a pejorative, as in dull and without-fun. They did not wear buttons, as the grinder-dust (bone and shell) destroyed the lungs of the workers.

Their houses and meeting-houses were plain-and-simple, although well-built. They believe in inward-signs, rather than outward-signs. Their direct-experience of truth-and-spirit was all they needed. This even reached baptism by water: In meeting for worship, Friends waited for baptism by the light to confirm belonging to Spirit. Friends found the experience of the Light, in meeting-for-worship, to be Truth. The inward experience of bathing in Light during corporate silence is reported to elders for clearness committee approval.

This simplicity extended to outward-relations with other cultures, to hold to one-truth. Their business-relations were trusted; they gave the same price to everyone. They kept good treaty-relations with the Seven Nations (Iroquois) and kept their mercantile-clients satisfied, through standing behind the goods and making-good on mistakes. These practical-spiritual ethics did not prepare traditional Quakers for the Modern commodity-consumer markets. They were makers and traders.

On the farms, the family was extended, and there may have been two-dozen folks, living-and-working on-farm. The only meal with all-present was dinner. Breakfast was served in waves, lunch-in-pails, and at dinner the whole-community got to hear what had gotten done, and what needed doing.

If a task required all-hands (harvests, burning, and animal-drives), everyone-that-could pitched in. Very low off-farm energy-inputs were required: Human-labor and simple-machines built to Human-scale. Animal-power (Horses and Oxen), dairy-Cows, and Sheep required half-the-fields for pasture, hay, and feed-grains.

Forestry and wood-lot management required a lot of labor. In the Adirondacks, we put up ten cords of hardwood (bucked and split) for the house-stoves and barn and ten cords of hardwood for the Sugar Bush (Maple-sap collection and boiling to syrup). This was done with two-person buck-saws and splitting mauls. Timber harvesting, milling, and seasoning was done with hand-tools, water-powered saw-mills, and animal-drayage.

The repeating pattern of work, decision-making, and social-arrangements holds *the good order of Friends* amongst the conservative Hicksite branch of the Society of Friends. The name Quaker was originally a journalistic-slander, following failure to doff a hat in the court-of-justice. "I do not quake before thee, magistrate, but only in the presence of the Light!"—some-such was spoken and the court-reporter re-named us.

We were originally self-named The Society of the Friends of Jesus, shortened to just calling ourselves Friends. We use the pronoun friend, as in *friend speaks my mind*. The small-book of our condensed-Truth is called *Faith and Practice,* and it contains unanimously processed *minutes-of-agreement*: social guidance called *Advices and Queries*. These are not fixed-rules, but works-in-progress. Etiquette is found in the *Advices*, and discipline questions (to meditate on) are the *Queries*.

> *Discipline is not punishment;*
> *it is a practiced-skill of discernment,*
> *alignment, and convivance.*

Conservative Quakers spoke *plain-speech,* which is a dialect of English, preserving the personal-pronouns thee, thou, and thy, and a phraseology-of-reminders. One does not know exactly whom is being spoken-to, if one only has the pronoun you. *If I am speaking to thee, friend, thee knows precisely whom is being spoken-to.* "You" has become singular or plural, indirect, accusative, and casual. Perfect for selling junk.

As a child I felt *spoken-to* when an Elder used *plain-speech* to give me advice. This etiquette-reinforcement allows-the-learning of a big-set-of-phrases that refer to stories-of-our-heritage (reinforced, deep-memory). Some Horticultural languages-of-Place have few verbs, or mostly use phrases-and-syntax, to be referential instead of authoritative.

"Soft-rain, gentle-rain, Deer-dance-time," is observation and association, not the causal-declarative, "It is raining." Multi-dimensional-relations flow in ecological-languages. Modern business-languages imply too-much-promise (hustle), or too-much-linear, or hierarchal-causal certainty, reduction with blinders. Modern English, now the global-business-language, has become unnatural and disconnected.

> *It is raining!*
> *Who is this it?*
> *God, right?*
> *And is means made-to.*
> *Raining is an effect?*
> *Crazy talk.*

In *Meeting for Worship,* Elders are honored with seats on facing-benches, raised up-front in the Meeting House, where they can hear and see more clearly. Worship is held in silence; *friends-who-are-moved* rise to offer brief messages. Worship can go an hour, without anyone speaking.

The room may be divided by gender or by clumping quiet-activities (sewing and knitting?). There may be *back-benchers,* who tend to the askance-attitude or have to come late and leave early. The meeting-room layout is adjusted to the social-relations.

All business and worship rises from quiet-centeredness and honors all contributing-members. Elders may have *weighty-advice,* but there is gender-equality, careful listening-to-dissent, and all-around sense-of-compassion and wonder (*gathered in the light*).

THE CIRCLE OF PRAYER

This word-poem
is about
Traditional Quaker practices.

Finding ourselves immersed in the great wonder, we can look about for the respectful practices of traditional and Indigenous people in home places. We find in First Nation sweat lodges and Quaker meeting-houses a common wisdom. The many shared stories map a pattern of prayer. Here is a way, a path of humility.

The wish rises unbidden, without effort and surprises us. We may hold the wish in the light without judgment and be amused and amazed.

The leading can then seize us. We are moved to quicken. An urge puzzles us to seek the clarity of true action. Our individual insight then benefits from counsel.

So we seek clearness. Best quality of discernment is found in silence and through careful ritual. As we live in cultures, we can ask others to sit with us and offer advice and reflection.

When clearness is reached there is Light. The spirit is palpable, present, and we are truly-moved. Here we enter alignment with the universal patterns and right-action is achievable. We move in dance, work, and practice with divine confidence, and yet we are not the authors of truth. Truth uses us.

Truly, mercy shines on us, the circle completes through careful practice, and the wish arises anew. Thus we find ourselves following the circle pattern of prayer.

There is no begging or promising on this path. We do not demean the mystery. Bargaining is not our privilege. We are all just parts of the infinite repetition. Patterns of patterns implicate and reiterated, as the beauty-magnificent that we are privileged to be present in.

We affirm our intentions and praise all. The prayerful ones say: amongst the we, standing open, blessed with humility, and thus willing to see, in solidarity with all sufferers and all relatives indwelling, find us here ready to dance the wonder-full burdens, elaborate and common to us all.

In the monthly-meeting-for-business, the *servant-clerk* and *recording-clerk* sit at a side-table to signal service, not leadership. The room is arranged so that all members can see and hear. A circle is common. The meeting starts with silence; the recording-clerk reads the minutes from the last monthly-meeting and business proceeds with reports, considerations, maintenance, calendars, nominations, and social-activities.

When a decision is apparent, with no dissent, the servant-clerk composes a minute that is approved by the gathered and recorded. This is the simple good-order-of-friends that is not always so easy. The clarity of the process and the care-taken by the members eases the challenge of difficult-business.

The traditional-practices of *Friends*, which can break-loose stuck-taboos and inappropriate-etiquette, are *testing-the-scriptures* and *continuing-revelations*. These are radical process-concepts that use *clearness-committees* to reach towards common-truths (and Truth).

Unanimous-consent procedures carry concerns-for-reform from local-clearness (monthly-meeting) through the visitations (carrying *traveling-minutes*) and many-silences, to an expanding-net of *monthly-meeting, quarterly-meeting,* and *yearly-meeting*, or even *five years meeting*. A larger and larger body of consideration weighs the concern and eventually records the changes in *Faith and Practice*.

Seasoning (holding a concern in-the-light), practice-and-process, good-order, and review take time and invest the communities in long-term goals and multi-dimensional harmony (*Peace*). Sometimes a strong-and-tested message is sent out on-walkabout to renew-the-concern, widely. Sometimes farmers are *released* from family/farm obligations to do this *visitation*; neighbors pitch in to cover the absence. This is a relations-net that is spokes-like, organic in its spread, and fractal. Back-and-forth, up-and-down, over-and-across.

ADVICE AND QUERIES ON SUSTAINABILITY AND SOCIAL ECOLOGY

Advice and queries amongst Friends are for meditative-guidance, to seek-clarity in our relationships and spiritual-endeavors.

A version of this proposal was originally submitted to South Mountain Meeting, Religious Society of Friends, "Quakers," Ashland, Oregon, early 21st century.

ADVICE

Careful considerations of sustainability should include study of the sciences of complexity and whole-systems mapping. An ecological-system consists of a minimum number of parts that are in right-relationship. In a resilient-system, relationships are more important than inventories.

The relationship-net amongst elements in a whole-system is more important than the diversity alone. In the long-story of humankind, many cultures have had some good and persistent times, living sustainably in place. Such cultures co-evolved spirituality and education that fit with ecological meta-stability. Others have stumbled through wastefulness and greed into darkness; they cut all the forests and used up all their soil. In times of catastrophic change, the ultimate goals may recede in probability but persist in Truth.

In the cultures with appropriate-fitness, local community reliance is the foundation of social ways. We are constrained to farm and live within the solar energy and water budgets. If we export more carbon from our field and forests than is fixed by photosynthesis and use too much water each year, our drainage-basins will decline in fertility. To fail to strive toward, and hold to, this peaceful-complexity and wholeness is to diminish the Light.

QUERIES

How do we as a meeting reach-out to the extended community of humans with a message of light, peaceable kingdom (caring universe), and stewardship with respect for the glory of diversity and interconnectivity in creation?

(continued)

As our held-in-common individual and community intelligence is the key to cultural sustainability, do we value our members' experience and keep record of our community story, so as to continue to learn and to persist in wisdom?

How do we nurture those of us that have witnessed brutality to creation and returned to our community? Do we provide debriefing and recovery to those whose ideals have led them to ecological and spiritual service?

Do we in our own lives live in the garden of peace and model right-sharing of world-resources? Do we constrain our needs to those fulfilled by local-production and do we grow a significant portion of our own food and medicine? Do we constrain our use of imported goods-and-energy?

Are we as a community of seekers ready to knit and support new right-livelihoods and to uphold traditional land-based cultures? Do we have real material and cultural relations with each other, within our meeting; do we barter, trade, and give gifts wisely?

With this modern information and economic speed-up and the industrial and climatic disasters, how do we remain centered in the wonders of creation and mystery, with enough time in our lives to grasp the vastness and wholeness of creation?

Do we practice non-violence with all of life, not just humans?

Do we have our own personal relationship with natural systems, which are the gifts of creation, manifestations of the Light? Do we take the time to observe, contemplate, and meditate on the lessons of complexity?

Do we hold our families, neighbors, and meetings in the light of cooperation, sharing, and care? Do we celebrate all the work-roles, genders, ages, cultures, extended-families, and elder-councils that lead to clarity and continuity for the work of living and love?

DISSENT AND RECALCITRANCE AMONG MEMBERS

Dissent is engaged by the meeting and listened-to carefully. Dissent may hold difficult or not obvious truths. Impasses are common and, as the meeting remains unsettled, some time is needed for seasoning. The clerks can ask for a minute, to form a *clearness-committee*, to sit with the concern or—with consent—send the concern to standing-committees, such as *Ministry and Oversight*, or *Hospitality*, or *Meeting for Sufferings*. The process of deep-consideration is called *seasoning-the-concern*.

Often when confusion seems to persist in *meeting-for-business*, the hesitant-silence is broken as an *Elder-rises, moved to speak-plainly:* "Let us hold all Friends in-the-Light, good-order here asks . . ." If the Elder's message resonates, all can palpably feel the gained *centeredness of the meeting*. The *gathered*-members sink silently into *reflective-clearness*. The clerks might recall (reread) the minute and find no dissent. Whether we see the *weight of Elders* as manipulation or as applied social-skills, the outcome is usually improved common-fitness and *the matter can be revisited*.

A consistently-recalcitrant member, who is disrupting the meeting, can be *read out of meeting* after a careful process. A not uncommon-discipline amongst introverted-cultures is shunning. The community stops paying any attention to the shunned ex-member. Shaming is also common; the member is held more-tightly by reinforcing taboo. These are hurtful to the object-of-scorn, but sometimes an extrovert-dissenter can disrupt *good-order* enough to cause general-harm and confusion. The body-of-attenders, the corporate-membership, feels *deeply-moved* (consent-minus-one) to expel or to harass-with-advice, if not-quite-ready to remove the disruptor.

TAKING A WIDER VIEW OF SOCIAL ORDER

When chiefdoms or principalities get consolidated into states, the central-government insists on a monopoly-of-violence and uses laws, courts, and police (or militia) to reinforce-the-edict, which reduces the local-freedoms for the bargain of fewer revenge-cycles and inter-tribal warfare (Diamond 2012). Enough peace to allow trade and crafts to flourish.

Did you know you agreed to this?

The Quakers benefited from the umbrella of the Modern industrial-state, which alternatively threatened and protected them, even as they practiced community non-violence, social-innovation, and settled-stewardship. The monopoly-on-violence gave Quakers a window to innovate socially.

One has to have a deep sense of irony to remember the centuries old story of Quaker social-activism. Many well-intentioned Modern liberal-reforms came from their efforts, appealing to governments and legislatures to pass laws that tried to modify the debilitating-effects of global-industrial extraction and Capitalism.

Many of these Quaker driven liberal-*retrofits* created legally compromised institutions and fudged natural principles-of-law, having been corrupted by clever-lawyers or just plain failed: public banking, penitentiaries, the Bureau of Indian Affairs, canals, conscientious-objection, and various industrial-efforts (including the slave-trade and Whale slaughter for a while!).

This is a complex story and there are many branches of Friends: some are evangelical-protestants, others are university-liberal silent-meetings. Very few working-class and farm-culture conservative-meetings persist. Everyday *Plain Speech* is rare and mostly used in meetings for business.

These insights and generalizations about Quakers come from personal, familial, and community experience (*walking in the light* and *seeking clearness*) and exploration in the wider-world (*released for visitation*). Any noticed local social-arrangements were site-specific at best and culturally-mediated; envisioning-council has a lot-of-patterns to visit. We as syncretic Ecotopians collect rumors off the street, ideas from study, travel, language, or customs to *bring-to-meeting*, as Friends would say..

For a Friends' Meeting, a local culture, to come to *good-order* and proper (*simple*) practices, there are achievements-of-abundance (basic ecological fecundity), cultural-maturity (elders as well centered systems-thinkers), and visionary-leadership (servant-facilitators) to accumulate from collective-experience.

Maslow's Hierarchy of Needs (1943) builds from base-levels of fitness-support, up to self-esteem and self-actualization. Finding our Modern selves poised on the brink of a phase-shift in this messy chaotic nest-of-nests, of post-peak-everything, we need to go post-petroleum with fast *quick-fix* re-localization planning.

The necessary sequence-of-tasks is carried by ecological-principles and universal-patterns. Envisioning is a process-of-discovery (observation) and adoption (brainstorm visions) that gets mapped (systems-theory-tools) and mulled-forever. We can find *light* in the *truths* of Nature if we are reminded by TEK.

Both the First Nations folks and Hicksites did not understand how secular modern movements could appropriate isolated cultural artifacts and social mechanisms without the deep, spirit-filled acknowledgements and affirmations of gratitude, repeated ritually in all circumstances of life. The practices of humility and the submission to the social order of Indigenous Horticulturalists is a challenging model for hyper-individualistic modern consumers.

We, the Human-species, are in-relation-to, not dominant-over, all Life, if we are to subsist within our local carrying-capacities. Mutualism not hierarchy: Where would we be without our resident, intestinal ecological-complexity? Humans co-evolved with uncountable other species and *biomes*.

As a constantly-evolving, and co-evolving species, in a complex mix-of-species, our social evolution-options could hold the key to mutual-thrivance, within-earthly-bounds. As Boolean Logic shows us (Kauffman, 1995), advanced-systems-evolution gets to the edge-of-chaos, with the highest-average-fitness, for the greatest-diversity of participants.

The Principles of Ceremony include

59.
Ceremony connects Humans to the Great Wonder. When we participate, we allow Spirit to whisper through us. Our daily busy-focus is diverted to enter the larger scope of Life. With our senses now open, we go forward to task with wider opportunities.

60.
A ceremonial bundle is a collection of symbols and sacred objects, wrapped up for storage and carrying. The bundle connects us to the greater Creation through sympathetic stories, represented by the contents. Mnemonic cues orient us. Gathering rods into a brindle reminds us how we are All, tied together.

CHAPTER 15

Carrying the Bundle

STATING OUR INTENTIONS

Let's start by cherishing-the-goal, that we might deeply-realize we are natural-beings on this landscape and we can belong-here. That is a big-healing for settlers; we're not necessarily only-taking. We have work-to-do-here and the land misses Humans, because co-evolution is deeply-embedded in our genetics and natural complexity. The essence of Life is relationship.

There have been First Nations peoples working on Turtle Island for anything from 13,000 to 250,000 years, depending on who you talk-to. It's only very-recently, 150–200 years ago, that a regular-cycle of seasonal-stewardship got broken in the Siskiyous. We can still see the remnants of that ethnobotanical-work. As we walk around landscapes, we find evidence of what the First Nations peoples were doing. So it's not that-big-a-jump to get back into-the-game, if you can see the long story.

Our intentions are first stated through prayers-of-gratitude. The ceremony of thanks-giving is the initial, primary-orientation of our councils. The reciprocation of gifts-of-Nature is foremost acknowledgement, a moment-of-explanation (what we think we are doing), gratitude, and alignment, before we dig into our tasks, culture, and sustenance. The inmost-bundle of our Spirit-ways is intentions, held deeply, and manifested through our craft and lifeways.

CHILDREN IN NATURE

Way out west near Ashland, Oregon, lived a cohort of three friends, who had magical-childhoods in the Adirondacks, upstate New York. We all experienced children's-pilgrimages. In our early teens, we took each-other to our childhood-pilgrimage Places, "Oh, you've got to meet this tree; oh, you've got to see this rock; you've got to climb this hill; you've got to jump in this ice-cold lake; you've got to squish in this mud." Now that we're grandparents, we think that kids are cute.

FRIENDLY CHILDREN IN NATURE

Childhood traverses the sequence-of-seasons, the stages-of-maturation. We pass through the portals of our life. We learn how Nature works, we share with new-friends, we raise our own children, and we care for Elders as they move-on. As we experience wonder, care-and-loss (praises-and-sorrow), we mature. And as the brain ages, it can make faster connections and see patterns that help us contribute, as Elders, to community-stewardship.

Paul Shepard, author of *Nature and Madness* (1998), says to be-sure that the children have full-support when they're infants: full-body-contact, at all times—it's called *continuum-parenting*. The baby is best carried face out-to-the-world or seeing over-the-shoulder in a carry-pack; the child sees what-is-going-on, and feels involved, arms and legs free to wave around. Then, from three to nine years old, they should have plenty, unsupervised nature-immersion.

As a child-in-Nature, we pick-up the metaphors-for-thinking, the archetypical-icons: Fox, mud, Lichen, dark-nights, bright-light, rain, cold, seasonal-flavors. The template to hold all those symbols is genetically built into our cognition. If we experience that full-complement of natural-symbology in our metaphoric-repertoire, we have a chance of becoming-adults, growing-up, making collective-decisions, and being-useful. A lot of us are missing parts of the full set of icons.

Without this symbolic, embedded, immersed-experience, we never grow-up and become adults, capable-of responsible-decisions. Guess what? Moderns live in a society of zombie-like, full-grown-children, who are acquisitive-dilettantes (meaning consumers). They are not adults— they're not taking responsibility for anything. They don't have a huge-understanding of how big Life can be, of multiple-generations together, of creative, regenerative-potential. That potential only blossoms when social-adults fully develop. There's social and brain-science that backs this up, helps explain the pathways.

The brain/mind/Nature connection changes through our whole lives. By the time we age, become Elders, our thoughts integrate creatively. Perhaps some loss of memory-capacity, but the corpus-callosum switch-board is kicking in. By the time we are Elders, masculine and feminine are fully integrated, and we are better-able to understand metaphor-and-poetry. Able to make connections in-a-snap that younger-people wouldn't see.

Everyone can contribute at all stages of life; attention is worthy; who knows where Truth rises? Every once in a while, a three-year-old girl can say the most outrageous truths. Or an eighty-nine-year-old woman reveals amazing-summations, because she makes-connections; she gets it instantly, even if she can't talk-fast or remember names as easily anymore. Our community members' long-lives are about coming-into the vast wholeness-of-being.

STORYTELLING

There are many types of stories and ways to tell them. We all are telling stories whenever we speak or move. There are ways of listening to stories: "What sort of story is this? How should we listen?" There are storytellers who seem unnoticed. They may be standing in a corner, dancing from foot to foot, singing an old language. Every sweep or brush of Nature is full of stories for us to hear and read—reading the landscape. What does the land want?

Human cultures are built on narratives. The power of story can glamor us. When the story fits the Place, the people feel the power of the land and are reconfirmed in their story.

When a traditional story is told, with set-and-setting considered, story-etiquette calls the gathered to listen. Perhaps the Elder-storyteller is already seated in the north-east, and the community files in to fill the circle. The prayers are said, and the room grows-quiet. In most cultures, when the children hear it is story-time, they circle-up and quiet-down.

Storytelling traditions vary with culture and Place, but universal-patterns are witnessed. Ceremony sets the stage. Drums or quiet instruments may accent the flow or drone as background, perhaps to remind us of rivers, winds, birds, and trees, outside the story-lodge. Dim, shifty hearth-light allows shadows and gestures to seem immense-and-compelling.

Storytelling is ceremony. Many cultures have Bard-guilds. The apprenticeship is long when a people have deep-epics and wide tales-to-tell. Bards study in sacred-groves of ancient-trees, or in natural mountain-cathedrals, or in temples-and-monuments. They may carry a carved staff or medicine bundles. The familiar public-spaces are filled with referential-shapes, icons, and flows, that act as the catalog, or pilgrimage, or visitation to the nature-of-the-universe, the epistemology-of-reality held-in-metaphor.

MEMORY

Deep, capacious-memory is becoming a lost-skill. Poets and orators used to be trained in the mind-storage (memorization) of epics and lists. From what we know of Northern European pre-Christian practices (Graves 1982) this was done in a sacred-grove of alphabet-trees (see "Gaelic Tree Alphabet Groves in the Pacific Northwest," below), or a cathedral or temple, that consisted of a series-of-altars around a floor-plan. The memorizing of an epic proceeds in sections or parts (stories) that associate with the familiar-architecture of the theater-of-learning.

To recall a long series of chapters, the orator imagines walking, with their well-worn-memory, through the spatial-sequence of the grove or plaza. This method of associative-memorization is called "mnemonic-architecture" (Illich 1988, Graves 1982, Sclain 1998). These methods are compatible with how our brain works (and the larger Mind). Spatial and tactile-memory, as well as olfactory-clues (smells), can trigger the retrieval of large chunks-of-words.

*The Mind
and individual brain
thinks in symbols,
not words.
Words are
the last thing done.*

The Gaelic Bards carried a rune-carved staff that could be fingered to remind. They often went blind when relatively-young, perhaps from the powerful-herbs used in the apprenticeship. With an apprentice-lad to lead-the-blind from tavern to tavern, the Bard could sit by the hearth and trace-the-runes with finger-tips, keeping the Elder in storytelling circulation.

The earliest writing was a string-of-consonants (no vowels, spaces, or punctuation), meant as clues to memory, not actual-text to read (Abrams 1996). Pictographic-writing, iconography, hieroglyphics—all trigger the memory of whole-stories, as the complex-symbol stands for a collection-of-words (a phrase).

The invention of alphabetic writing-and-reading reduced wholesale-memorization but enabled bureaucratic-imperialism (laws and memos). The oral-languages of Indigenous people-of-Place usually consists of phrases, not atomized-words. The familiar-phrase, when repeated in conversation, triggers the memory of a longer-story, learned-in-childhood.

Referential-phraseologies allow for ecologically-complex knowledge to be patterned and remembered culturally. Oral-storytelling is traditionally very-exact. Precise-memorization is taught, to make-sure that a people does not forget its stories, the TEK, and the Indigenous Ecological Knowledge (IEK). IEK is Place-based, and TEK has been known to travel (Martin, 1978). Scrolls-of-parchment or scraped-hides with pictographic sequences-of-symbols; strings of colored and/or shaped-beads; shells (wampum); and columns-of-ideograms alongside ink-illustrations—all these formats trigger deep-memorization, the preservation of ancient-wisdoms of Place, and stories-of-wandering or displacement. Symbol-bundles allow a culture to stay-intact, keeping some inventory of origin-stories, even with the loss of Elders or during times of change-and-trouble, to remember who-we-are.

The Druid-priesthood in Gaelic Europe walked in a forested northern-latitude that was relatively impervious to the grassland-based Pony-horde invasions. The forest-itself became the cathedral-of-memory that holds the epic-memory of a people who came themselves from-the-steppes, but now lived in villages, surrounded by forests, bogs, and lakes. The corridors of travel and trade, from east-to-west below the Baltic and North seas, allowed the rapid-spread of tools and Horticultural varieties into northern-Europe.

We have some ragged-memory of long-times with the Druids, from collected-fragments (monk's journals) of poetry-and-songs. The story goes that sacred-groves hold the widest biodiversity. In these northern forests, the alphabet-of-trees groves were connected by pilgrimage routes: students traveled from forest-shrine to tree-shrine and memorized the epic-stories associated with each tree and its ecological-implications.

After memorizing the epic alphabet-of stories, the now bearded-and-mature Bard would receive a walking-staff with the rune-marks for the trees, carved on the heft (the top of the stave), where the Bard, even if blind from too much *Amanita muscaria* potions, could finger the carvings and trigger-the-memorization. These traveling-repositories of the long-story then answered questions, found apprentices, and adjudicated-conflicts by finding the appropriate-story for the occasion-and-situation (set-and-setting).

*The collective memory
carries survival and flexibility,
entertainment and elaboration,
all tied to Place and way.*

Places that are familiar-to-us, from long-appreciation, get hung with stories. If it is our-valley, every hill, hole, and flat has a story and a name. The Greek Pantheon (imitation-forest?) columns tell tales that are now forgotten; they served as a mnemonic-monument. The Christian cathedrals of the Middle Ages contain the stations of the cross, the niches of the saints, and many-other mnemonic-symbols (murals and statues) that refer to accumulated-wisdom.

Writing allowed for bureaucratic extension and the ruling-of-empires but it also undermined memory. What written-fragments of oral-memory that we have from late-medieval-feudalism were written down by monks and missionaries, who translated and adulterated the eco-lessons embedded in-the-story.

GAELIC TREE ALPHABET FOLLOWS US TO THE PACIFIC NORTHWEST: A STORY, OF COURSE

In the late 1970s, I was familiar with only Robert Graves (1982), when my editor, Gwion, asked me to go out into the Pacific Northwest forests and collect the woody-twigs of nineteen or twenty of the sacred-grove tree-species. He wanted to construct the "Boca" (also know as "Ogham"), a bundle of twig-bundles where each bundle-twig is rune-marked, so that he could toss themed species-bundles and—laying the four-twigs in the shape of the dolmen-stones (two or more cap rocks on two pillars)—he could read the four or five runes exposed by-the-toss, in sequence as the dolmen is built, of alphabet-trees' poems.

These runes recalled the poems or stories of each tree-rune, as answers to the question held-in-ceremony and in the choice of twig-bundle-theme. One chooses the twig-bundle that best represents the question and then reads-the-results from the roll-out of the four bundle-twigs onto the ceremonial-cloth.

*The Boca
is a bundle of bundles
of twigs,
and each twig bundle
tells a bundle of stories.*

I was skeptical. "This is not Europe," I complained. "Get out of here, I gave you an assignment," he said. So I looked-out and by-goddess, there were close-relatives, if not the same-species, right-here on the other-side of the world. Yes, I did take some liberties-of-identification by seeing ecological niche-equivalents as substitutes and, actually, I found the original European-species in yards and parks.

After all, I knew that the vast-majority of plant-species, now found on the coasts of North America, were from Europe. We of settler-descent also brought-with-us, in the belly-of-ships with the livestock, our very-familiar weeds, many of which we know—from our own traditions—how to use medicinally.

Next, looking at the alphabet-tree matrixes on-record, especially the Irish layers of fines to-be-paid for damage-done, in four ranks (see just below), I began to think maybe there was a set of ecological-clues here. Perhaps the ranks represent plant-guilds with shared fungal-hyphae, connecting the member-plants through mycorrhizal-interpenetration in the soil. The ranks of trees could represent stages of forest-succession, enabled by different sets-of-mushrooms.

In the early 1980s, leading plant-walks at Country Faire near Eugene, Oregon, I looked at the species-inventory and found the floodplain fairgrounds (the essential Ecotopian-event) were blessed with almost the whole-sacred-grove complement. Since then, the one or two missing archetypes have been insinuated. The Hoedads tree planting coop has been at the faire from the beginning, and they know how to plant trees.

Thus the implication of this adventure is that we never escaped our ancient-inspiriting of the landscape by sailing away-west. The northern-hemisphere forest is trans-boreal, or circum-boreal. We might as well get used to this as *home*. Here we are, and there are the clues to live with the forest and how to take-care of our medicine and our sources of epic-poetry. The language settlers speak (Modern English) is rooted in these trees. Our poetic-metaphors have their origins in alphabet-grove lore.

The epic "Battle of the Trees" poem (Graves 1982, Matthews 1991) is a set of the fragment-sources for relearning the feel of the alphabet-carved rune-staff. As we re-deepen our languages, we will re-elaborate these myths. We have the woods-lore of utilitarian-reputations to guide us. Alder as the tree of Bran, the hero who bridged the English Channel with an Alder-bridge. Alder posts are known to resist-rot if kept underwater (pilings). The myth follows the practical use.

So here is the localized-version of southwest Oregon, so we can begin to remember the ecological-significance of key trees and shrubs.

NATIVES OF OREGON:

Goddess
Garry Oak, Oregon Crab Apple, Red Alder, Spoon Willow, River Birch.

Fairy
Quaking Aspen, Oregon Ash, Ponderosa Pine, Douglas Haw, Pacific Yew.

Witch
Modoc Plum, Mt. Ash, Blue Elder, California Hazel, Oregon Bearberry.

Brownie
Lacquer Bush, Fox Grape, Giant Reed, Deer Brush, Greenleaf Manzanita.

(Oregon Bearberry is *Mahonia aquifolium*, and Lacquer Bush is *Rhus diversilobum*)

British Version:

Chieftain
Oak, Apple, Alder, Willow, Birch.

Peasant
Aspen, Ash, Pine, Hawthorne, Yew.

Shrub
Blackthorn, Rowan, Elder, Hazel, Holly.

Bramble
Ivy, Vine, Reed, Furze, Heather.

To make the rune-sticks, one carves short, fat twigs—all of one species—into either three-sided or four-sided sticks. One rune is carved in each face of the four-siders and in each edge of the three-siders. Thus it would take five four-sided sticks to have twenty rune-marks and six three-sided sticks to have eighteen runes. The tree-alphabets that we have recorded, have various numbers of letters. Notice the matrices above have twenty species listed.

Refer to the texts mentioned at the beginning of this discussion and to *The Twenty One Lessons of Merlin* by Douglas Monroe (2002) for suggested-equivalencies between

Tarot-cards, Gaelic gods-and-goddesses, and associated poetry-fragments. We, as newly arrived settlers, have a-lot-to-learn. As we take more-and-more of these inherited-gifts to heart, we will find the characteristics and uses that would give a sacred-tree its metaphors. Leonard Shlain (1998) suggests we might better do with speech, symbols, and icons to communicate-with, rather than the written-word (such as this book).

May the new/old epic
re-emerge
as we re-localize.
Greenward Ho!

CEREMONY

Humility is rare in the global-hustle. Growth beyond carrying-capacity is folly. Moderns are people without appropriate-culture, especially when they chase greed as an organizing-goal. Ceremonial Societies in Horticultural People-of-Place remind, repeat, reinstate, and reiterate The Original Instructions to keep-sane. This concept of traditional-philosophical-essentials has been lost in the sea-of-greed.

We as the Humans do not know what we are doing, but there are old-stories that tell us what-will-happen if we mis-behave relative to our small-planet and its local carrying-capacities. These old-stories also remind us that there are times when, no-matter how much humility and precaution we apply, the planet goes through big-changes on geological-time scales and we will be lucky to survive. Ceremony reconnects us to humility as it tells the stories of loss-and-recovery in a centering-way.

Perhaps moderns are always ready to move? In this global-network of empires, there is no longer anywhere to move to. Humans (and the Wild) are going to try to escape big-changes, but we will run into each-other, not into "new lands." Ceremonial Societies that persist with tribes-and-peoples on-the-move allow continuity that supports community-cohesion (as we have discussed, above) and eventual re-settlement, relocalization, and eco-restoration..

Going deep-local is an appropriate-strategy, if the carrying-capacity supports it. This only works on a *fecund*-landscape, as opposed to a *brittle*-landscape where eco-systems can collapse under-stress (Diamond 2005). Martin Prechtel (2015) speaks of feeding-the-past. The past is hungry for our memories, so that the cycles and balance in-Place can be maintained. Grief is both loss and praise. Ceremony facilitates our deepening.

The rise and fall of ancient empires in Eurasia rested on agriculture and trade-routes connecting fecund, irrigated-flatlands, and these cradles-of-civilization were easily invaded by highly-mobile and well-organized nomadic-tribes. Most persistent local-cultures have held remote mountainous-valleys and defendable-basins. The wider-geography (Kaplan 2012) of the northern-hemisphere as the context of local-culture, largely-determines the survival of ceremonial-continuity.

There are exceptions to this geographic-determinism of wandering-peoples with strong language-and-story roots that allow a nationless culture to infiltrate and cooperate with more-settled cultures. The common aspect of cohesive-cultures is a ceremonial-memory. Ceremonial Societies can feed-the-past, All in beauty, with perfect timing and elegant-etiquette, keeping-the-calendars and watching-the-skies. The work of ceremonial-facilitation is both a privilege and a calling; the pageantry is skilled-and-honored, but the training can be exacting.

The Gundestrup Caldron was found in a Danish bog. This is a ceremonial-cauldron that consists of a pile of hammered silver curved plates that fit together like a puzzle, with wire or string, to make a big-bowl. Each of the plates shows a ceremonial-figure in poses ranging from yoga-postures to shaman-costumes. Careful-examination of each plate shows different-origins, including the mining source of the metals and the tool-marks of the crafts-workers.

This appears to be a traveling ceremony-bowl that could be carried by a nomadic-group with deep-ceremonial-practice and multi-generational-cohesion. There is gender-fluidity and cross-cultural symbology in the images. Perhaps the carriers-of-the-cauldron were able to serve a wide-range of cultures with special-celebrations and ceremonial-passages. Perhaps they brought wide-knowledge to scattered-villages, who learned-a-lot and appreciated the cross-pollination.

The olfactory and tactile senses tickle the memory, surprise the entranced. We remember clearly, almost like a vision, and we may not know how we were glamored.

The storyteller uses cues-and-referrals to weave the trajectory-of-memory and lead the listener without too much talking. The listener on the next and following-day just makes a gesture or a face to mean-so-much, as brought-out by story. The People of a Place, who eat many plants, fungi, and animals, have a big-memory need. The intensification of Wildness by Horticulture requires deep-training and memory.

Knowledge of Place is multi-sensory and multi-modal; it takes the whole-people to carry as much story as possible. Lots of decorative-pattern clues, odiferous smudges, tones, and rhythms. A set-of-villages (drainage-basin) see/hear/feel way beyond any personal-knowledge. The collective-memory carries survival-and-flexibility, entertainment-and-elaboration, all tied to Place-and-way.

The Shaman and Herbalist use memory-training extensively. They put-on a great-show to help bring-us into-focus and include-us All in the great-immensity. Reminding the lost where and who they are is powerful-healing; if the distracted-one can come Home, whatever-outcome will be improved.

Personal-healing in crises begins with smokes-and-color and the showing-of-bundles. The one-in-need is shown who-they-are, where-they-are, and sometimes the re-integrative effect of compassion-and-concern is enough to bring-back health. *Holding the Friend in the Light* of identity-and-community. Demonstrating the wonders-of-Life to keep us in Place with All-of-our-Relations.

The wisdom of Human-nature, gained by Elders and kept in story-lines, is a map of Human-cultural variation. The highly flexible (opportunistic) Human-species seems to have developed a wide-range of tendencies, orchestrated in social-groups for the greater-good. Some of these behavioral-characteristics can also be ascribed to social-groups, as in clan-totems, symbolic-animals (with all the associated stories), and spirit entities.

VERNACULAR WAYS

A drainage-basin repertoire of colors, materials, land-forms, denizens, and Place-spirits suggests building-styles, clothing, diets, art, and philosophies. These tastes and feels of Place are celebrated through ornamentation and elaboration, using iconic-representations of myths and stories.

When we visit over the mountains to the east, our hospitality host feeds us those stories and we are enriched. The vernacular-way is a style, decorated as Being Home. Our crafts-guilds are happy to move with these local-forms and allow their materials to teach them. We end up with lovely garden-cottages; sturdy barns and council-lodges; and colorful mile-posts. *Prayer-flags flutter in the summer mountain passes; faint bells sing the herds.*

In the tipi, no one goes-around the center-hearth the wrong-way (widdershins, or counterclockwise) for both practical and symbolic reasons. There are stories about seating, hearth-tending, sleeping, and guests; living in a tipi is living on-the-land is living in-spirit-space. Never pass anything across the line between the hearth and the anchor-peg, where the altar and smudge sits. The seats and chores of the lodge-hold—on their mats, tucked into the liner-shed—are all relative to the micro-climates, the flow-of-food, crafts-and-story, and the continuous-feeding of the small but effective hearth-flames.

The grandmother, sitting in the southeast, can pass food to or from a storage-side (northeast) helper, across the east-door, to the matron and cooks on the south-side mats. Loads coming in the east-door are set north or south, depending on the flows: food north to cool, brush-wood south to dry. Folks occasionally go a short way, around to the north-side and back-out the east-door (same on the south-side) but if one goes south to the west back-side, they go all-the-way around the north to the east-door exit.

The anchor-peg and hearth-altar are symbolic of centeredness and are referred to by gestures, story-phrases, thanksgiving-recitations, and reciprocations (gifts). So-much meaning in such a diaphanous nest of thin shells of canvas, that it needs an anchor-rope to hold it all-together. The well-worked-out and storied-pattern of light-living in close-quarters, for semi-nomadic ecosystem interlocutors, is graciously decorated. Not hung with signs-of-wealth, as much as connected to-the-whole through a wrap-of-icons and sacred-patterns. Entering, socializing, or sleeping in the deep and well-defined-space nurtures the community in multi-dimensional-ways that celebrate an epic-story of land-stewardship, social-cohesion, and Human-worth.

Ornamentation can be full of symbolic-representation of story-cycles and myths. There might be colorfully woven fiber strips running up the inside of the liner as stripes, where the north, south, and door poles (the tripod, for the Sioux), or the north, south, east (door pole), and west poles (the quadra-pod, of the Crow) run down between the liner and the cover. There might be materials-bundles hung up in the smoke. There are ceremonial herb-bundles and tools gathered at the altar and all the baskets, clothing boxes, hair-braids, and mats are woven with meaning-and-representation. This is different from pretty glitter (consumer-glamor): The tipi as a portable story-loom, weaving library, cathedral, and medicine.

Simple-truth, plain-clothes, plain-language, and simple appropriate-elegance are stewardship-landmarks. Can one guess what the Ecotopian meeting-hall looks like? Solid timber-frame with kiln-baked tile-roofs in the coast-ranges. Re-localization usually takes generations to build deep, but we have many clues from previous conserver-societies and from a world-of-stories and visits.

A syncretic culture-building sociology is experimental and without much institutional back-up. This is the bench of our alchemy: Place, People, and Gift-sharing. The social evolution is best held-and-matured in council, in a round-lodge, with All Species representation and whole-story ideogram-boards backing us up. Surrounded by icons of the whole-work, we can collectively-glimpse the options for consensus-and-restoration.

The return of Salmon, Beaver, and Fire.

PART IV
Visioning

CHAPTER 16

A Year in Wagner County

*In the Ashland library basement,
on a cold December night,
we gather to hear a story
about the eco-future
in longed-for Wagner County.*

FIRST WE HEAR SOME BACKSTORY THAT SETS THE FORE-STORY

OK folks, we're going on an adventure and it's going to require all of us to do this. I'm going to start out with how this series of stories is happening. There's different types of storytelling that I do. One type is the practiced story and I'm going to start with an example of that. These talks that I've been giving in Ashland are where I begin to tell a story. So this is going to be the first time I've told this story tonight and you're going to help, because you'll either go to sleep or wiggle a lot; you-all will teach me. I have my teenage sweetheart and senior partner Elizabeth, sitting here by me to keep me going.

There's a big difference between a story that's been practiced a lot and a story that's extemporaneously unfolding right before you. This is in a series of talks I've been giving and it goes way back to the 90s and the 70s; I've been telling stories of this town for decades. One of the favorite story lines that got repeated back to me was the tradition of promenade. If we were real people that lived someplace, we'd get dressed up in the evening, just a bit, walk around the neighborhood, admire the flowers and say hello to people; a home-centered economy might go this way. I will tell the story of what it might look like to live someplace.

I tell this story about what Place could be. One of the most important parts of that is the Return of the Salmon. Once we have Salmon back, it's a big sign of success, and there's a lot that needs to be done before Salmon are running up the creeks regularly, several times a year.

HAZEL ORGANIZES NOTES AND SETS THE STORY

Tonight what we will do is walk through an entire year of festivals, work-guilds, and social arrangements. What does that look like, how does that work, and what works on this landscape? I've been here and heard local stories for so long that I made this one up, and our collective-part is to follow the thread.

Every person, as I see it, has four basic social attachments. One is family; the second is clan, which might not be your genetic family, but is your Raven clan or Wolf clan or Elk clan; the third is guild: what work you specialize in (Everybody knows how to do everybody-else's jobs generally, but each of us has specialties. We're in an herbalist guild, the Mothers, or a woodworker's guild, Bodgers, or Herder's guild. We're all in craft-work associations.); and the fourth is through the connectivity web of all those things, our council situation—set-and-setting—where decisions get made. We may be observers, or spokes council delegates, or overseeing Elders, or ceremonial society space-holders. We're all involved in considerations.

I'm also going to mention people who come though here at different times of year. Some of us may want to identify with Trader-travelers, the news-bringers, or council delegates off on discussions.

*I'm only going to be able
to fill this vision
with side-comments,
readers fill unseen with sensed,
the waves of my hand
and the faces of emphasis.*

Let me set the scene one more way. We can still see the remnants of that ethnobotanical work as we walk around, and that shows what the First Nations were doing on this landscape. So it's not that big jump to get back into the game. I want to be encouraging.

> *Let me repeat
> that my experience
> of the wildlands,
> mostly the mid-elevation
> hills and gulches,
> where I can still see
> the ethnobotany,
> is that the land
> is lonely for us.*

Stories can be remembering, revisiting, and it's good if we remember the Indigenous stories of Place. My favorite local source is Thomas Doty—he is Dhalgelma/Karok and studied the settler's English. Then he stepped up to put on plays with Phoenix High School, staging the Dhagelma myths, and traveled to other Oregon schools with his tales. His published story collection is *Doty Meets Coyote* (2016).

> *There are storylines
> to remember forever.
> Then there are stories
> of envisioning,
> and that's what we're going
> to be in tonight.*

The culture that Elizabeth and I came out of is Quaker culture, and the Friend I want to first recommend is Elise Bolding. She went around to progressive cities and did positive envisioning with people. The first thing she had to do was get people to let go of their but, but, buts. She was very powerful in releasing people's imagination. Her skills are very important. Then there's all these science fiction writers—Ursula Le Guin, *Always Coming Home* (1985), Kim Stanley Robinson (a favorite of mine), Octavia Butler, Margaret Atwood—these are writers who are telling future-stories.

Now the story of Wagner County begins. As you are reading this, remember that literature is a derivative of oral storytelling. This transcript has been edited to turn it into written text. Try to imagine sitting at a campfire, listening to a well bundled old storyteller.

HERE IN THE SISKIYOUS, WE FIGURE OUT HOW TO BE

We have come to live here, and the Siskiyous are one of the more fabulous places on the planet. They're so complex, and we have generally mild winters; the valley-bottoms are below snowline most of the time. The hills and mountains above were glaciated in the last Ice Age and held snow-fields not that long ago. So we have all these elevations and climate zones.

This complexity supports an incredible repertoire of different pleasant-places-to-be at different times of year. We have all these geologies. We have different waters; we have all these strange soils; we have all these seasonal creeks and pools; we have valleys that point, opening aspects in every direction. So you want a north-facing valley, no problem. You want a southeast-facing valley? Sure. It means that we have a very convoluted and complex landscape to tend, and our new-societies will grow and gather with that landscape's whims. How does this look year-round?

What can we do
to help
All Sentient Beings?

WELCOME TO WAGNER COUNTY, THE HEADWATERS OF THE EASTERN SISKIYOU CREST

I propose a new County called Wagner County. Wagner County is the south half of the older Jackson County, Oregon, and the new line goes basically between Talent and Phoenix; so we're thinking the Greensprings, Ashland, Talent, and the Upper Applegate and Little Applegate canyons, all the way downstream to Thompson Creek. It's mostly the high-ground, a set of drainage basin headwaters, the snow-pack melt-grounds, and narrow valleys with some open bottomlands.

The early settler Mr. Wagner was run out of Talent for being too nice to the Indians. I was up at Grouse Gap shelter one day decades ago, and this woman and her tall lanky partner got out of a small car and stopped to talk to me. I asked, "Where are you going?" and they pointed at the bushes. So I said, "You obviously know your way around here," and she said, "I'm Wagner's great-great-granddaughter." There's a plaque in Talent that shows where his trading post was and there's some written history of who he was. The other Dahlgelma-friendly settler-family in Talent was the Beesons, and they were Quakers.

What story I tell tonight, about adaptation after the collapse of globalism, is a visionary future in Wagner County. We've moved out of the flat valley bottoms. Ashland is mostly perched up on north facing slopes, with old irrigation canals crossing the upper, sometimes southeast facing slopes above the fogs. All the valley bottoms were smothered in nasty winter smogs during the industrial times. Almost all the Pacific Northwest First Nation winter villages were up at mid elevation, between 2,000- and 4,000-foot elevation, not in the valley bottoms.

Ashland in the early settler days, when everyone was doing some farming, built irrigation ditches stacked on the hillsides, diverting canyon streams. After World War II, bigger, feeder-canals and reservoirs were built, distributing Klamath River Basin water down through giant head pipes to electric generation turbines, and then to the new big feeder canals.

In this story, Wagner County still gets some water power and distribution from these engineered networks. We really depend on the Bodgers and Rangers to figure out the salvage work and retrofits. Wagner County is far more judicious and careful with water distribution, using local sources with well-sealed ditches and pipes. We have also found the settler-built irrigation distribution pattern is useful in other ways.

Imagine that the big canals have narrow-gauge railways on the bank-berm. The canal path is improved enough that only where we cross steep canyons and the canal goes through an inverse siphon did we have to build timber trestles to carry the train over to the canal bank again.

During the time of year when there's plenty of water and where the canals do not cross deep canyons, we're actually using those canals for transport, for moving materials: logs, hay, rocks, sand. We're floating special canal barges and boats, narrow and shallow, ganged together like a train, pulled and polled, on those canals.

Our villages are connected by this scaled down railroad that goes in and out of the side-valleys. The train visits all these good village locations on the southeast facing slopes, out of the big storm winds and below snow line. That old industrial irrigation grid has been repurposed and is connecting the perimeter of the big valleys together.

To get to the Little Applegate, we wagon up Wagner Creek to a cog-railway that winches gondola cars up to Wagner Gap. We've got steam-engine wagon lines, running on salvaged roads. We've taken old vehicles and retrofitted them for steam and they're running up and down the remnants of paved roads. This is how we connect this whole long, skinny, east-west Wagner County together.

Let me read this:

> *Thus we are becoming Indigenes all,*
> *semi-nomadic Herbalist,*
> *sedentary Miller,*
> *Horticulturist,*
> *Ranger and Teamster*
> *guilds all working together*
> *to Keep It Living.*
> *Everyone has praise, work, family,*
> *and is deeply connected to the land.*
> *Everyone, because we-all live here.*

JANUARY IS COLD AND DARK

I'm going to start the year-round tale with the old Roman calendar month of January and end with December. We've all just been enjoying the Winter Solstice festivals in Wagner County, and we're now in early January; the valley fogs are thick. There are really cold night freezes. We're up in our south-east aspect canyon-side garden villages. Hardly anyone is out wandering around, except some Rangers and some hardy mountain-people who love to go up into the snow, but pretty much the clans are tucked in. That means in the saunas, drinking Apple Jack. There's sort of an alcohol thread through this story: what we're drinking, how we preserve some nutrients.

> *How to make Apple Jack:*
> *take Apple cider that's fermented*
> *and freeze it down in the canyon bottom;*
> *every morning take out the floating ice;*
> *after a month of that,*
> *pour off what isn't frozen.*
> *Apple Jack.*

It's very tasty. We're eating nut cakes—acorn-berry cakes made with the mountain berries from the fall and late summer gathering—dry pressed berry cakes, and berry soup. We're firing the ovens, we're firing the pottery kilns, we've got medicinal wines bubbling through air locks. The creeks have risen and the water-power mills are turning. Craft-shop time. Crafts materials fill the warehouses, waiting for nimble fingers.

Out on the landscape, we're downhill cool-burning some mid-elevation south-aspect slopes that we can't burn any other time of year without harming the soil-seed-bank from burning too hot. But in January we can do cool-burning, where the ground actually stays wet and you burn dry standing understory, the slow line of burning creeping over wet ground.

WINTER COOL BURNS

We have been learning to do this at Wolf Gulch Ranch. Guided by similar traditional First Nations downhill burning, in the middle of January. The fire passes over still green emergents and leaves a lace of char stretched between burnt bunchgrass stems.

We're also basket weaving, making handcrafts, and fiber processing with storytelling. The lot of our nomadic/itinerant friends are resting in winter-camp, and we're all listening to reports from the Rangers and Mothers, feeding us information that they've gathered about the greater landscape. Everyone's noticing complex relational patterns about what's happening to whom, in order to plan the next year. We're listening to the watchers; guild-folks are reporting to us and telling us about edge-effects, while we're working; then we retire to another sauna. Most folks are in town, sunning above the fog and below the snow.

IMBOLC SURPRISES US, HIPPETY-HOP

As dark and cold as it has been, the quickening sneaks up on us.

Then we get to our mid-winter festival day, which is February 2. This story is hung on a thread, using the Gaelic cross-quarter days calendar, because it's sensitive to animal patterns, celestial events, and Horticulture. Since so many of us settlers are from Northern European removals, we carry stories about Imbolc, the first cross quarter day after the Winter Solstice. In January, the Sun doesn't seem to be doing anything; it's just always cold and the days are always short.

As we approach February 2, we can feel the acceleration of change—it's called the quickening—it almost makes us dizzy because the rate of change of the rate of change is strong around the cross-quarter days. Day-length either speeds up fast or slows down fast. The first lambs are born, and at the same time the last Herd culls are taken. So there's necessary feasting; we're selecting and slaughtering. Some of the Herd-animals won't make it, as we don't have enough hay for all the Herds to get them through the rest of the winter. So we're thinning, but we're also seeing birth at the same time.

*The Alder
and the Hazel
are tasseling,*

And we're out looking for the Grouse Flowers (Snow Queen); these little purple flowers at the end of January and beginning of February. We also call Imbolc the First Flowers Festival. Stories are told of the harbingers of spring. At the festival crafts-faire, we're looking at winter-woven woolen and linen cloth; musical instruments; and hand tools, (booths offer immediate sharpening and smithing). We're having a repeat of our favorite shows from the holiday season, because the temporary stage-troops are still in town; they haven't split up and gone off with their guilds on their Hoop walking. We're gathered in the great council-halls; we're cozy in the side-drainage villages.

In just the right weather, cloudy at night and not too warm during the day, the Maple and Alder saps rise. Where the sap run is strong, in trees alongside water, we can collect buckets of sugar water that can be boiled to syrup or fermented to beer. A good sap-sugar year is cause for further celebration and praises.

If we can safely run the train, there's a rolling celebration party. A nice day near Imbolc: the train is on tour and we're stopping at the villages where there's a little bit of trading and lots of gifting. We build bonfires and we're throwing the old stuff out; it's ritualistic, we're getting ready for this year's abundance.

We're building and firing our charcoal kilns, those below the snowline; we're busy that way, as we can at mid-winter. This is when we gather Hazel withies for our very best baskets. They all need to be peeled; in the Siskiyous we have a lot of bark and boring beetles. They do a lot of important ecological chores, but we need to plan for them to preserve our baskets. We heat the Hazel over fires and it smells really cool, like popcorn, and we're building up a stash of de-barked and split Hazel bundles, all white and smooth.

The quickening of Imbolc is the cue for the last patch burns; when March arrives, the birds start to set up territories. We shouldn't be disturbing bird-culture by working in the woods anymore. The birds are looking for nesting-places and the mother Deer are chasing away last year's adolescents, while searching for a nook to have their new fawns in.

There's a bunch of exciting green growth about to happen in March. We have a lot of early-emergent medicinals: Cresses, Violets, *Lomatiums*, and Cranesbill greens. All of these are spring tonics, as all winter we've been low on fresh greens. We have dried things, but this is when we get to praise the fresh new emergents, which are important to us health-wise. We feed these special treats to the children and Elders, mostly.

> *We're looking
> for early greens,
> while we're out staging
> the year's tasks.*

We're compost building, planting some greenhouse flats, and we're working in the valley-bottom tree-crops, orchards, and pastures doing clean up, clearing irrigation ditches, pruning and thinning, getting ready for a good year.

This is when the craft-guild council meetings happen, during early March, to make very specific plans for the sequence, staging, sorting, and stashes. Remember, in January, the Rangers brought down the reports; now the guilds meet to mull the Rangers' news and start making their plans. While conspiring, they are working on their packs, the pack-animal harnesses, the tools they need—only what they can carry; nothing heavy, we're walking.

SPRING EQUINOX CLIMBS BY FAST

Then comes the Equinox, March 21. At the Equinox, the day-length is changing exactly the same from day-to-day, and we are halfway between Solstices. The sun travels the mid-arc across the sky. At Imbolc you feel the quickening, but at Equinox changes are moving steadily (schwoosh); that's the rising sensation of Spring Equinox. There's a specific time-point of the precise day, and most cultures that watch for the Spring Equinox have a cave where, on a certain morning, the rising sun-beam touches the altar at the back of the cave.

This is the first spring children's-pilgrimage since taking them up on the snow at Imbolc to tell some snow-stories. Spring Equinox offers rushing rivers, and verdant cliffs, and caves for the sunrise-moment. We're looking for the first flowers, and the early migrant birds, and telling stories of all those awakenings.

> *Outdoor activity
> becomes more attractive.
> Time to get out
> of the buildings.
> We get excited.*

We're going to be outside for the rest of the year until the snows come back again. Our last chance comes in late March, to divert excess water flows (snow-melt floods) safely, without being overwhelmed, and before we need to leave all stem flow in channel for the fish. We're diverting into ponds, we're diverting into soak-aways, we're water-table building. This is the last of our winter water harvest. We're also doing a lot of ditch work, cleaning out debris and repairing leaks in the distribution web, while we're impounding surpluses in reservoirs and perched water-tables. This is also the spring Salmon run, so we have a Salmon ceremony: "Look, spring is coming up the rivers."

It's especially exciting for the children of Place.

Now we're in April. We're doing our first spring-sowing of pelleted-seed—seeds encased in clay. We're doing our first sowing of early seeds to get fields growing: the fastest Barley, the short season Daikon, the varieties that we've worked out that will get us food the fastest on fields we will not irrigate, trusting in spring rains. We're using pelleted-seed so that just the right conditions will germinate them, the Mice will eat less of the seed, and the seed will lie dormant waiting for strong germination after natural rain. This gets us our earliest crops, on lands we can't irrigate later in the summer. We've really dialed that in.

We let out animal-guilds on our earliest pasture; the grasses are coming up in the orchards in the tree-crop savannas. We're releasing our newly birthed animals from confinement with their moms, onto fresh grass, and we're very closely facilitating. We need to move them continuously, because it's pretty soft ground and the emerging grasses are tender. The new grass is just coming up, and we want to give our animals super-food; they're like, "Oh, that barn was really ugly." The deep-rooted native bunch grasses in our pastures respond to a light early grazing and grow back strong for cutting hay later, while the Shepherds and Hoop crews work the elevations.

The new lambs frolic; Herders move up slope, following the season.

Our relationships with animals are complicated. We are trying to act as good stewards, and we praise the powerful feedback through animal spirits and dreams. Our Hunters-guild has learned much from Wolves—who are eminently social—in their relationships with prey. With taboos, etiquette, and cautionary tales, our springtime Human councils consider much and approach decisions and protocols with respect, humility, and hopefully wisdom.

The animal societies that we live closely with and work with are facilitated to hold generational range (youth and elders), language, ritual, and social arrangements. Many of the animal societies we share with are also enabled to work with other totemic species in multicultural swarms, doing land tending with us.

We Humans live in clan clusters, our homes. As we all belong to nature-based totem clans, we empathize with and notice coincidence and complexity. Many of our social arrangements are patterned and informed from careful observation of the immense wonder of being alive together, adapting to our Place in loopy relationships. Much art emerges; we learn to converse with Others; important help is rendered, with perfect timing, as we dance.

5 Animals in Ecosystems

We're doing a lot of our turning, milling, and manufacturing, because we still have water-power, even if we are not any longer diverting to reservoirs and ditches. April is busy for everyone; we're moving more work outside to porches and pavilions. The Millers are busy; the Bodgers are busy; the Sawyers are busy; the winter logs were skidded down to the water-power and the sawmill is screaming. We've got the grinding stones turning, the machine tools cutting, the floppy belts fluttering overhead in the shops and the water wheels spilling water back into the stem flows. It's kind of noisy: the creeks are noisy, the machines are noisy, and many of us are still in town, feeling bumpy and grumpy.

"When do we get out of here?
Been in here all winter.
Now do we go on Hoop?"

There's spring bird-nesting and the early morning-chorus. This is the season of the most extensive birdsong. Time to tell the story of the origins-of-music and how music came to earth. Get the children up, bundled warm, and out for this early-morning pilgrimage: to hear all those dawn-chorus songs. Tell stories about each one of those birds as they get active and sing. It's a sequence, there's first one bird and then another one, and there's a story about why they're in that order.

Now, there's lots of wildflowers. April has a lot more wild-greens. We're antsy after being kept cooped all winter, although it was entertaining. Now we're getting some good fresh food, we're starting to feel our blood thinning and warming, as we're eating wild-greens. We've taken the last Cabbages and made them into sauerkraut; the last of the winter storage vegetables are weird looking. We're fermenting the winter leftovers. We've got our fresh wild-greens now, so forget it, we're fermenting it all—setting big crocks of sauerkraut. There are multiple birthings, both wild-animals and animals living closer to us. The busy seasons of growth and harvest are aborning.

BRIDGET, BELTANE: THE SECOND CROSS-QUARTER DAY GIVES US A SPIN!

Here comes Bridget, May Day. We can't contain ourselves anymore; we want to frolic. On May Day, as we put on the Wild, with flower chains in our hair, and circle up to dance, we might take our clothes off, because it's May Day. The days are not getting longer so fast. The day length is already long and stays long until early August.

> *We can feel the slowing-change and need some wild dancing, to get us ready for the long summer.*

The Herders and many Herbalists have already left the villages to move up slope with the blooms and verdancy. The Rangers are ahead of the Herds and gatherers, setting up summer camps and clearing trails of winter log-falls, rock-rolls, and slope-slips. Stashed and cached building materials are readied for Bodgers and Charcoaliers to move to higher camps. Kiln burns, and some pile burns, are ongoing. The Sawyers and the Bodgers will be building bridges, tools, cabins, and sheds from the winter's forestry harvests.

There is still a significant contingent of villagers and workers waiting their turn to go up-country. Many children, Elders, Field-tenders, Orchard-watchers, and systems-engineers are tending and gardening the hamlet cottages and keeping the winter-camp support-base solid. Travelers and Traders are coming through, heading north, bringing goodies from the warmer southlands.

The hedges are full of things happening, including children. It's time for May-baskets. Making these little rough baskets with flowers and treats, hiding in the hedges, sneaking up to put them on people's front stoops, running and hiding again. This, the time when secret gifts are being given: "Oh no, that wasn't me, I didn't . . . " It's always flowers and under the flowers in the May-basket is some Maple sugar-cake. The excitement of the season keeps everybody busy, just like the insects, birds, and small animals in the hedges.

We've got Spring Fair. This is when all the new growing season starters come to market: hot-house plants to set out, baby Chickens, baby fowl of all sorts, lots of cheeping, and bouquets of the first garden flowers. The summer clothes (made over-winter) are displayed. It's May 1st and it may still be cool, but we need to get ready to be off to summer-work. People are getting all their tools together; the scythes need sharpening; it's the last-minute packing for the last guilds to be off. First summer field planting is at the end of May, and the last frost is expected at the end of May, depending of where we speak, because it's the Siskiyous.

There's a big barn dance, the hay lofts are empty, and we sweep big indoor floors that were stacked with hay that we burrowed into last fall (that was nice, too). It's the first cutting of fresh hay on just the right slopes. There are special scythes that come out. The cutting of hay is going to happen during the next few months at rising elevations. So this is the warm-up, getting your feet and strokes to dance, the moves smoothed.

We have had barn dances with fancy steps, so we're tuned-up to take long up-country walks or waltz through the grass with the scythe swinging itself, without excess effort, gracefully (swish-swish). Down comes the fresh, tall grasses, laid over delicately, not raked too rough, dried briefly, carefully brought up, green and fluffy, to the hungry Herds and immediately fed to the animals that have been in quarters all winter. Get some energy up; we're about to move to high summer pastures. All the subsequent cuttings will be turned and dried and stacked to preserve nutrition into the winter. Curing the hay.

We will be cutting many sorts of hay. on a variety of slopes and meadows.

Then it's early June and the first guilds who get all the way up to the high country are the Mothers, the herb guild. These gatherers left town ahead of everybody else because they want to get to the herbs, roots, and flowers before the Herd-animals do. The Mothers are the lead-guild; they cover a lot of ground, and the Rangers work closely with the herb gathering guild. They relay information to the gatherers, who are moving upslope; the Rangers are often ahead, on their rounds, scoping out the best meadows.

Then the Herders-guild head out of winter camp and valley savannas with the Herds. They've already mob-grazed the early pasture in the orchards and Oak meadows at the low elevations, and now they're moving up the mid-slopes. That south face of Grizzly Peak would be the first that would be good summer pasture. The Rangers and Sawyers are clearing the trails for the Herders to move up. The Rangers re-stack cairns and reset guide posts, knocked over last winter.

Rangers are marking the way.

The Herders are coming up behind the Mothers, slowly moving up slope, looking carefully at the grasses and how the Shepherds disturb/tend the ecology of the meadows, working with the animals to enhance ecological functions. The Herds stimulate fertility with a walk-through light graze, then move on without compacting or taking too much. The Shepherds also do some reseeding.

We've really become effective, topsoil building with animals.

Early summer is when we cut, cure, and stack a lot of hay. Early June in the villages, the summer resident-tenders weed, set out successive crops, trellis, and prune. Village folks harvest the over-winter fields. This is an opportunity on the valley-bottom that we have dialed in. We process the over-winter polyculture guilds and green manures, such as Barley-Daikon-Fava or Lentil-Oat-Salsify. These crops were sown last fall, germinated by early fall rains, sat under the snow and frost, and flushed early in spring.

On the southeast aspect slopes, at mid-elevation, above irrigation ditches, winter-sown grains and beans are harvested in mid-June, bringing in the first field crops. The early Barley is dried, lightly toasted, and brewed up into the new summer beer. One of the things we absolutely enjoy while we cut a lot of hay is this summer beer, to keep the dance step right. It's light and thin, not brewed from dark roasted grain.

Out on the Hoop, the first root camps, lower slope Lomatium (Biscuit root), and some late upslope Camas.

When everybody's in town over-winter, there's the Rangers' report to drainage basin council— everybody hears it. Guild plans are made, and people start to leave on Hoop. The Mothers wander off, the Rangers stride off, the Shepherds round up the flocks, and who's left? Some people don't leave: the Gardener-guild, the folks who have work at lower elevations—Millers, Weavers, Warehouse keepers. Older and convalescent folks, grandparents and small children who are not up to the long walks, stay near the villages and enjoy the summer.

The local Traders-guild holds the market space in town, and the Warehouse-guild women guard the stores, and guesthouse keepers decorate and present vernacular-ambiance to get ready for the summer Travelers coming through. There may be trades for obsidian, grinding rocks, pipe stone, and for salvage material and tools from that long ago empire we heard of in the winter epic storytelling.

SUMMER SOLSTICE: FLAT OUT

Then everything at the base camps seems to settle down. Here comes the Summer Solstice; summer feels like it's going to last forever. Every day seems just as long as the next: get up in the early morning, start to get lazy early afternoon. It's summer, everything's good. It's the festival of flowers along the valley basin circuit: dried bouquets and wreaths are arranged; Willow-basket withie harvesting proceeds in the hedges; and Hazel splint weaving seems endless, with all of last spring's split and barked bundles.

This is when certain rites-of-passage, weddings, and council meetings take place in the villages, just before the last guild-workers go off to summer high-country camps and work sheds. Some young folks are ushered into guilds after apprenticeship. Arrangements happen at the solstice that are proper to the season and workflow.

Most importantly, the Clan-Elders council meets almost daily; they gather in the shade by the creeks under big trees, after midday, and hold Salon. There are enough younger adults to watch the young children while the Elders circle up. In the cooler hours of the summer days, the special bond grandchildren have with the parents of their parents can shine.

The Iroquois model, along the Mohawk River, has a seasonal cycle, a gardening village culture. The smallest children, the elders, and the women who run things, that's who's around all summer. They are visited by traders and scouts, cycling on the canoe routes.

After Midsummer, in Wagner County, narrow-gauge steam train and canal tours take the Elders and the children around the big valley on contour. They pass lovely garden cottage villages, all swimming in flowers. There may still be enough water to float boats in the canals as irrigation proceeds; trains are running one round trip a day—wherever we get on, we can return to. The summer nights are short and warm; the train runs in a valley scale circle round the clock. Old friends meet, make village-to-village contact, perhaps stay up all night and take naps.

After Midsummer, the last unirrigated valley hay is cut; then we fallow the fields. We're not going to get another valley hay-cut. The Mowers are cutting at higher and higher elevations, making a second cut, following the Herders, as the highest pastures persist and hold the Herds. As the guilds moved upslope, the highest hay cutting is on meadows below the crest. The cutting is sparse but cures well and contains medicinal plants: very special hay. The upslope hay is dried on racks and stacked on platforms under roofs. We will be hauling a lot of it down to the valley, from the hay sheds near the upper meadows to the winter-barns. As the Herders and animals move down, we harness wagons and tie packs to carry hay down. The last loads will come down on sleds to barns just below snow line.

Many folks are traveling on Hoop, staying at cool lodges in the mountains, or trading over mountain passes. There's hardly anyone down in the hot villages. The small kids love it nonetheless, flowers and butterflies everywhere, your grandmother and ancient uncles are telling stories.

There is a lot of playing in creeks and ponds, napping in the deep shade of canyon trees.

This is also when we have the summer-camps for older children at the lakes. There we vacation. There's not as much hay to be cut, the Herds are being moved across the crest meadows, the Mothers are at a pause. This is when folks meet to swim and float at the lakes.

Moonlight canoe flotillas.

It's mid to late July; in the villages we weed, make compost, cut forage, take care of the smaller animal-cultures that don't travel far, do nursery work and plant propagation. Interestingly, all of the field crops and all of the biological systems around the villages are being watched and guarded, because a lot of wildlife is moving down to fatten up on late summer berries.

In *Keeping It Living*, the book by Nancy Turner (2005), she talks there about how the First Nations Peoples didn't just walk across the landscape and pick a little of something here, and a little of something there. They had intensively gardened patches; they magnified the naturally productive stands of herbs, roots, berries, and nuts. Every one of those patches are marked with carved totem poles, not those mask-like totem poles; these had rings or diamond patterns on cedar staves. Every clan had marked patches and those patches were guarded 24/7 mostly by young boys learning to become hunters. They all had small bows and arrows—slingshots—these young men and the women, because the girls and the boys want to do the same things at that age.

Meanwhile back in Wagner County, the interesting people start coming through: the Travelers, the nomads, the wandering tribes, the Traders. And they're weird looking; we don't initially recognize their guild and clan marks. Most of us Wagner County folks have names that are like, "I'm Hazel from the Fish Clan from the Village of South Glens Falls," and so you know my name, my clan, and my home-place. But the Wanderers come though, and we ask, "Where are you from?"

> *"We're not from anywhere,*
> *but from everywhere."*
> *"Where are you going?"*
> *"Oh, to the next place"*
> *"How do you make it?"*
> *"Note we're headed south;*
> *the summer is turning."*

And perhaps a couple of kids run away with the Wanderers; villagers want to make sure, via respectful stories, that potential runaways know what they're getting into, if they're going to pull this off and come back with great stories and wisdom.

There's a lot of local information, many different categories of information, that the Rangers and Hoop guilds are bringing to council—landscape-scale information. Now, in late summer, the camps and villages have long distance, regional-scale news coming through, and all that story-flow is filtering down to summer Elders Salons to get ready for the storytelling in the fall, when the Hoop walkers almost all come back down. The Women's Warehouse-guild is stocking-up long-lasting supplies, repairing storage bunkers, and getting ready for the new crops and materials to come into the platform caches, caves, and storage cellars. There's much bustling making sure things are swept out and cleaned.

19 Wagner Councils

There is ongoing summer learning, outdoor skills; lots of conversational growth happens in the middle of summer for pre-teens, not during the "school year" as it was known before. Some gardeners and crafts-folks have been freed up: this is when the barn raisings hatch; roofs get repaired; pottery is shaped and dried to be put in the kilns after the first fall rains; road work is being done. A whole bunch of infrastructure maintenance is done in late July, just ahead of Llamas.

The Eco-Pilgrims, up on the Siskiyou Crest, are looping through on the Mount Shasta circuit pilgrimage. It is summer high-country pilgrimage time. The villages send supplies, catering to the pilgrims, up to the mountain pass lodges. Some of the capable Elders have spent the summer hosting teahouses, and tell the local stories to the pilgrims as they go through. The beads of our Place—special woods, pottery, glass, and stone—are added to the prayer necklaces and bracelets. The pilgrims come from other Places, and the story and song exchanges are generative and overlapping, gifting confirmation of wisdom.

LLAMAS: FAT AND HOT

Then August 2 arrives, another season of cross quarter days. All of sudden we all notice that summer's waning, because the day length is changing. The rate of change of the rate of change is dizzying. We think of the cold-times again. Time now for a big early-harvest festival; we all have a big appetite; we can eat the bounty that isn't going to store well. Eat all the cucumbers and the first tomatoes, basil, peppers; the salads are so delicious. The first guild-folk are down from the high-country. The Mothers are back by Llamas, and they carry a great breadth of harvests.

The Warehouse Keepers are ready for this; the children are ready for this. Here comes baskets full of seeds, fragrances, flowers, roots, herbs, and stories: "We're the early Hoop returns, and we can tell you how things are." This is when the Harvest Crews form and plan; this is it—the harvest. Some of us sober up by drinking summer beers and indulging in our last summer nap. There are August campouts in shady glens not far from the villages, near the tree and field harvests. We build summer shelters, brush shelters.

It's still comfortable living outdoors, and there are cold pools in the canyons.

The first taking of Deer is mostly done around the crops. At this season, the Hunters are working with the village folks. They've already helped with the infrastructure work in town. Now they're the ones going uphill, not far, when the other Hoop guilds are coming downhill, to all meet up at the mid-slope Oak and nut groves for Deer dance and Acorn dance. Some seed crops, in mid-elevation pocket fields with good solar, also have excellent frost drainage and can be harvested later, after the valley fields freeze. The Hunters-guild folks, while helping the Harvest Crews, watch the wildlife, making observations about what to do and how to do it.

*"Let's go this way this fall;
we hunted over there
last year."*

Plans are being made. Guild councils are conspiring in the Hoop camps. In the middle of August we plant the over-winter crops; September can be too late. Grains, roots, and beans polycultures go in the ground, when it's dry and hot. We mulch the pelleted seeds, all in their clay jackets, ready for the first rains. To get that winter crop established, we get ahead of the very first rain, seed already in the ground. When we have early September rains, we might have a very cold winter; it's a good thing to have snow cover on the germinated seeds in early December.

Which slope aspects and topographies we plant on is critical. The winter field slopes that are gently northeast facing are where the snow holds the best, insulating the tender sprouts, and also where the sun is sufficient in the early summer for ripening the grain. We have knowledge about aspects of Place. In the lower elevation savannas, meadows, and orchards, where the Herds did that quick light grazing in the spring, the Mothers are now collecting seed crops. Grasses and perennial roots were stimulated and have rebounded strong; grass, mustard, *Lomatium*, *Madia*, and sunflower seeds are beat into baskets in that late August window.

Lightning-lit fires are burning in the mountain forests. We have smoke settling down valley in the night and washing back upslope with the valley pattern winds. Our camps and villages are in safe terrains where fuel reduction from intensive tending with Horticulture and Cultural Burning keeps our most lived-in zones, near home and work, safe from most big fires. We gather to burn our woodlots, savannas, and canyonsides in September, after some good rain. The small animals are nimble, the burning is quick and in patches, the smoke does not puddle in the right wind moments, our homes and camps have some smoke filtering, and we have done these landscape-scale controlled burns for generations. The old decrepit industrial forest mess has been converted to open old-growth trees with diverse understory mosaics, full of food, medicine, and craft materials. Special spot burning, for many reasons, happens all year round.

*Fire has spirit,
we dance
with respect.*

In early September, winter prep is proceeding in the high-country. So much has been ongoing in the high country, and now the Herds are still up there, the Mothers have left, the Hunters haven't come up yet. The people who are up-country, the laggards from the summer camps and mountain pass lodges and the Rangers, are putting in wood for the Hunters and late-travelers, making sure all the trailside guild-shacks are tight, and converting the tea houses to emergency winter shelters. They're shuttering the open-view summer tea houses by putting big boards and bark shingling up around them and rigging the ladder access door at the top of the stove pipe tower, so the Rangers, Hunters, and travelers lost in storms can find these shelters, snow drifted around, and climb down in there.

The early fall seasonal changes happen fast in the high-country. The Rangers and late summer Pilgrims are doing watershed, lip of the drainage basin, circuits: drainage basin pilgrimages.

WATERSHED CIRCUIT PILGRIMAGES

For many decades, we've been doing the pilgrimage around the Ashland watershed. We know where the hidden springs are, and the sheltered camps. A three- or four-day pilgrimage that completely walks around the Ashland Creek drainage-basin and starts and ends in Lithia Park. We've done this ten to twelve times and it's so amazing to start in town at 2,000 feet and go up onto the 7,000 foot range, and we circle our heart, along the watershed range, circling the water. and It's our last high-country excursion, just before the snows. The colors are incredible. This helps us get focused on who we are, where we are. The children start with side-drainage day loops, perhaps, or walk with us a half first day and go back down to town.

Then the fall-run Salmon ceremonies take place, the special ones. We follow the Dahlgelma traditions and have continued the Salmon ceremony on Dragonfly's River just below the Table Rocks. They take the first caught Salmon, and before any more are taken, the sacred Salmon bones are carried down into the deep hole, by divers, and put under the rocks. This is when really sacred prayers are said and songs sung; winter is coming, and what would we do without the Salmon, and the berries, and all the tending that we've done on this landscape to bring it back to life and to Keep It Living? It's the core of who we are; we're helping the Salmon people on these river-landscapes, all together.

*It's late fruits
and nuts time;
we're starting to wrap up
our harvests.*

FALL EQUINOX: SLIPPING DOWN THE SLIDE

Then we're at September 21, the fall equinox, and we feel the free-fall. Instead of that feeling at Llamas, of looking over the edge, now at fall equinox we're on that slope—we're headed down, every day is shorter: fast changes. We have the farm festivals, the special athletic events, ballgames on the cleared fields. We're strong for the winter, we're strong from the summer; we're healthy, eating the best. We get to play with our bodies, to see what we can do, to push ourselves, while it is still easy to be outdoors. Long races take place; the runners go out over passes to other drainage-basins. This is when invitations are sent; the long distance runners carry satchels. This is the last chance to get over the passes and come home again, without heavy storm gear at ready.

Early October, the Herds are down on valley pastures. We can hear the Shepherds coming singing; we hear the Herd bells. We freed up a great deal of our people to go away for the summer on Hoop. Those of us who are still here are getting ready to welcome everyone back. I'm a fan of Morris Berman, who is a sociologist/historian thinker. He wrote a book called *Wandering God* (2000). In the last millennia of history, the nomad-wanderers, the Shepherds, have been not welcomed by farmers and the townspeople. It's been, "Please go away, you're crossing fences."

We really need to all work together; there's this big healing we all welcome. So when the Herds are coming down, and the guilds are coming down from their summer Hoop, there's parades. We parade into town and share a feast.

*Welcome back,
we're glad
to see you.*

Remember all those work-identities? The people are coming back together and sorting. We've been in our guild-identity on Hoop, but now we're re-uniting with our clans and our family. We-all comes back-together. All these social-arrangements are re-knit for the winter, we be in close-quarters, and we-all get-along. This is all done ceremonially; it's all beautifully laid out, the way we-all be welcomed back. Everybody is re-integrating—everybody carries gifts, stories, treats from far Places, animal-cultures all fattened up. Then we get busy processing, storing, stashing under cover: acorns, chestnuts, mushrooms. We're busy, busy, busy.

Roots, berries, seeds, barks, flowers, essences, fungi, and craft materials galore.

SAMHAIN: THE DARK CROSS QUARTER, WRAP THE SCARF

At the end of October the last cross-quarter day arrives, Samhain, previously called Halloween. Day length change is breaking: it's a decelerated slump into long nights. At this inevitably dark season, we notice winter is not far away. The days are short from then on, and the freezes are icy, on from the end of October. It's so evident—we get all dressed up, we light all these shielded lights. We have another feast, because we have to eat everything that won't last in storage. We celebrate the past year, and we celebrate our ancestors who taught us how to be here, who brought back Salmon, Beaver, and Fire.

It's the day of the Dead, the feeding of Ancestors, and we've got hard Apple cider to share. We-all and the children pilgrimage down and out into the flat valleys to the ruins; we-all go down to what's left of the cities. We visit the broken-down walls and tell stories of these abandoned Places. We all troop to the ancient cemeteries, honor those pieces of the past, tell stories about collapse.

ANCESTRAL CEMETERIES

Where Elizabeth and I came west from, on the west Vermont border, there's lots of cemeteries. My family has been there for 400 years, and there are plenty of cemeteries and ruins that involve ancestors. We used to go to the ruins of a canal and a steep set of seven small locks from the 18th century. On our ponies, we got perspective from visiting those ancestral story-Places.

Then, it's November. This can offer a late mountain-pass opening, to travel out of the valley. If anyone is going to cross the pass, this is last chance; they better have snowshoes and a big parka because a storm could happen at any time. The Rangers are up above the passes on skis, touring the winter-camps and lodges, making sure winter-prep got done and taking care of last-minute adjustments, making sure the latch strings are out and firewood is in, ready for any travelers.

At the end of sufficient pasturage, is another time when we take from the Herds and Flocks of animals. It's cold enough now that we can hang the meat in special tight sheds and it will keep. We have negotiated this taking with our companion cultures, offered ceremony and blessings. Some guidance has come our way, through traditional wisdom, close observation, prayer, sweat lodge, and dreams. In return for the sustenance we receive, we give protection, cover, stored feed, and appreciation. The Hunters are moving upslope, with small dogs, to work with the wild-land animal-cultures.

There,
the spirit events
are rich and humbling.
They go quietly and quickly
to high camps.
Sit in sweat lodges,
lose ego and expectations;
be open to the gifts.
The early snow and wet ground
are good for tracking;
much is learned of wild-ways.

We're getting ready for the first big engineering feats of the winter. We're rebuilding canals, trails, roadwork, rail beds, and other earthworks. We dredge our ponds, dig new ponds, clear ditches, check culverts, and reinforce bridges. There hasn't been so much rain that things are mucky, but the dust has been knocked down and the topsoil is soft. There's just enough moisture in the subsoil that it moves well and packs well.

We're using the oxen to pull bucket dredges and drag blades, re-shaping our empty reservoirs that get refilled in March. Almost everybody is back from Hoop-work, so we have competent work-crews. We're all strong, been eating well, and moving cross-country all summer. We can do big stuff: earthworks, roads, trails, ditches—rewarding hard work, with our precious salvaged iron equipment.

And we're putting up firewood, charcoal, and kindling in sheds and clever piles. Our charcoal kilns are smoking; the Charcoaliers are up in their ridge-pad camps, processing the wood left behind by Sawyers and Bodgers. Logging season begins as soon as the ground's frozen, up above 4,000 feet the snow isn't too deep, and the weather is comfortably cool for the heavy work-clothes. No birds nesting, no animals giving birth, a good time to head up there, drag logs, stash materials, and selectively thin the forest. We can see the forest layers clearly; the leaves are down, the conifers stand out starkly between the hardwoods.

WINTER LOGGING

In the Adirondacks, that's what my brothers did: the logging in the winter on the snow. It's easier to skid logs on winter snow than in any other season; that's when the logging season is, the upslope logging season. Trees are selected, felled, limbed, and the trunks are slid down hill in well established chutes. Very little soil compaction or disturbance.

The branches are bundled and skid on toboggans, if not cut up small enough to be left as carbon mulch. The woods are well known and the way of work is familiar and complicated. Small-time winter logging. Down to the homestead sawmill.

Here come the late-fall feasts. A great deal of eating fatty food, because oily foods—like certain acorns we're feasting on in November, like White Oak acorns or Live Oak acorns—don't last in storage, they go rancid. Black Oak and Tan Oak, with more like 5% oil, last in storage for several years in some cases. And even Acorns can be fermented, in stream-side pits. We're feasting on the rich foods that aren't going to last.

Reunions breed exchange and recall, news, and reviews: "Did you hear the news? Did you know? Do we remember when? And meet your new cousin." Catching-up happens before the more ceremonious season's celebrations. With enough time, perhaps while doing useful social tasks, oral story-traditions can be kept full and rich. The storylines are repeated, and new branches are sprouted. This is a social sequence to also portray, as a reinforcement of oral traditions.

The Smithies (another term for Blacksmiths that can include other metal workers—refers to "keeps a hot fire") and Bodgers are busy; there's no water power, but there's charcoal and there's dense Manzanita wood; and the forge and craft sheds are covered—folks are ready to work in hot places. Fresh-dug ore and carving stone has been brought down in the fall, from the quarries, caves, and mines. There's a lot of wagon repair; the wagons that have come down from the high-country are beat up. The Smithies and Bodgers are on it. They're also turning and assembling toys, tools, stools, boxes, and fancy frames, for the gifting and crafting seasons.

All the craftspeople are keeping busy. The craft shops, bakeries, wineries, creameries, basket sheds, fabric looms, wool processors; it's all happening, as winter closes in and we are indoors on the rainy and cool mornings. We take the materials we warehoused and make what is needed to be comfortable and entertained. These are the craft support bundles that we've brought down from our ecological restoration work on the whole landscape.

Larders, cool caves, log bunkers, and cellars are cleaned out, smudged, sung to, and stuffed with crocks, barrels, shelves, bins, and pole racks. Some areas in the Siskiyous are better than others; the mining areas, as messed up as they are—stripped, down cut, and blown away—allow us some opportunities for long term storage.

> *Gulch bottom pits,*
> *benches on steep north slopes,*
> *and tunnels,*
> *deep into the mountain.*

We want to thank all those poor, miserable, starving miners, most of whom were desperate men, who'd come a long way looking for gold and had very little to eat. We have ceremonies to thank them for what good they did and to settle their hungry ghosts. They built us some really good storage facilities that could last forever.

So the engineers are in the tunnels, shoring up the old mines; they have been stabilized, not to mine but to put cheeses, wines, beer kegs, mushroom beds, and dry goods when we find a dry, ventilated chamber. We use some remote, deep chambers to have initiation ceremonies and vision quests; cave painting is back, but we have a lot to learn about the arts of natural pigments on dimensional surfaces.

We can work at any hour in the cool, dark places, where the temperature is steady. Although the afternoon outdoor temperatures are relatively mild, the short days hurry our winter preparations. We have social time indoors with big breakfasts, outdoor basket lunch picnics, and early light dinners by fires that drift into Salon and song.

The children are celebrating the bright falling leaves and dressing themselves up in the colors of the season. Now, it is post-Samhain, almost Thanksgiving, but not the colonial Thanksgiving, something else: a celebration of abundance, the long presentations of acknowledgements and gratitudes. At the end of November, we feast with traditional foods of Place, foods we learn to appreciate through our tending work, guided by the ecological knowledge of the First Nations, long before our time but on the multi-dimensional Hoop of Creation.

Then we get to December. The Charcoaliers are busy again gleaning left-behind wood, after the loggers, who have left a lot of material to stack and dry. Last year's seasoned culls are pyrolyzed in kilns, if we can still get up to the ridge benches. The Charcoalier huts and campfires have great views of the stars, see the weather coming in, and host several-day sessions of story and song.

The snowpack is not too deep yet at the 2,000 to 4,000 feet elevations, so some guilds can still be up-slope. The Sawyers are wrapping up the selective thinning, and they're decking the logs on road sides to bring them down to the sawmill in the spring, when there's water power for ripping.

All our forestry work is sequence/stage/sort/stash, and every guild knows their seasonal repertoire.

The millers are busy, Ox-power turns the stones, the demand for grits and flour is growing with the season's promise. The last-forcing mushrooms are found and the Hunters are down from the high-country; the snows grew too deep and the campfires barely warm enough to hold the circle talk. It's not the best time for Human ecological Wild-tending. We would need heavy clothes and snowshoes. The ancient predator/prey winter-dance carries on. The Hunters yield the high-country to the Wolves; the Wolf-culture is strong. We can hear them sing and toast.

Here comes the festivals—we-all are excited. Winter's not so bad; we get to play in the snow and come in to fires to warm up again. We enjoy all these social activities. There's visiting, sharing, and distributing the surplus, and folks promenading in their new winter clothes, while they are still clean and bright. It's less convenient to do laundry in the winter, even though we have steam baths; pretty much we wear one set of outer clothing all winter, and in the spring, we carefully clean and pack the thick woolens away in moth-proof chests. It takes a long time for wool clothes to dry.

Now we remember the winter-routines, making sure the chores are all covered, especially caring for the animal-cultures that support us. Many animals are clumped in close quarters: they're going to have parasite challenges—it's dry hay (high quality herbal) not pasture and sun, they're not frisky, huddled for the long hunker-down. The Herds and the Shepherds are remembering the high-country.

It's not just Humans telling stories about that glorious storm on the Soda Mountain Crest. The very old epic stories tell of Shepherds dragging themselves out of winter camp, heading back upslope with who's left of their companion Herds and after a long sparse winter, walking over the first hill—where they can at last see the high-country—and as soon as they can see the new green below the retreating snow-line, they burst into song. They sing the song-lines, the rest of the way, all summer.

Now, back here in winter-camp, they kind of forget how to sing certain songs. This is sad for nostalgic wanderers.

Townspeople want to make everything nice for the Hoop nomads. We're mending the fences. There is good hay in the barn lofts. We have a well-stocked apothecary. Our great-lodges allow Hoop-crews and guilds to meet comfortably, and hold report and plan councils.

The Sawyers are putting up next year's firewood, so it will be well-seasoned and dry for the winter after this. We're putting up firewood two years ahead. We're cutting up the hardwood logs skidded to the woodshed yards; we're sorting the tree types and values—some of it's going to furniture-turners (Bodgers), some of it's going to kilns (Charcoaliers). The rest is bucked up for firewood. Many categories of sorted and bundled materials are cached. The craft-guilds are focused on preserving the highest quality woods. Winter logging leaves the least damage of soils and drainages to allow an even more elegant harvest in following years.

WINTER SOLSTICE, FLAT-OUT DARK, WITH A MIRACLE OR FEW

Seasonal learning opportunities—the performance schools—are organizing. In lofts that are not full of hay, or empty council-lodges, there's sequestered practice going on, working up the dances and plays for the Solstice festivals.

We want surprises;
we're in the season of surprises;
we seldom spy
on the stage rehearsals.

And the biggest surprise of all: the Sun turning up—we hope it is coming back again. We've got the sky watchers at the great stones or cave shafts, we await that touch of light, we've ochered the mark. The Sun-rise beam shines through and stays there at the mark for way too long, almost twelve days, and then starts to move. We helped it, our expectations did not fail, the Sun's up coming back, we-all get excited about that.

There's enough lamp oil, the Warehouse-guild announces: "There's this much surplus; we can burn this many gallons of oil." So the great halls are lit, the homes are lit, we set way more lights than later in the winter—that's the festival of lights. Also we're trying to encourage that Sun to turn around and come back, with our meager sympathetic gesture.

The very last long-distance intrepid travelers return: the people who had gone the furthest, those messengers, the people who were off on a long trip, they've returned. At last everyone is back or accounted for. Then starts the season's gifting, all those surprises that we-all were crafting, that the secretive artists didn't talk about, come out into the lights.

We indulge in social-dances, song, gifts, and music. Elaborate staged plays, epic storytelling, the long tales. We're all here in the great-lodges, there's plenty of lights, there's wide dance floors, we've got plenty of firewood and we have lots of distilled-spirits that have been cooked and condensed out of the extra beer that didn't get drunk during the hay cutting season. Wine has been turned into brandies. And Apple-jack from the canyon freeze-out decks.

We've stashed a lot of nuts and seeds; we've roasted and parched them to reduce the seed coat acids and to delay rancidity. In this cold and dark, we love our fats and carbohydrates. With Hazelnuts, we need to toast and store them in oil, or confections, or brandies. One of the favorite ways we like to have Hazelnuts stabilized in oil is in Bay-nut "chocolate;" that's why the Swiss and Germans, mountain people, loved these Hazelnuts in Chocolate. We have accumulated all this understanding of how to have the highest nutrition from this beloved Place, where we live and steward.

We're working on restoring ecosystem functions, the return of Beaver, Salmon, and Cultural Fire.

We tell very long stories (much longer than this one); when the epic stories get told, the stories take several days to wrap up. The children are eager to hear them—they've heard them before. The early winter sports beckon to the adventurous and cabin-bound; we can use the cog railway to get to Wagner Gap from Talent. We get up to Greensprings Pass via a wagon up Tyler Creek Road, then up 1,000 feet higher on the other cog railway we build there. This Solstice season is also when we might get the big floods. We are monitoring the bridges and culverts; a handy task for those expeditions to the snow.

Back before the long drought that accompanied the fall of the empire, there were floods in '73, '64, '54, '36 and '24 of the previous century. By counting tree rings on very old trees, we learn of big cycles; we seem to have a 60-year cycle, a period of floods followed by a period of drought. Now, in Wagner County, we keep counting sticks to try to map or recognize what has happened since '76, when a long drought exacerbated the industrial-charged climate-change challenge. That global-weirding was drastic, and thank goodness the whole craziness collapsed. The long stories tell of a string of many civilizations failing by idiocy, so we are interested in what happens next. May we have wisdom and flexibility.

THE CRECHE

With my village family, back east, we'd go out to the snowy winter woods to get plants of the forest, such as Running Cedar (*Lycopodium sps.*). White Cedar boughs and Balsam Fir trees. Elizabeth's family got other plants, in the Polish tradition, such as Rose hips to string up for the Fir trees, Mistletoe, and Wintergreen. My family would go to special woods we knew about and look for special plants.

Even though we were Quakers, a sort of Christian that rejects outward signs, we brought the forest into our homes to remind us of where we lived. As we arranged the crèche, following the biblical sequence, we'd put new characters in every weekend. We'd also be bringing in different parts of the forest into the house every weekend. It was delightful for the children to go out, get a special piece of the forest that's decorative, bring it home, and sing the songs that celebrated life.

Even though we might be trapped indoors by the weather and the bitter-dark cold, and are enticed by the goings-on, we're always remembering where we are; we're on this meaningful landscape, and we have brought home icons of that landscape—green, shiny, evergreen icons, such as Tannenbaum, the Yule tree, the multi-floral dried wreaths, the fresh aromatic conifer wreaths and ropes, the shiny red berry strings, the Mistletoe clumps hung over doors—bringing these treasures out of the woods, bringing them home. All this connects us with this storied Place, reminds us of who we are, that we belong here, that we're people of this Place. This Place loves us, and we love this Place.

Happy
High
Holidays
Folks!

WE ARE THE PEOPLES OF FORESTS

These stories go on and on. Forests are a crucial part of any possible future for life on this planet. If we pull together and free up the space for working with Nature, a green, pleasant world stretches out long before and long after these times on the fulcrum.

In this moment in the epic of trees on Earth, we Humans are front and center. We have the means to rise to the challenge of our perceived situation and adapt to change and opportunity.

Our cultural imagination leverages deep knowledge in story cycles, towards healing and persistence. We are moving forward with the inheritance of evolution and natural diversity.

Taking our part in the immense company of continuous creation, Humans will come to a more comfortable perspective. The problem is the solution. Step up and dance.

BIBLIOGRAPHY

This is a list of some of the reading that has contributed to this book.

Abbey, Edward. *Desert Solitaire: A Season in the Wilderness*. New York: McGraw-Hill, 1968.

Abrams, David. *The Spell of the Sensuous: Perception and Language in a More-than-Human World*. New York: Pantheon Books, 1996.

Adkins, Jan. *Moving Heavy Things*. Brooklin, ME: Woodenboat Publications, 2004.

Akwesasne Notes, ed. *Basic Call to Consciousness*. Summertown, TN: Native Voices, 1978, 1981, 2005.

Alexander, Earl B. *Soils in Natural Landscapes*. Boca Raton, FL: CRC Press, 2014.

——, Robert G. Coleman, Todd Keeler-Wolf, and Susan P. Harrison. *Serpentine Geoecology of Western North America: Geology, Soils, and Vegetation*. New York: Oxford University Press, 2007.

Anderson, Kat, and Thomas C. Blackburn. *Before the Wilderness: Environmental Management by Native Californians*. Menlo Park, CA: Ballena Press, 1993.

Anderson, Kat. *Tending the Wild: Native American Knowledge and the Management of California's Natural Resources*. Berkeley: University of California Press, 2005.

Arnold, Mary Ellicot, and Mabel Reed. *In the Land of the Grasshopper Song*. Lincoln: University of Nebraska Press, 1980 (1957).

Ashley, Clifford. *The Ashley Book of Knots*. New York: Doubleday, 1944.

Ashworth, William. *The Carson Factor*. New York: Hawthorn Books, 1979.

——. *The Encyclopedia of Environmental Studies*. New York: Facts on File, 1991.

——. *The Left Hand of Eden: Meditations on Nature and Human Nature*. Corvallis: Oregon State University Press, 1999.

Austin, Mary. *The Land of Little Rain*. New York: Penguin, 1988 (1903).

Backlund, Gary and Katherine Backlund. *Bigleaf Sugaring: Tapping the Western Maple*. Ladysmith, BC: Backwoods Forest Management, 2004.

Bager, Bertel. *Nature as Designer*. Translated by Albert Read. New York: Van Nostrand Reinhold Company, 1976 (1955).

Baker, Richard St. Barb. *My Life My Trees*. Moray, Scotland: Findhorn Press, 1981 (1970).

Bakker, Elna. *An Island Called California: An Ecological Introduction to Its Natural Communities*. Berkeley: University of California Press, 1971.

Barber, Elizabeth Wayland. *Women's Work: The First 20,000 Years: Women, Cloth, and Society in Early Times.* New York: W.W. Norton and Company, 1994.

Barlow, Connie. *The Ghosts of Evolution: Nonsensical Fruit, Missing Partners, and Other Ecological Anachronisms.* New York: Basic Books, 2000.

Bauer, Peter Michael. "Playing with Fire: Social Forestry with Hazel." https://www.petermichaelbauer.com/playing-with-fire/. October 10, 2016.

Bean, Lowell John. *Temalpakh: Cahuilla Indian knowledge and Usage of Plants.* Banning, CA: Malki Museum Press, 1979 (1972).

Beard, Daniel Carter. *Shelters, Shacks, and Shanties: The Classic Guide to Building Wilderness Shelters.* Ottowa: Algrove Publishing, 2000 (1914).

Beckham, Stephen Dow. *Requiem for a People: The Rogue Indians and the Frontiersmen.* Corvallis: Oregon State University Press, 1996 (1971).

Beinhart, William and Peter Coates. *Environment and History: The Taming of Nature in the USA and South Africa.* New York: Routledge, 1997.

Bender, Tom. *Environmental Design Primer.* Minneapolis: Self-published, 1973.

———. *The Heart of Place.* Minneapolis: Self-published, 1993.

Bergström, A., Stringer, C., Hajdinjak, M. et al. "Origins of modern human ancestry." *Nature* 590, 229–237 (2021). https://doi.org/10.1038/s41586-021-03244-5.

Berman, Morris. *Coming to Our Senses: Body and Spirit in the Hidden History of the West.* London: Unwin Hyman, 1990 (1989).

———. *The Reenchantment of the World.* New York: Bantam Books, 1984 (1981).

———. *Wandering God: A Study in Nomadic Spirituality.* Albany: State University of New York Press, 2000.

Berry, Wendell. "Manifesto: The Mad Farmer Liberation Front." *Fellowship* 52, no. 4–5 (1986): 16.

Blandford, Percy W. *The Woodworker's Bible: A Complete Guide to Woodworking.* Cincinnati: Popular Woodworking Books, 2007 (1976).

Bleything, Dennis, and Ron Dawson. *Poisonous Plants in the Wilderness.* Manning, OR: Life Support Technology, Inc., 1971.

Bookchin, Murray. *The Rise of Urbanization and the Decline of Citizenship.* San Francisco: Sierra Club Books, 1987.

———. *Toward an Ecological Society.* Montreal: Black Rose Books. Montreal, 1996 (1980).

Boyd, Robert T., ed. *Indians, Fire, and the Land in the Pacific Northwest.* Corvallis: Oregon State University Press, 1999.

Brady, Lynn R., James E. Robbers, Varro E. Tyler. *Pharmacognosy.* 8th ed. Philadelphia: Lea & Ferbiger, 1981 (1936).

Brinker, Francis J., ed. *An Introduction to the Toxicity of Common Botanical Medicinal Substances.* Portland: National College of Naturopathic Medicine, 1983.

brown, adrienne maree. *Emergent Strategy: Shaping Change, Changing Worlds.* Chico, CA: AK Press, 2017.

Brown, Joseph Epes, ed. *The Sacred Pipe: Black Elk's Account of the Seven Rites of the Oglala Sioux.* 176pp. New York: Penguin, 1971 (1953).

Brown, Lauren. *Weeds in Winter.* New York: W. W. Norton and Company, 1976.

Brown, Mildred L., and Chloe Ann Rounsley. *True Selves: Understanding Transsexualism–For Families, Friends, Coworkers, and Helping Professionals.* San Francisco: Jossey-Bass, 1996.

Budworth, Geoffrey. *The Complete Book of Knots.* London: Hamlyn, 1997.

Burt, William Henry. *A Field Guide to the Mammals: Field Marks of All North American Species Found North of Mexico.* Boston: Houghton Mifflin Company, 1976 (1952).

Cameron, Anne. *Daughters of Copper Woman*, Madeira Park, BC: Harbor Publications, 1981.

Camp, Orville. *The Forest Farmer's Handbook.* Ashland, OR: Sky River Press, 1984.

Catton, William R., Jr.. *Bottleneck: Humanity's Impending Impasse.* Bloomington, IN: Xlibris, 2009.

———. *Overshoot: The Ecological Basis of Revolutionary Change.* Urbana: University of Illinois Press, 1982.

Charnley, Susan and Melissa R. Poe. "Community Forestry in Theory and Practice: Where Are We Now?" *Annual Review of Anthropology* Vol. 36 (October 21, 2007): 301–336, https://doi.org/10.1146/annurev.anthro.35.081705.123143

Chatwin, Bruce. *The Songlines.* New York: Open Road Media, 2016. E-book.

Conant, Roger. *A Field Guide to Reptiles and Amphibians of Eastern and Central North America.* Boston: Houghton Mifflin Company, 1975 (1958).

Corbett, Jim. *Goatwalking: A Guide to Wildland Living, a Quest for the Peaceable Kingdom.* New York: Viking Penguin, 1991.

Dahlberg, Frances, ed. *Woman the Gatherer.* New Haven: Yale University Press. 1981.

Deloria, Vine, Jr. *Red Earth, White Lies: Native Americans and the Myth of Scientific Fact.* Golden, CO: Fulcrum Publishing, 1997.

Densmore, Francis. *How Indians Use Wild Plants for Food, Medicine and Crafts.* New York: Dover Publications, 1974.

Deur, Douglas, and Nancy J. Turner, eds. *Keeping It Living: Traditions of Plant Use and Cultivation on the Northwest Coast of North America.* Seattle: University of Washington Press; Vancouver: University of British Columbia Press, 2005.

Devall, Bill and George Sessions. *Deep Ecology: Living as if Nature Mattered.* Salt Lake City, UT: Gibbs Smith., 1985.

Diamond, Jared. *Collapse: How Societies Choose to Fail or Succeed.* New York: Viking Penguin, 2005.

———. *The World Until Yesterday: What Can We Learn from Traditional Societies?* New York: Viking Penguin, 2012..

Diaz, Nancy and McCain, Cindy. *Field Guide to the Forested Plant Associations of the Northern Oregon Coast Range.* Portland, OR: USDA, Forest Service, Pacific Northwest Region, 2002.

———. *Field Guide to the Forested Plant Associations of the Westside Central Cascades of Northwest Oregon.* Portland, OR: USDA, Forest Service, Pacific Northwest Region, 2002.

Dodge, Jim. *Fup.* New York: Simon and Schuster, 1983.

Doty, Tom. *Doty Meets Coyote.* Ashland, OR: Blackstone Publishing, 2016.

Douglas, J. Sholto and Robert Hart. *Forest Farming: Towards a Solution to Problems of World Hunger and Conservation.* Emmaus, PA: Rodale Press, 1978 (1976).

Dove, Michael R. "The Theory of Social Forestry Intervention: The State of the Art in Asia." *Agroforestry Systems* 30, no. 3 (January 1995): 315–340.

Drengson, Alan R. *Doc Forest and the Blue Mountain Ecostery.* Victoria, BC: Ecostery House, 1993.

Farb, Peter. *Man's Rise to Civilization as Shown by the Indians of North America from Primeval Times to the Coming of the Industrial State.* New York: Dutton Books. 1968.

Fasenfest, Harriet. *A Householder's Guide to the Universe: A Calendar of Basics for the Home and Beyond.* Portland, OR: Tin House Books., 2010.

Fleming, David. *Lean Logic: A Dictionary for the Future and How to Survive It.* White River Junction, VT: Chelsea Green, 2016.

———. *Surviving the Future: Culture, Carnival, and Capital in the Aftermath of the Market Economy.* White River Junction, VT: Chelsea Green. 2016.

Fowler, Connie and J. B. Roberts. *Buncom: Crossroads Station: An Oregon Ghost Town's Gift from the Past.* Lincoln, NE: iUniverse, Inc, 2004 (1995).

Fukuoka, Masanobu. *The One-Straw Revolution: An Introduction to Natural Faming.* Emmaus, PA: Rodale Press, 1985 (1975).

Garner, R. J. *The Grafter's Handbook.* New York: Oxford University Press, 1979 (1947).

Gibbons, Ann. "Experts question study claiming to pinpoint birthplace of all humans." *Science*, October 28, 2019. https://doi.org/10.1126/science.aba0155.

Gilman, Charlotte Perkins. *Herland.* 1915.

Gleick, James. *Chaos: Making a New Science.* New York: Viking Penguin, 1989 (1987).

Graves, Robert. *The White Goddess: A Historical Grammar of Poetic Myth.* New York: Farrar, Straus and Giroux, 1982 (1948).

Gunther, Erna. *Ethnobotany of Western Washington: The Knowledge and Use of Indigenous Plants by Native Americans.* Seattle: University of Washington Press, 1977 (1945).

Haggard, Peter and Judy Haggard. *Insects of the Pacific Northwest.* Portland, OR: Timber Press. 2006.

Halpern, Daniel, ed. *On Nature: Nature, Landscape, and Natural History.* Berkeley, CA: North Point Press, 1987.

Hard, Roger. *Build Your Own Low-Cost Log Home.* Charlotte, VT: Garden Way Publishing, 1977..

Harlow, William M. and Ellwood S. Harrar. *Textbook of Dendrology.* New York: McGraw-Hill, 1958 (1937)..

Harrington, H. D. *How to Identify Plants.* Chicago: Swallow Press. 1957.

Harrison, Robert Pogue. *The Dominion of the Dead.* Chicago: University of Chicago Press, 2003.

Hart, Carol and Dan Hart. *Natural Basketry.* New York: Watson-Guptill Publications, 1976.

Hasluck, Paul N., ed. *Rustic Carpentry: An Illustrated Manual.* Ottawa: Algrove Publishing, 1998 (1907).

Hemenway, Toby. *Gaia's Garden: A Guide to Home-Scale Permaculture.* White River Junction, VT: Chelsea Green Publishing Company, 2009 (2000).

Hickman, James C., ed. *The Jepson Manual: Higher Plants of California.* Berkeley: University of California Press, 1993.

Hill, Christopher. *The World Turned Upside Down: Radical Ideas During the English Revolution.* New York: Penguin Books, 1991 (1972).

Hitchcock, Charles Leo. *Flora of the Pacific Northwest: An Illustrated Manual.* Seattle: University of Washington Press, 1987 (1973).

Hobbs, Christopher and Kathi Keville. *Women's Herbs, Women's Health.* Loveland, CO: Interweave Press, 1998.

Holmgren, David. *Future Scenarios: How Communities Can Adopt to Peak Oil and Climate Change.* White River Junction, VT: Chelsea Green, 2009.

———. "Retrofitting the Suburbs for the Energy Descent Future." *Simplicity Institute Report 12i,* 2012. https://simplicityinstitute.org/wp-content/uploads/2011/04/RetrofittingTheSuburbsSimplicityInstitute1.pdf

Howe, Carrol B. *Ancient Modocs of California and Oregon.* Portland, OR: Binford and Mort Publishers., 1994 (1979).

Hunn, Eugene S. with James Selam and Family. *Nch'i-Wána "The Big River": The Mid-Columbia Indians and Their Land.* Seattle: University of Washington Press, 1990.

Hutchens, Alma R. *Indian Herbalogy of North America.* Boston: Shambala Press, 1991 (1969).

Hyde, William F. and Gunnar Köhlin. "Social Forestry Reconsidered." *Silva Fennica* 34, no. 3 (2000), 285–314.

Illich, Ivan and Barry Sanders. *ABC: The Alphabetization of the Popular Mind.* New York: Vintage Books, 1988.

International Analog Forestry Network. "Analog Forestry." Accessed October 17, 2022. https://www.analogforestry.org/about-us/analog-forestry/.

Jackson, Wes. *New Roots for Agriculture.* San Francisco: Friends of the Earth, 1980.

James, David G. and David Nunnallee. *Life Histories of Cascadia Butterflies.* Corvallis: Oregon State University Press, 2011.

James, Wilma Roberts. *Know Your Poisonous Plants: Poisonous Plants Found in Field and Garden.* n.p.: Naturegraph Publishers, 1973.

Jason, Nancy, Dan Jason, Dave Manning, Robert Inwood, and Tom Perry. *Some Useful Wild Plants.* Vancouver, BC: Talonbooks, 1972.

Jeavons, John. *How to Grow More Vegetables than You Ever Thought Possible on Less Land than You Can Imagine.* Berkeley: Ten Speed Press, 1979 (1974).

Jensen, Edward C., Warren R. Randall, Robert F. Keniston, and Dale N. Bever. *Manual of Oregon Trees and Shrubs.* Corvallis: Oregon State University Book Store, 1994 (1958).

Jensen, Edward C. and, Charles R. Ross. *Trees to Know in Oregon and Washington.* Corvallis: Oregon State University Extension Service. 1999 (1950).

Johnson, Nathanael. "How One Town Put Politics Aside to Save Itself from Fire." *Grist*, September 1, 2021. https://grist.org/extreme-weather/how-one-town-put-politics-aside-to-save-itself-from-fire-ashland-oregon/.

Kaplan, Robert D. *The Revenge of Geography: What the Map Tells Us About Coming Conflicts and the Battle Against Fate.* New York: Random House, 2012.

Kauffman, Stewart. *At Home in the Universe: The Search for the Laws of Self-Organization and Complexity.* New York: Oxford University Press, 1995.

Keator, Glenn, PhD, and Ruth M. Heady. *Pacific Coast Fern Finder.* Rochester, NY: Nature Study Guild Publishers, 1981.

Kennedy, Joseph F., Michael G. Smith, and Catherine Wanek. *The Art of Natural Building: Design, Construction, Resources.* Gabriola Island, BC: New Society Publishers, 2001.

Kesey, Ken. *Sometimes a Great Notion.* New York: Penguin Books, 2006. First published 1964 by The Viking Press.

Kimmerer, Robin Wall. *Braiding Sweetgrass: Indigenous Wisdom, Scientific Knowledge, and the Teachings of Plants.* Minneapolis: Milkweed Editions, 2013.

Kingsbury, John M. *Deadly Harvest: A Guide to Common Poisonous Plants.* New York: Holt, Rinehart and Winston, 1972.

Kirk, Donald R. *Wild Edible Plants of the Western United States.* Happy Camp, CA: Naturegraph Publishers, 1975 (1970).

Knobel, Edward. *Field Guide to the Grasses, Sedges and Rushes of the United States.* Revised by Mildred E. Faust. New York: Dover Publications, 1977. First published 1899 by Bradlee Whidden, Boston, as *The Grasses, Sedges and Rushes of the Northern United States.*

Knuth, Priscilla and Thomas Vaughan, eds. *Oregon Historical Quarterly*, Volume LXXVI, No. 1 (March 1975). Portland: Oregon Historical Society.

Kourik, Robert. *Designing and Maintaining Your Edible Landscape Naturally.* Santa Rosa, CA: Metamorphic Press, 1986.

Kroeber, Theodora. *Ishi in Two Worlds: A Biography of the Last Wild Indian in North America.* Berkeley: University of California Press, 2002 (1961).

Kruckeberg, Arthur R. *Gardening with Native Plants of the Pacific Northwest.* Seattle: University of Washington Press, 1996 (1982).

LaChapelle, Dolores. *Sacred Land, Sacred Sex: Rapture of the Deep: Concerning Deep Ecology and Celebrating Life.* Durango: Kivaki Press, 1988.

LaLande Jeff. *From Abbott Butte to Zimmerman Burn: A Place-Name History and Gazetteer of the Rogue River National Forest.* Medford, OR: Rogue River National Forest, 2000.

Lambert, Frederick.. *Tools and Devices for Coppice Crafts.* London: Evans Brothers, 1957.

Lampe, Kenneth F. and Mary Ann McCann. *AMA Handbook of Poisonous and Injurious Plants.* Chicago: American Medical Association, 1985.

Laubin, Reginald and Gladys Laubin. *The Indian Tipi: Its History, Construction, and Use.* Norman: University of Oklahoma Press, 1970 (1957).

Lauck, Joanne Elizabeth. *The Voice of the Infinite in the Small: Revisioning the Insect-Human Connection.* Mill Spring, NC: Swan Raven Books, 1998.

Law, Ben. *The Woodland Way: A Permaculture Approach to Sustainable Woodland Management.* East Meon, UK: Permanent Publications, 2008 (2001).

Le Guin, Ursula K. *Always Coming Home.* New York: Harper and Row, 1985.

Lenentine, Miku. "The Brave New World of Forestry: Designing a Participatory Strategy for Successful Community-Based Collaboration in Packwood, Washington." Master's thesis, Simon Fraser University, 2009.

Leopold, Aldo. *A Sand County Almanac and Sketches Here and There.* New York: Oxford University Press, 1968 (1949).

Levi, Herbert W. and Lorna R. Levi. *Spiders and Their Kin.* New York: Golden Press, 1990 (1968).

Link, Russell. *Living with wildlife in the Pacific Northwest.* Seattle: University of Washington Press, 2004.

Little, William, H. W. Fowler, and J. Coulson. *The Oxford Universal Dictionary of Historical Principles.* London: Oxford University Press, 1955 (1933).

Logan, William Bryant. *Oak: The Frame of Civilization.* New York: W.W. Norton and Company, 2005.

Lopez, Barry. *Horizon.* New York: Penguin Random House, 2019.

——— and Debra Gwartney, eds. *Home Ground: Language for an American Landscape.* San Antonio, TX: Trinity University Press, 2006.

Ludlum, David M. *National Audubon Society Field Guide to North American Weather.* New York: Alfred A. Knopf, 1991.

Macfarlane, Robert. *Landmarks.* United Kingdom: Penguin Books Limited, 2015.

Macy, Joanna. "The Council of All Beings". The Rainforest Information Centre, n.d. https://www.rainforestinfo.org.au/deep-eco/Joanna%20Macy.htm.

Maitland, Sara. *From the Forest: A Search for the Hidden Roots of Our Fairy Tales.* Berkeley, CA: Counterpoint Press, 2012.

Mann, Charles C. *1491: New Revelations of the Americas Before Columbus.* New York: Vintage Books, 2006.

Margolin, Malcolm. *The Earth Manual: How to Work with Nature to Preserve, Restore, and Enjoy Wild Land–Without Taming It.* Boston: Houghton Mifflin Company, 1975.

———. *The Ohlone Way: Indian Life in the San Francisco–Monterey Bay Area.* Berkeley, CA: Heyday Books, 1978.

Marianchild, Kate. *Secrets of the Oak Woodlands: Plants and Animals Among California's Oaks.* Berkeley, CA: Heyday Books, 2014.

Martien, Jerry. *Shell Game: A True Account of Beads and Money in North America.* San Francisco: Mercury House, 1996.

Martin, Calvin Luther. *In the Spirit of the Earth: Rethinking History and Time.* Baltimore, MD: Johns Hopkins University Press, 1992.

———. *Keepers of the Game: Indian-Animal Relationships and the Fur Trade.* Berkeley: University of California Press, 1982 (1978).

Martinez, Dennis. "Coyote, Science, Fire & Indians". Self-published. 1993.

Maser, Chris. *The Redesigned Forest.* San Pedro, CA: R & E Miles Publishers, 1988.

Maslow, A. H. (1943). "A Theory of Human Motivation." *Psychological Review*, 50 (4), 430–437. https://doi.org/10.1037/h0054346.

Mason, Howard C. *Backward Glances.* 3 vols. Hudson Falls: Warren County Historical Society, 2014.

Matthews, John. *The Song of Taliesin: Stories and Poems from the Books of Broceliande.* London: Aquarian Press, 1991.

———. *Taliesin: Shamanism and the Bardic Mysteries in Britain and Ireland.* London: Aquarian Press, 1991.

McHarg, Ian L. *Design with Nature.* Garden City, NY: Natural History Press, 1971 (1969).

McKnight, Kent H. and Vera B. McKnight. *A Field Guide to Mushrooms of North America.* Boston: Houghton Mifflin Company, 1987.

McPhee, John A. *Assembling California.* New York: Farrar, Straus and Giroux. 1993.

———. 2011. *The Control of Nature.* Farrar, Straus and Giroux, New York.

———. *Encounters with the Archdruid: Narratives About a Conservationist and Three of His Natural Enemies.* New York: Farrar, Straus and Giroux. 1971.

Meadows, Donella H. *Thinking in Systems: A Primer.* White River Junction, VT: Chelsea Green Publishing, 2008.

Metzner, Ralph. *Green Psychology: Transforming Our Relationship to the Earth.* Rochester, VT: Park Street Press, 1999.

Miller, Howard and Samuel Lamb. *Oaks of North America.* Happy Camp, CA: Naturegraph Publishers, Inc. 1985.

Milne, Lorus and Margery Milne. *National Audubon Society Field Guide to North American Insects and Spiders.* New York: Alfred A. Knopf, 1980.

Moerman, Daniel E. *Native American Medicinal Plants: An Ethnobotanical Dictionary.* Portland, OR: Timber Press, 2009.

Mollison, Bill. *Permaculture: A Designers' Manual.* Tyalgum, Australia: Tagari Publications, 1988.

Monroe, Douglas. *The 21 Lessons of Merlyn: A Study in Druid Magic and Lore.* St. Paul, MN: Llewellyn Publications, 2000.

Morsbach, Hans. *Common Sense Forestry.* White River Junction, VT: Chelsea Green Publishing, 2002.

Moulton, Phillips P., ed. *The Journal and Major Essays of John Woolman.* Richmond, IN: Friends United Press, 1971.

Mullens, Linda and Rachel Showalter. *Rare Plants of Southwest Oregon*. Grants Pass, OR: Bureau of Land Management, Grants Pass Interagency Office, 2007.

Müller-Ebeling, Claudia, Christian Rätsch, and Wolf-Dieter Storl. *Witchcraft Medicine: Healing Arts, Shamanic Practices, and Forbidden Plants*. Translated by Annabel Lee. Rochester, VT: Inner Traditions, 2003 (1998).

Nearing, Helen and Scott Nearing. *Continuing the Good Life: Half a Century of Homesteading*. New York: Schocken Books, 1979.

———. *Living the Good Life: How to Live Sanely and Simply in a Troubled World*. New York: Schocken Books, 1970 (1954).

———. *The Maple Sugar Book: Together with Remarks on Pioneering as a Way of Living in the Twentieth Century*. New York: Schocken Books, 1970 (1950).

Neihardt, John G. *Black Elk Speaks. Being the Life Story of a Holy Man of the Ogalala Sioux, as Told through John G. Neihardt (Flaming Rainbow)*. New York: W. Morrow & Company, 1932.

Nickell, J. M. *J. M. Nickell's Botanical Ready Reference*. Beaumont, CA: Trinity Center Press; Lakemont, GA: C. S. A. Press, 1976 (1892).

North Pacific Yearly Meeting of the Religious Society of Friends. *Faith and Practice*, 3rd ed. Corvallis, OR: Friends Bulletin Corporation, 2018 (1986).

Olcott, William T. *Field Book of the Skies*. Revised and edited by R. Newton and Margaret W. Mayall. New York: G.P. Putnam's Sons. 1954 (1929).

Orlov, Dmitry. *The Five Stages of Collapse: Survivors' Toolkit*. Gabriola Island, Canada: New Society Publishers. 2013.

Palmer, Dan. "Holistic Management and Permaculture." *Permaculture Design Magazine*, no. 119 (February/Spring 2021): 11–17.

Paschall, Max. "The Lost Forest Gardens of Europe: People of the Hazel." *Shelterwood Forest Farm* (blog), July 22, 2020. https://www.shelterwoodforestfarm.com/blog/the-lost-forest-gardens-of-europe#People-of-the-Hazel.

Paterson, Jacqueline Memory. *Tree Wisdom: The Definitive Guidebook to the Myth, Folklore and Healing Power of Trees*. London: Thorsons Publishers, 1996.

Paul, Frances. *Spruce Root Basketry of the Alaska Tlingit*. Lawrence, KS: Department of the Interior, Bureau of Indian Affairs, 1944.

Peacock, Doug. *Grizzly Years: In Search of the American Wilderness*. New York: Henry Holt and Company, 2011.

Pearce, Joseph Chilton. *Exploring the Crack in the Cosmic Egg: Split Minds and Meta-Realities*. New York: Julian Press, 1974.

Peat, F. David. *Blackfoot Physics: A Journey into the Native American Universe*. Boston: Weiser Books, 2006 (1994).

Peterson, Roger Tory. *A Field Guide to Western Birds*. Boston: Houghton Mifflin Company, 1961 (1941).

Petrides, George. *A Field Guide to Trees and Shrubs*. Boston: Houghton Mifflin Company, 1958.

Pilarski, Michael, ed. *Restoration Forestry: An International Guide to Sustainable Forestry Practices.* Durango, CO: Kivaki Press, 1994.

Plevin, Julia. *The Healing Magic of Forest Bathing: Finding Calm, Creativity, and Connection in the Natural World.* Berkeley: Ten Speed Press, 2019.

Prechtel, Martín. *The Smell of Rain on Dust.* Berkeley: North Atlantic Books, 2015.

Pyle, Robert Michael. *The Butterflies of Cascadia: A Field Guide to All the Species of Washington, Oregon, and Surrounding Territories.* Seattle: Seattle Audubon Society, 2002.

Randall, Warren R., Robert F. Keniston, Dale N. Bever, and Edward Jensen. *Manual of Oregon Trees and Shrubs.* Corvallis: Oregon State University Bookstores, 1994.

Ray, Paul H., PhD, and Sherry Ruth Anderson, PhD. *The Cultural Creatives: How 50 Million People Are Changing the World.* New York: Harmony Books, 2000.

Rhodes, Frank H. T., Herbert S. Zim, and Paul R. Shaffer. *Fossils: A Guide to Prehistoric Life.* New York: Golden Press, 1990 (1962).

Ridley, Matt. *The Origins of Virtue: Human Instincts and the Evolution of Cooperation.* New York: Viking Penguin, 1996.

Riotte, Louise. *Nuts for the Food Gardener: Growing Quick, Nutritious Crops Anywhere.* Charlotte, VT: Garden Way Publishing, 1975.

Robbins, Chandler S., Bertel Bruun, and Herbert S. Zim. *Birds of North America: A Guide to Field Identification.* Revised by Jonathan P. Latimer, Karen Stray Nolting, and James Coe. New York: St. Martin's Press. New York, NY. 2001 (1966).

Robinson, Kim Stanley. *Aurora.* New York: Orbit, 2015.

———. *The Ministry of the Future.* New York: Orbit, 2020.

Roscoe, Will. *Changing Ones: Third and Fourth Genders in Native North America.* New York: St. Martin's Press, 1998.

———. *The Zuni Man-Woman.* Albuquerque: University of New Mexico Press, 1991.

Ross, Anne and Don Robins. *The Life and Death of a Druid Prince: The Story of Lindow Man, an Archaeological Sensation.* New York: Summit Books, 1989.

Ross, Herbert. *A Textbook of Entomology.* New York: John Wiley and Sons, Inc, 1965 (1948).

Ruediger, Luke. *The Siskiyou Crest: Hikes, History and Ecology.* Jacksonville: Self-published, 2013.

Sahlins, Marshall. *Stone Age Economics.* Hawthorne, NY: Aldine de Gruyter, 1972.

Savory, Allan with Jody Butterfield. *Holistic Management: A New Framework for Decision Making.* Covelo: Island Press, 1999.

Schopmeyer, C. S. *Seeds of Woody Plants in the United States.* Washington, DC: US Department of Agriculture, Forest Service, 1974.

Schwenk, Theodor. *Sensitive Chaos: The Creation of Flowing Forms in Water and Air.* London: Rudolph Steiner Press, 1996 (1962).

Seymour, John. *The Forgotten Arts and Crafts: Skills from Bygone Days.* New York: Dorling Kindersley Publishing, 2001 (1984).

———. *The Lore of the Land.* With illustrations by Sally Seymour. New York: Schocken Books, 1983 (1982).

Sheeran, Michael J. *Beyond Majority Rule: Voteless Decisions in the Religious Society of Friends.* Philadelphia: Philadelphia Yearly Meeting of the Religious Society of Friends, 1996 (1983).

Shepard, Paul. *Nature and Madness.* Athens: University of Georgia Press, 1998 (1982).

Shlain, Leonard. *The Alphabet Versus the Goddess: The Conflict Between Word and Image.* New York: Viking, 1998..

Sims, Lorelei. *The Backyard Blacksmith: Traditional Techniques for the Modern Smith.* Gloucester, MA: Quarry Books 2009 (2006).

Sky, Gino. *Appaloosa Rising: Or, the Legend of the Cowboy Buddha.* Garden City, NY: Doubleday and Company, 1980.

Sloane, Eric. *A Museum of Early American Tools.* New York: Ballantine Books, 1973 (1964).

———. *A Reverence for Wood.* New York: Ballantine Books, 1973 (1965).

Smith, J. Russell. *Tree Crops: A Permanent Agriculture.* New York: The Devin-Adair Company, 1953 (1929).

Smith, Richard P. *Animal Tracks and Signs of North America.* Mechanicsburg, PA: Stackpole Books. 1982.

Snyder, Gary. *The Practice of the Wild: Essays.* Berkeley, CA: North Point Press, 1990.

Stamets, Paul. *Growing Gourmet and Medicinal Mushrooms.* Berkeley: Ten Speed Press, 2000 (1993).

Stebbins, Robert C. *A Field Guide to Western Reptiles and Amphibians: Field Marks of All Species in Western North America.* Boston: Houghton Mifflin Company, 1966.

Steen, Edward, B. *Dictionary of Biology.* New York: Barnes & Noble, 1971.

Stegner, Wallace. *Angle of Repose.* New York: Penguin. 1971.

Storl, Wolf D. *Culture and Horticulture: A Philosophy of Gardening.* Illustrated by Midge Kennedy. Milwaukee: Bio-Dynamic Literature, 1979.

Tainter, Joseph. *The Collapse of Complex Societies.* Cambridge, England: Cambridge University Press, 1988.

Tilden, James W. and Arthur Clayton Smith. *A Field Guide to Western Butterflies.* Boston: Houghton Mifflin Company, 1986.

Tilth. *The Future Is Abundant: A Guide to Sustainable Agriculture.* Arlington, WA: Tilth, 1982.

Tavris, Carol and Elliot Aronson. *Mistakes Were Made (But Not by Me).* Orlando: Harcourt Books, 2007.

Tresemer, David. *The Scythe Book: Mowing Hay, Cutting Weeds, and Harvesting Small Grains with Hand Tools.* Brattleboro, VT: Hand and Foot, 1982 (1981).

Udvardy, Miklos D. F. *The Audubon Society Field Guide to North American Birds, Western Region.* New York: Alfred A. Knopf, 1977.

U. S. Department of Agriculture Forest Service. *Silvics of Forest Trees of the United States of America.* Agriculture Handbook 271. Washington, DC: U. S. Printing Office, 1965.

U. S. Navy. *Tools and Their Uses.* New York: Dover Publications, 1973 (1971).

U. S. Department of Agriculture Soil Conservation Service. *Conquest of the Land through 7,000 Years.* Agriculture Information Bulletin No. 99. Washington, DC: U. S. Government Printing Office, 1978.

Verdet-Fierz, Bernard and Regula Verdet-Fierz. *Willow Basketry.* Loveland, CO: Interweave Press, 1993.

Vogel, Virgil J. *American Indian Medicine.* Norman: University of Oklahoma Press. 1970.

Wallace, David Rains. *The Klamath Knot.* San Francisco: Sierra Club Books, 1983.

Ward, Tom. "The Chicken Story, Part I and II." *Permaculture Drylands Magazine,* 1990.

———. *Greenward Ho! Herbal Home Remedies, An Ecological Approach to Sustainable Health.* 5th ed. Ashland, OR: Wild Foods and Flora Ltd, 1990.

Watts, Tom. *Pacific Coast Tree Finder: A Pocket Manual for Identifying Pacific Coast Trees.* Rochester, NY: Nature Study Guild Publishers, 1973.

Weschcke, Carl. *Growing Nuts in the North.* St. Paul, MN: Webb Publishing Company, 1954.

Wiggington, Eliot, ed. *The Foxfire Book: Hog Dressing, Log Cabin Building, Mountain Crafts and Foods, Planting by the Signs, Snake Lore, Hunting Tales, Faith Healing, Moonshining, and Other Affairs of Plain Living.* Garden City, NY: Anchor Book Publications, 1972.

Wilber, Ken. *Sex, Ecology, Spirituality: The Spirit of Evolution.* Boston: Shambala Press, 1995.

Williams, Terry Tempest. *When Women Were Birds: Fifty-Four Variations on Voice.* New York: Picador, 2013.

Woelfle-Urskine, Cleo, ed. *Urban Wilds: Gardeners' Stories of the Struggle for Land and Justice.* 2nd ed. Oakland, CA: water/under/ground publications, 2002.

Wolf, Crossing [Deh-Gah-Nn-Dee Wah-Hyah]. *True Hunting Things I Make Known to You (Gaw-Naw-Hee-Lee-Daw Deh-Gunh-Yeh-Yaw Hunh-ss-Gah Yah!).* N.p.: Chill-A-Witt Rare Press, 1983.

Wolff, Virginia. *Orlando: A Biography.* London: Wordsworth Editions, London, 1995.

Wright, Amy Bartlett. *Peterson First Guide to Caterpillars of North America.* Boston: Houghton Mifflin Company, 1993.

Yocom, Charles and Vinson Brown. *Wildlife and Plants of the Cascades.* Healdsburg, CA: Naturegraph Publishers, 1971.

Yoemans, P. A. *The City Forest: The Keyline Plan for the Human Environment Revolution.* Sydney, Australia: Keyline Publishing, 1971.

Zim, Herbert S. and Robert H, Baker. *Stars: A Guide to the Constellations, Sun, Moon, Planets, and Other Features of the Heavens.* Illustrated by James Gorden Irving. New York: Golden Press, 1956 (1951).

Zuckerman, Seth, ed. *Salmon Nation: People and Fish at the Edge.* Portland, OR: Ecotrust, 1999.

Zwemer-Margulis, Kim. *Shadowchaser of the Siskiyous.* Illustrated by Elizabeth Zwick. Jacksonville, OR: One Sky Press, 2009.

INDEX

acorn mush, 193, 279
adobe plaster, 305, 308-310
Adirondacks, 452
air flow(s), 9, 202-203, 209, 232, 245-246, 264, 267-268, 277, 282, 285-286, 288, 290-292, 297, 312-314, 316
All Sentient Beings, 5, 8, 19, 25, 67, 69, 145, 147-148, 153, 159, 162, 199, 248, 249, 280, 359, 376, 398
Ashland, 429, 432, 448
assumptions list, 373, 375

basket basics, 252
Barley-Daikon-Fava, 442
Beltane, 172
biochar soils 14, 76, 167, 174, 188
block and tackle, 252, 267, 298-299
Boaz branch of anthropology, 43
Bodger Guild, 254-255
bodger shed, 283
brindles, 253, 259-260, 262-264, 303, 305, 308, 314
broad knife, 269
broadscale burning, 4, 43, 76, 150, 227, 242
brown gas, 173, 232
brush bundles, 295, 305
brush field conversion, 178, 228
brush wall, 259, 283, 289, 308, 313-314
brush wood cabin, 268, 291-292, 294, 424
Buckbrush, 15, 208, 225, 228, 237-238, 258, 260, 262, 264, 294, 297, 304-305
Burrell, Bill, 325

carbon fixation, 135, 168, 174, 220, 233, 238
carbon monoxide, 7, 216, 245, 286

carbon sequestration, 83, 116, 123, 133-134, 145, 173, 177, 225, 241, 247
ceremonial bundles, 12, 412
chaparral brush fields, 220
charcoal stove, 77, 277, 316
chimney thermometer, 287-288
clans, 24-26, 69, 71, 110, 112, 315-316, 356, 380-383, 395, 398, 434, 439
colonists, 43, 71, 192, 206, 242, 258
community inventory, 14, 32, 74, 347-350, 372
consensus, 15, 343-344, 359, 372, 395
convenience, license and privilege, 77, 214
coppice forestry, 253
coppiced thickets, 259
coppicing, 59, 133, 171, 174, 240, 253, 258, 260-263, 280, 391
council, 14-19, 25, 27-28, 30, 32, 34-35, 37, 42, 46-47, 54, 62-63, 65-67, 70, 88, 125, 134, 140, 151, 158-160, 162-163, 171-173, 175-178, 180-181, 187, 189, 212, 249, 282, 314, 316, 324, 328, 331, 333-335, 343, 347-350, 355-359, 364, 366-368, 370, 372-376, 381-383, 389, 391, 394-395, 397-400, 408, 410, 413, 424-425, 433
cross diagonal cribbing, 258, 301, 303, 305
cultural reciprocation, 8, 248
curing clay, 306

Deerbrush, 90, 149, 260, 262, 264, 278, 304-305
detention structures, 169
dew point, 285-287
disturbance regime, 13, 66, 86, 97-98, 137, 144, 151, 192, 258, 355, 369

Dixie Delta, 302
Doty, Tom, 51
double retort kiln, 7, 216, 236
Douglas Fir, 52-53, 109-110, 120, 140, 142, 146, 165, 178, 181, 184, 206, 258, 278-279, 286, 294, 297, 299-300
downdraft tube, 243-246
downhill burn, 167
D-Q University, 45, 52, 393
draft control, 286, 288
Dragonfly's River, 448
drama diet, 332-333, 379
dynamic airflow, 291

ear cuffs, 227, 272-273, 275
ecological mosaic, 14, 29, 42
Ecostery, 338
ecosystem functions, 3, 13, 17
ecotopian culture, 355
Elders, 12, 26, 45-46, 66, 173, 175-176, 212-213, 268, 279, 292, 316, 329, 332-333, 343-344, 359-361, 368, 381, 388, 393, 398, 400, 403-404, 409-410, 414-415, 417, 423
Elise Bolding, 431
embedded logs, 210, 228, 233
energy descent, 322-323
Equinox, 437
etiquette, 5, 26, 30, 34

fall equinox, 172
fiber basics, 250
field loom, 255-256
fire breaks, 42, 183, 195-197, 200, 203, 207, 210-212, 214, 228
fire control district, 214
fire dollies, 196, 210, 212-213
Fire Pig, 229, 232, 234, 236-239, 245
fire sector, 154, 160, 197, 200, 203, 214
First Nations Horticulture, 41, 214
forest bathing, 8, 14, 26, 67, 98, 149, 248, 436
forestry camp, 74, 76, 95, 101, 247, 262, 285, 294, 296-297, 299, 301, 304, 315, 324-325, 339, 364, 367

fuel ladder, 6, 117, 167, 183, 190, 197, 220, 222
fuel reduction, 27, 53-54, 58, 76, 157, 166-168, 194, 230, 237, 242-243, 252, 262, 272, 304

gender diversity, 386, 392
ground fire, 117, 167, 195, 198, 203, 262
grouse Flowers, 436
Guardian Oak, 110, 221

hand log-arch, 299
Hazel baskets, 276
holistic management, 42, 377, 435-436
hoop camp, 27, 127, 162, 177, 346, 355, 369
hoop culture, 32, 278, 287, 315, 359, 364, 368-370, 375
hot center, 222, 224-227
householding, 345

Imbolc, 172
infiltration earthworks, 144, 169
inner bark, 273, 279

keyline trails, 241
Klamath River Basin, 433

lamas, 172
learning styles, 11, 362, 376, 378, 392, 394
Lomatiums, 436

Manzanita, 53, 90, 119, 139, 177, 208-209, 225, 228-229, 234, 237-238, 243-244, 247, 262, 264, 280, 291, 294, 302, 326, 370-371
Manzanita charcoal, 243, 247, 326
Martinez, Dennis, 44-45, 151, 434
mast year, 89, 279
metamorphic soils, 72, 278
mid elevation slopes, 170, 175, 205-206
mopping up, 213
Mothers Guild, 176, 188-189, 212, 368, 391
mountain side burn piles, 224
multiple genders, 384, 389

natural building, 52, 76, 157, 189, 229, 260, 296, 310
natural plaster 217
natural succession, 221
net, sink, and loop, 64, 118, 133, 218, 325
nutrient-dense foods, 278

Oak Pine savanna, 76, 86, 88-89, 94, 127, 143, 163, 166, 205, 214, 219, 242, 273, 275, 294, 304, 331
Old English Law, 213
oxygen quenching, 224, 232, 238

perfect timing, 5, 9, 54, 79, 88, 102, 104, 140, 150-151, 168, 170-173, 175, 187, 212-213, 229-230, 244, 280, 282, 333, 339, 342, 364, 391, 422
pile burning, 183, 194, 247, 262
Pinole, 279
pit and mound topography, 116, 228
pit oven, 279
pit saw, 170, 265
Place, 14-16, 19, 25-30, 34-35, 43-44, 46-50, 59, 61, 63, 65-71, 74, 76, 83-84, 88, 90, 92, 98, 103, 105, 125, 140, 142, 151, 153, 156, 159-160, 162, 164, 172, 178, 192, 211, 249, 252, 266, 315, 322, 324, 330, 333-335, 337-338, 341-342, 347-348, 354-355, 357, 370, 380-381, 386, 397-400, 404, 416-417, 423-425
plate glass, 301
pocket deserts, 74, 232
post and beam frame, 264, 299-301, 314
proper clothing, 226
purposeful simplicity, 249

queries, 15, 70, 403, 407

ramial tissues, 174, 188, 223, 286
refugia, 30, 32, 40, 121, 123, 131, 136, 184, 228, 233, 357-358
relative location, 157, 206, 293
retro feudalism, 84, 362-366, 368

Return of the Salmon, 430
rewilding, 204
rigid, 11, 84, 340, 378, 402
riparian zones, 136, 138, 144, 168, 170, 183, 304
Robinson, Kim Stanley 431
rocket stove, 40, 239, 247, 277
rod mat, 250, 255-256, 258, 260-261, 263-264, 287, 293
rubble foundation, 297
rural land holders, 232

Samhain, 172
scenario practices, 326, 348, 353, 372
senescent vegetation, 137, 220, 233
settlers, 24-26, 28, 30, 35, 41, 48, 70, 72-73, 104, 125, 146, 159, 181, 194, 324, 327, 355, 359, 399, 413, 419, 421
sharpening saws, 270
sheep herders stove, 304, 324
Silvo pastural systems, 203
simple living, 76-77, 151, 345
single retort kiln, 234-235, 239-241, 245
six genders, 389
skid chutes, 179
slash pile, 222-223, 231-232
smoke flaps, 251, 267, 269, 290-291
social attachments, 430
social fire tending, 211
social order, 12, 34, 48, 65, 331, 335, 342, 344, 355, 359-362, 396-399, 402, 409
Soda Mountain Crest, 454
soil micro life, 7, 216-218
soil seed bank, 114, 168, 199, 222
spokes councils, 5, 14, 34, 343, 348, 373
spring-equinox, 172
stewardship contracts, 55, 327, 339, 374
stone boiling, 252, 279
strewn floor, 285-287
summer solstice, 172, 262, 297

taboos, 5, 26, 30, 34, 89, 103, 140, 160, 163, 333, 341, 354, 360, 362, 370, 399-401, 406

TEK, 2, 35, 38, 45-47, 67, 88, 199, 206, 280, 315-316, 325, 330, 353-355, 358, 360-361, 369-370, 375, 384, 399, 410, 417. See also Traditional Ecological Knowledge
top lit pile burn, 231
totems, 67, 380, 382, 423
Traditional Ecological Knowledge, 2, 26, 38, 41, 45-46, 350, 397
transition, 10, 16, 18, 29, 32, 35-37, 39, 52, 55-56, 59, 96, 105, 158, 160, 181, 194, 196, 214, 219, 231, 240-241, 250, 298, 313-315, 321, 327-328, 334-335, 338-339, 348, 352-353, 355, 357-359, 361, 363, 365-369, 371, 373-375, 377, 386, 402
tree alphabet, 416, 418, 420
tripod, 255, 265
of poles, 267-269
stool, 291
tyloses deposits, 273, 276

ultimate goal, 36, 159, 206, 248, 387, 407
ultimate restoration, 144, 242

vernacular living, 77, 269, 315

Wagner County, 371, 456
wall tent, 285-287
water detention, 169, 233, 284
water tables, 4, 49, 69, 118, 125, 131, 133, 136, 141, 146, 150, 175, 278, 290
wattle walls, 255
White Oak, 53, 86, 88-90, 92-95, 103-104, 129, 135, 145-146, 166, 205, 207, 208, 228, 234, 263-264, 272-273, 275, 279, 294, 296-298, 452
wildfire, 49, 75, 115, 117, 135, 143, 166, 170-171, 176, 178, 190-192, 194-200, 202-208, 211, 213-214, 219-221, 228, 231-233, 240, 242, 244-245, 258, 260-265, 283-284, 294, 296-299, 302, 304-305, 310, 315, 330, 347, 350, 363-364, 374
wildlife corridors, 27, 169, 376
Willow basket, 282-283, 293
winter solstice, 172, 297